Food quality management

a techno-managerial approach

P.A. Luning
W.J. Marcelis
W.M.F. Jongen

Wageningen Pers

CIP-data Koninklijke Bibliotheek
Den Haag

ISBN 9074134815 paperback

Subject headings:
Quality assurance
Agri-food chain
Food safety

First published, 2002

Cover design:
H. Kunst

Wageningen Pers, Wageningen
The Netherlands, 2002

Printed in The Netherlands

Contents

Preface

In the last decades quality has become of utmost importance to society. Consumers have become more conscious of quality, and organisations are now judged more on their overall quality performance instead of their financial performance alone.
In the fifties, when thinking about quality management started to develop, it could not have been foreseen that it would have such a big impact on today's quality management. The most drastic change in quality thinking is probably the change from production-oriented to customer-oriented concepts. Moreover, integrative approaches, system thinking, the focus on advanced technologies, and belief in human capacities have had a considerable impact on current quality management. Whereas before the focus was mainly on technological (physical) product quality, it is now extended to additional aspects like organisation flexibility and reliability. In fact, nowadays quality is equal to total business performance in the context of superior relationships with all stakeholders of the company.

This trend in increased quality consciousness is also evident in the agribusiness and food industry. In the last decade, tremendous progress has been observed in the food industry, especially with respect to technological product quality and food safety. Much effort has been put in the development of quality assurance systems to guarantee product quality and safety. In the agribusiness and food industry, the change from production-orientation to customer-orientation continues gradually but undeniably, with chain collaboration playing a central role. These developments are only feasible when they are continuously supported with up-to-date technology and modern management methods. Therefore, the modern food industry is underway to become a knowledge intensive business, with completely different characteristics when compared with the situation of twenty to thirty years ago.
It has become increasingly dependent on technological developments (such as packaging, processing, post-harvest technology but also biotechnology) and on management knowledge, in order to comply with the wide range of continuously changing consumer demands. Logically, food quality management has become a major and interesting topic of research and education. Food quality management is a scientific field in which integration of the essential disciplinary knowledge domains is a key issue. Only hence can it assist the food industry in making the jump from good quality to superior quality.

In this book we introduce the techno-managerial approach, which aims at integrating the different disciplines to contribute to achieving superior quality. The basis of the techno-managerial approach is that quality issues are simultaneously perceived and analysed from a technological and managerial perspective.

The basis for this concept was established in the mid-eighties at Wageningen University when the first course in food quality management started. Dr. ir. W.J. Marcelis designed this course according to the principle of integrating technological and management sciences. This resulted in a close collaboration between the Management Studies Department and the Food Technology Department (the late Mrs. ir. J.C. de Wit and dr ir. J.P. Roozen). Subsequently, the education programme on food quality management was extended with a course on quality systems and MSc thesis projects. During this period, Prof. M. van den Berg, who had a comprehensive experience in quality management in the dairy industry, has significantly contributed to the

development of food quality management education at Wageningen University, which is reflected in his book "Quality of Food products".
Till his leave, far more then a hundred students have finished their MSc-thesis in food quality management with Prof. M. van den Berg and Dr. ir. W.J. Marcelis as supervisors. Most of these thesis projects were carried out at international reputable companies in agri-business, food industry and institutions involved in food quality. The food quality management courses were subsequently carried out and improved by Ir. A. Jellema and Dr. ir. G. Ziggers.

In 1998, education in food quality management received a new urge by Prof. dr. W.M.F. Jongen and Dr. ir. P.A. Luning, who have taken the initiative to develop an integrated MSc programme "Quality Management in Food Chains" and to write this book.
This programme offers an interesting curriculum for all types of students in the new Bachelor-Master system. It consists of technological, management and consumer courses and of food quality management courses that integrate the technological and managerial knowledge. Besides, there are supporting courses on methodology, modelling, and the role of government in food quality and food safety and health. Students with backgrounds in different disciplines, such as food technology, bio-product management, animal science, plant science, veterinary sciences, management and consumer sciences can follow this MSc programme. Built on a profound disciplinary knowledge, the MSc programme offers students from various countries the opportunity to study Food Quality Management using the techno-managerial approach as a basis.

The book as presented to you is the basis of the educational programme on Food Quality Management at Wageningen University and Utrecht University. We have developed a total concept for food quality management, which is described and underpinned in this book with the techno-managerial approach as starting point. We have attempted to make no appeal on detailed foreknowledge and therefore some parts of the book have an introductory character.
To our knowledge, such concept has not yet been published and we hope that this book will contribute to the further development of food quality management as a scientific research area. At the same time we believe that this book provides a basis for further integration of all disciplines that are necessary to achieve superior quality.

Last but not least, we are grateful to several people that enabled the completion of this book. We thank Mrs. ir. M. Breithbarth and Dr. ir. G. Ziggers for their contribution in collecting and compiling literature for this book. We also thank Prof. dr. ir. M.A.J.S. van Boekel for critical reading of the manuscript and thank Dr. ir. P. Overbosch for the stimulating discussion.
We also thank the participants of the Erasmus Curriculum Development Advanced project "Total Food Quality Management", co-ordinated by Prof. R. Verhe, for their financial support. The partners are from Gent, Vienna, Athens, Bonn, Copenhagen, Murcia, Valencia, Wageningen, Olstzyn, Cluj Napoca, Nitra, Maribor, Riga, Kaunas, Burgos and Godollo.

Wageningen, Maart 2002

P.A. Luning
W.J. Marcelis
W.M.F. Jongen

Authors

Mrs. Dr. ir. P.A. Luning

Pieternel Luning studied Food Chemistry and Microbiology at Wageningen University (Netherlands). After she finished her study in 1988, she was involved in different projects in the area of food flavour research. She worked at the University for two years on bread flavour. From 1990-1995 she was employed as project manager at the Agrotechnological Research Institute ATO-DLO, where she finished her PhD work on flavour of fresh and processed bell peppers. She was then involved in a flavour project on vegetable oils for Unilever Research. Since May of 1996, she started at TNO Research and Nutrition Institute as product Manager "Innovative Packaging", where she initiated research on new packaging concepts for food. Since July 1998, she is employed as assistant professor in Food Quality Management at Wageningen University. She is now co-ordinator of the new educational programme "Quality Management in Food Chains" and is responsible for research in sensory science and product quality.

Dr. ir. W.J. Marcelis

Willem Marcelis finished his study on Mechanical Engineering at Wageningen University (Netherlands) in 1972. Thereafter, he did research on maintenance management and published two books on that topic. He finished his PhD in 1984 and was management consultant in the field of maintenance management for many years. He collaborated with Professor A.A. Kampfraath in developing a management theory based on administrative processes. A book about this topic was published in 1981. In the mid-eighties he started with the initial educational programme on Food Quality Management. Since that time he collaborated with the different technological departments on extending and improving the food quality management programme. He is now involved in the development of the new curriculum on "Quality Management in Food Chains" at Wageningen University.

Prof. dr. W.M.F. Jongen

Wim Jongen studied analytical chemistry at the Technical University Heerlen (Netherlands). He graduated in Food Toxicology at Wageningen University and worked as a post-doc at the WHO centre for cancer research IARC in Lyon, France. In 1989 he joined the management team of the newly founded agro-technological research institute ATO-DLO in the Netherlands, as head of the division on post-harvest technology and product quality. In 1994 he took up the chair of Integrated Food Technology at Wageningen University, later renamed into Product Design and Quality Management. In May 2000 he became part-time scientific director of ATO bv and since November 2001 he is director of business development of the newly formed Science Unit Animal Production of Wageningen-UR. In addition he is professor at Wageningen University in Product Design & Quality Management, with special attention to chain aspects. Wim Jongen is an advisor for the food industry and has been involved in a large number of scientific advisory committees both for government and industry.
Currently he is also member of the executive board of the European Federation of Food Science and Technology (EFFOST).

1. Introduction: food quality management

Quality has become a vital distinctive feature for competition in the world market of food products. To obtain a good quality end product, quality is more and more managed along the whole food chain from the supplier of raw materials to consumption. Striving for quality is not a free choice. Customer understanding of food quality and the ultimate concern for health and food safety force actors in agribusiness and food industry to use quality management as a strategic issue in innovation and production.

1.1 Food quality

Especially in agribusiness and food industry quality management is an issue of importance. The consumer has an emotional relationship with food products. Food is something you put inside your body every day during life , which affects the wellbeing and health of the individual directly and indirectly. Safety and healthiness of food products are public issues of primary importance. Traditionally quality management in the food has primarily been focused on issues of food safety. Nowadays the concept of quality is widened to all other aspects of the food product.

One consequence of the industrialisation and the associated scale enlargement of food production systems are that, *e.g.* by accident a contaminated food product appears on the shelf of a supermarket and a large number of consumers can be at risk. Quality should be built into the production system to prevent, rather then to correct, failures.

Safety and healthiness of food products is often not a visible feature and the consumer has to believe in the information provided by the producer. Nevertheless, one of the major achievements of the current situation is that in most western countries there generally is a relationship of trust between the consumers and the producers of food products.

Hoogland and co-authors (1998) have mentioned several features that are characteristic for food quality management, such as

- Agricultural products are often perishable and subject to rapid decay due to, amongst others, physiological processes and microbiological contamination. These processes can be hazardous for human health and therefore profound product knowledge is required for adequate quality control;
- Most agricultural products are heterogeneous with respect to desired quality parameters, such as content of important components (*e.g.* sugars), size and colour. The kind of variation is dependent on *e.g.* cultivar differences and seasonal variables, which cannot be fully controlled;
- Primary production of agricultural products is performed by a large number of farms operating on a small scale, which renders control more complicated.

Besides these typical aspects for agri-food production, there are several trends and facts that also underpin the need for appropriate quality management.
Despite progress in medicine, food science and technology of food production, illness caused by food borne pathogens continues to present a major problem in terms of both health and economic significance. In 1990, an average of 120 cases of food borne illness per 100.000

population was reported from 11 European countries. Estimates based on another study indicated that in some European countries there are at least 30.000 cases of acute gastro-enteritis per 100.000 population (Notermans and Van der Giessen, 1993). Nevertheless, only a small part of actual cases of food borne illness are brought to the attention of food inspection, control and health agencies. This means that the real number of cases due to contaminated food products is much higher than registered data suggest. This fact illustrates the urgency of proper attention to control and assurance of food safety.

With respect to consumer trends, it should be noticed that there is an increasing demand for convenience foods, requiring minimal handling or preparation in user-friendly packaging. In addition, consumers are seeking foods which are more 'fresh' and with enhanced natural flavours that inevitably challenge the industry to use less harsh processing and production regimes. Whilst there is little evidence that such trends have led to increased food borne illness, it must be noticed that these foods will require greater care in their production, distribution, storage and preparation prior to final consumption. In fact, the food industry faces the challenge to achieve high sensory quality while assuring a safe product with an appropriate shelf life.
Moreover, there is a trend towards pharma-oriented food products, *i.e.* functional foods. Functional foods claim disease protection and health progress, they penetrate the zone between food and medicine. These functional foods also provide new challenges to quality management. Significant changes are taking place in food production systems such as *e.g.* in animal husbandry. The focus is changing from quantitative towards qualitative production, putting more effort in *e.g.* animal welfare and reduced use of veterinary medicines. Another trend that requires special attention is the extended use of a range of raw materials and products originating from countries all over the world. This has increased the potential for a geographical spread of diseases associated with particular contaminants and pathogens. In addition, many new processing techniques have been (or are being) introduced which alone or in combination with each other, offer distinct quality advantages, such as 'fresher' sensory properties by using milder thermal processing, microwave heating, ohmic heating and high pressure processing techniques. However, a lower (heat) treatment provides products that are more sensitive towards microbial spoilage, therefore proper control (*e.g.* temperature) in the production chain is a basic need.
Anticipation to specific demands of consumers on the one hand has resulted in product diversification and a large assortment on the shelf of the supermarket. On the other hand the strong and increasing market competition makes it necessary to produce at minimal costs. These developments have been resulted in the concept of mass customisation. In other sectors of industrial production this concept has already been implemented and the food industry is following now.
Last but not least, consumer expectations regarding taste, fat content, variety, package and so on, make food quality a complex but fascinating field of study.

1.2 Quality and business performance

The relationship among quality, profitability and market share has been studied in depth by the strategy Planning Institute of Cambridge, Massachusetts. The conclusion, based on performance data of about 3000 strategic business units, is unequivocal (Buzzell and Gale, 1987; Ross, 1999): "One factor above all others - quality - drives market share. When superior quality and large market share are both present, profitability is virtually guaranteed.

Even producers of commodity or near-commodity products seek and find ways to distinguish their products through cycle time, availability, or other quality attributes. In addition to profitability and market share, quality favours growth and can reduce costs. Return of investment will increase due to better productivity. Moreover, by increasing quality it is possible to drastically decrease the need for intermediate stocks in logistic supply chain". The linkages between these correlates of quality are shown in the "quality spiral" of Figure 1.1 (Bergman and Klefsjö, 1994; Ahlmann, 1989).

The above mentioned research supports the statement that customers are prepared to pay more for a product of higher quality than the costs required to achieve higher quality. In food production systems quality pays. Quality does not cost extra money, it is non-quality that creates additional costs. Costs arise when defective products are manufactured in such a way that rework in different forms is necessary: when process quality is so uncertain that special inspection has to be performed, or even worse, when recalls have to be carried out. The way to create a number one position is to invest in the development of new products and in quality improvements resulting in increased margins of profit.

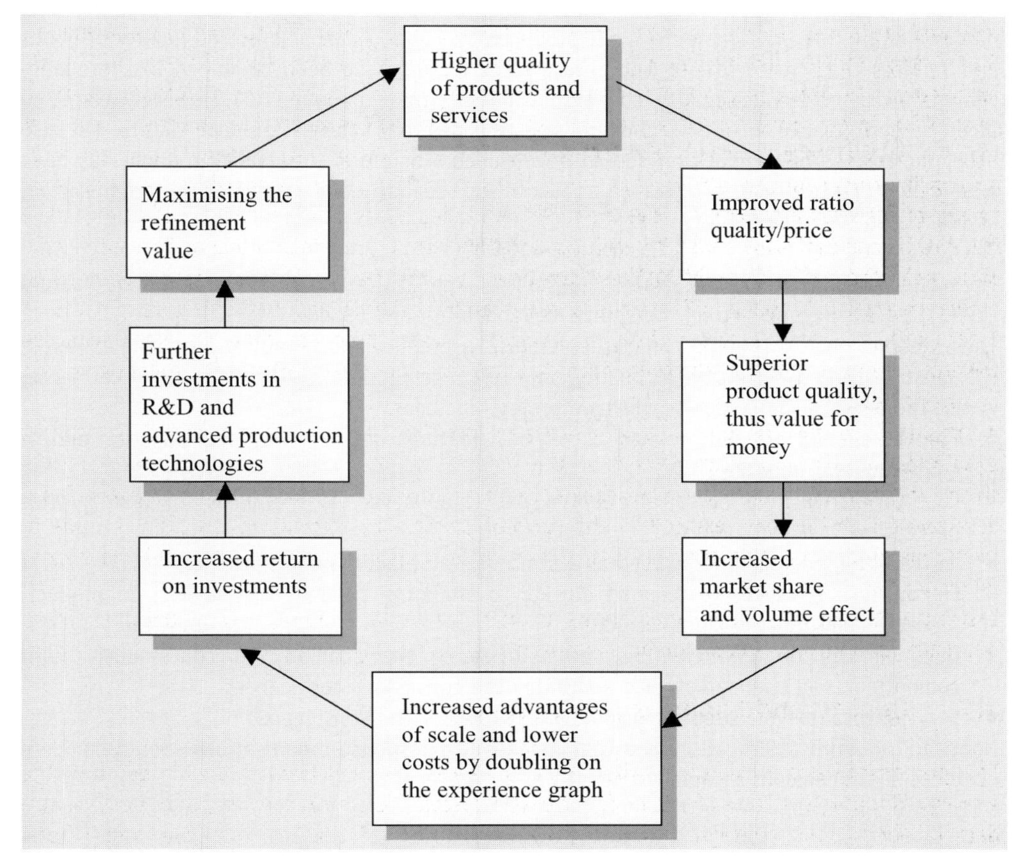

Figure 1.1. The quality spiral illustrating long-term maximisation of refinement value instead of short-term maximisation of profit.

This orientation gives the quality department an important role not only in production but also in innovation processes. It also requires quality consciousness throughout the whole organisation. Feigenbaum (1961), one of the quality gurus stressed that "quality is everybody's job". In particular, active participation of the senior management in QA issues is a prerequisite for successful implementation and operation of quality management systems.

Quality has been defined in many ways (see also Chapter 2). Some link quality as superiority or innate excellence, whereas others view it as a lack of manufacturing or service defects. Today most managers agree that the main reason to pursue quality is to satisfy customer demands. A common definition of quality is "the total of features and characteristics of a product or service that bears on its ability to satisfy given needs." The view of quality as the satisfaction of customer needs is often called *fitness for use.* In highly competitive markets, merely satisfying the needs of customers will not achieve success. To beat competition, organisations often must *exceed* customer expectations. Most progressive organisations now define quality as follows, and we use this definition in this book:

"Quality is meeting or exceeding customer expectations"

Already in 1887, William Cooper Proctor noted that profitability of an organisation is determined by three critical factors, *i.e.* productivity, costs of operations, and quality of goods and services that create satisfaction. These three factors have been combined in the quality triangle including product quality, cost and availability (part of Figure 1.2). However, related to business performance, quality cannot only be considered as physical product quality, but also other dimensions of competition should be included. Therefore, the quality triangle has been extended with additional factors such as flexibility, reliability and service of the organisation, as illustrated in Figure 1.2.

The model is mainly based on the six dimensions of competition as defined by Noori and Radford (1995), *i.e.* quality, cost, time, flexibility, dependability and service.
The extended quality triangle can be depicted as a product and organisation triangle. Herein the most restricted definition of quality only refers to product quality in a technical sense, whilst the broadest definition - which we prefer - includes:

- Product quality, not only related to physical features but also way of production and/or environmental impact;
- Cost, primarily in a customer-oriented net value sense, is the basis for quality/price perception;
- Availability, wherein time is used as a strategic value-adding concept;
- Flexibility, *i.e.* ability to respond quickly to changing situations in respect to product, processes and organisational structure;
- Dependability or reliability, *i.e.* possibilities of the company to continuously fulfil commitments in the marketplace and enjoy the consumers' confidence;
- Service, total service support for products, suppliers and consumers.

Noori and Radford (1995) discussed that modern fast response organisations should compete on these six dimensions in order to reach world-class quality.

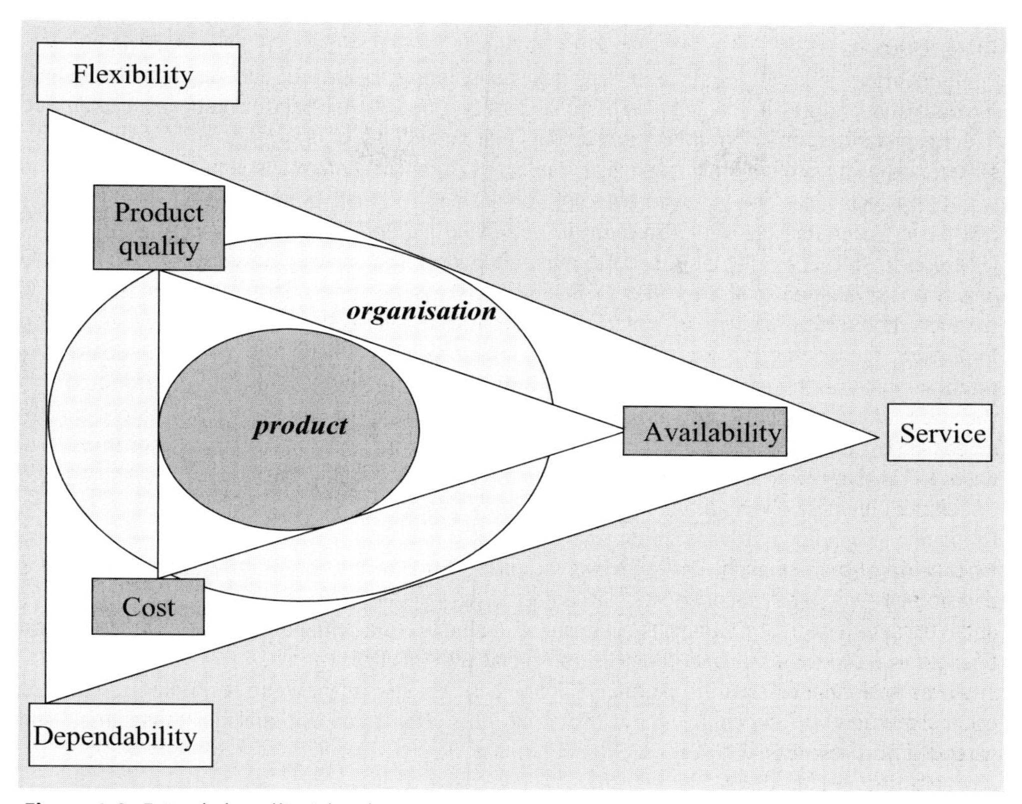

Figure 1.2. Extended quality triangle.

1.3 Techno-managerial approach

W. Edwards Deming (1900-1993) is - together with Joseph M. Juran - generally accepted as the father of quality management. His theory, summarised in Deming's 14 Points of Management (Chapter 8), was based on what he called profound knowledge (Deming, 1993). Profound knowledge consists of four parts (Bergman and Klefsjö, 1994; Dean and Evans, 1994):

1. *System thinking.*
   System thinking plays an important role. A system is a network of interdependent components trying to fulfil an aim. Deming emphasised that without an aim, there is no system, and that the job of the management is to optimise the system.
2. *Understanding variation.*
   It is not easy to understand complex interaction of variations in materials, tools, machines, operators and environment. Variation due to any individual source appears at random. However their combined effect is stable and can usually be predicted statistically.
3. *Using theory.*
   Deming emphasised that management is very much about prediction. In order to plan, a comprehension of what will be the consequence of different actions is needed. One should predict. To make a prediction it is necessary to know present and past history and to have a theory from which past behaviour may be coupled to future behaviour.

4. *Psychology.*
   Psychology is important since in human systems, human behaviour and decision making will determine future. Psychology helps to understand people, interaction between people and circumstances, interaction between leaders and employees, and any system of management.

This concept of profound knowledge is also relevant in food quality management. However, food quality analysis faces one big difference with analysis of "normal" products. Food, food products and basic materials from agricultural origin are *living materials* that change continuously over time due to (bio)chemical, physical and microbiological influences.

The main consequence of this fact is that variation is huge and interdependencies in food products and processes are of great complexity. This makes technological knowledge very important. Technological domains, such as microbiology, chemistry, process technology, physics, human nutrition, plant science and animal science, are needed for understanding this variation and complexity and in controlling it. Technological theory and research provide managers with, for instance, predictive models, which contain cause-effects relationships.

Thus in analysing food quality management are needed both the use of psychology to understand human behaviour and use of technology to understand behaviour of living materials. Of course psychology is very important, but not the only science to be used in analysing management problems. In management studies there is an integrated use of - besides psychology - sociology, economics, mathematics and legal science.

As a consequence, food quality management embraces the integrated use of technological disciplines as well as the integrated use of managerial sciences.

Three different approaches are possible (Figure 1.3): the managerial, the technological and the techno-managerial approach. They differ in the extent of integrating managerial and technological sciences.

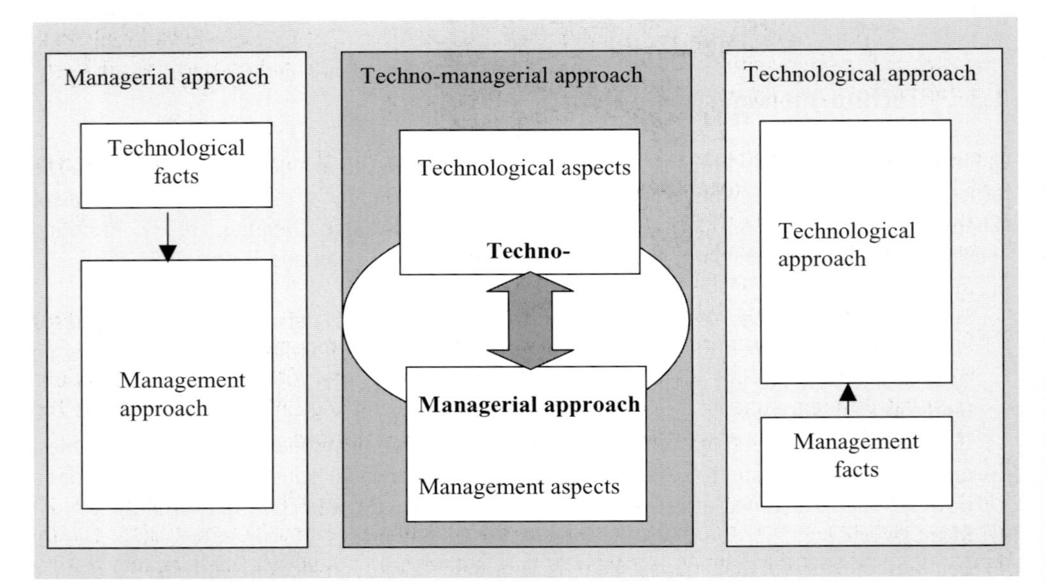

Figure 1.3. Different approaches to food quality management.

Typical for the managerial approach is that technological aspects are contemplated as facts: "We can make everything we want to make, in fact there are no technological restrictions". In the traditional technological approach, management aspects are considered as boundary restrictions: "They want everything finished yesterday and never provide the appropriate budgets".
In contrast, the techno-managerial approach encompasses integration of both technological and managerial aspects from a system's perspective. Quality problems are considered interactively from both a technological and managerial viewpoint.
In a first attempt to describe the innovation process from the techno-managerial perspective, Jongen (2000) used the DFE concept as a simplified approach that uses three questions in a specific order to put both managerial and technological aspects in the right perspective. The first question is that of the *Desirability (D)*, which product concepts for which markets should be developed. The second question deals with the *Feasibility (F)* of the concepts, which technological capabilities are required to make the product. Do we have them or do they have to be developed? The third and final question is that of the *Effectiveness (E)*, how do we organise the production process from a chain perspective in order to be cost-effective.
In this book, the techno-managerial approach is introduced and detailed in different chapters. The core element of this approach is the contemporary use of technological and managerial theories and models, in order to predict food systems behaviour and to generate adequate improvements of the system. Integration of both sciences should be developed to such an extent that:

a. technological behaviour and changes can be judged on the effects they have on human behaviour and,
b. human behaviour and changes can be judged on the effects they have on technological behaviour.

A good example of techno-managerial thinking is the HACCP-system (see Chapter 7), wherein critical technological hazards are controlled by human control and monitoring systems, and Quality Function Development (QFD), wherein consumers' wishes are translated into technological requirements through intensive and organised collaboration of different departments in the company.

Figure 1.4 shows how the techno-managerial approach resulted in the food quality management model, which is the basis for this book. The model includes:

a. the organisation in its environment, wherein
b. management and technology interact, striving for
c. product quality that meets or exceeds customer expectations,
d. wherein technology is perceived as a technological system, with complex interactions fulfilling different functions in order to meet product quality requirements, and
e. wherein management is perceived as a management system with complex interactions fulfilling different functions in order to activate the technological system, to give it the right direction and to ensure that it meets customer expectations.

The quality functions form the main structure of the book. After an introduction on technological and managerial aspects of food quality (chapter 2 and 3), the typical quality functions, quality design, quality control, quality improvement, quality assurance and quality policy and strategy are elaborated in subsequent chapters (4 to 8).

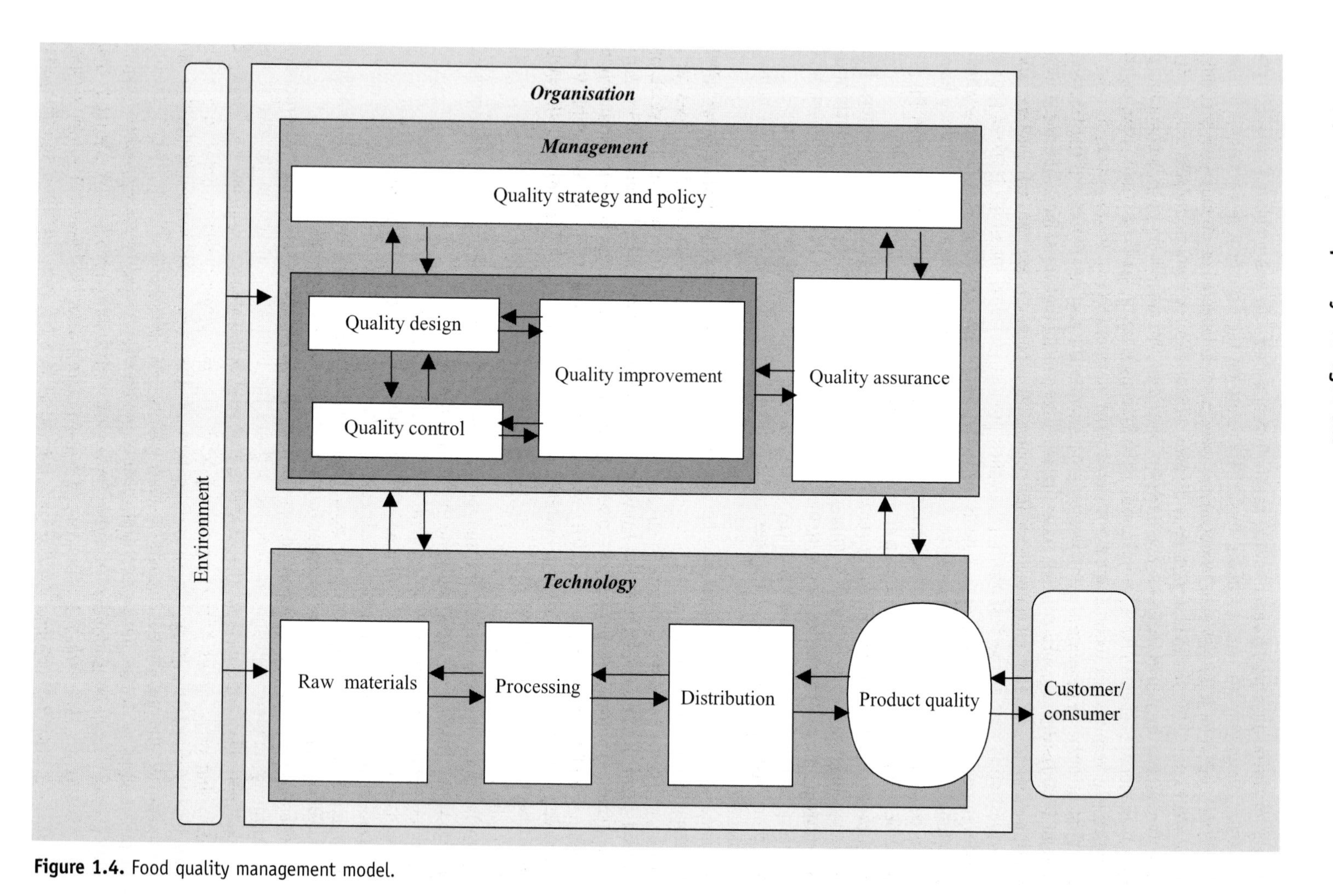

Figure 1.4. Food quality management model.

1.4 Chain perspective

From the quality management perspective, every company is part of a complex network (actually many chains and sub-chains) of customers and suppliers. Each company is a customer to its suppliers and a supplier to its customer (Dean and Evans, 1994). Sometimes a company must focus on both their immediate customers and those customers next in the chain. Food companies certainly work hard to satisfy the needs both of the people who use or consume their products, and the retailers who sell them, labelling the former as "consumers" and the latter as "customers". Food companies do have an additional objective in assuring food safety and health aspects. They are for that partly dependent on what their supply chain can deliver. More and more companies at the end of the food chain feel obliged to broaden their management attention to former parts of the chain in order to succeed in meeting customer expectations.
In this book attention to chain aspects is paid throughout the whole book:

- technological aspects and interactions through agri-food production chain (Chapter 2),
- supply chain management and partnerships (Chapter 3)
- supplier - and customer - relations (Chapters 4, 5 and 6)
- chain quality assurance (Chapter 7) and
- strategic alliances in food supply chains (Chapter 8).

1.5 Food quality management research

Food quality problems always have been problems of a special kind. Apart from the fact that food quality problems require the integrated use of both technological disciplines and managerial sciences, they are rather complicated due to the complexity of food as a product, as well as quality management as an element of the social system. Boulding's hierarchy of systems is helpful in explaining this (Boulding, 1956). Boulding developed a hierarchy of nine system levels. Each higher level includes the features of all lower levels, but adds a new dimension that was not present before. So each higher level is more complex and more differentiated. The nine levels are, starting with the lowest:

1. frame-work system: a static structure;
2. clock-work system: a simple dynamic system with predictive motion;
3. cybernetic system: equilibrium through built in control;
4. open self-maintaining system: the cell, life and self-reproduction;
5. genetic social system: differentiation in organs and blue-printed growth (plants);
6. animal system: consciousness and gathering of information;
7. human system: self-reflection, knowledge and anticipation;
8. social system: social relationships and (conflicting) interests, culture;
9. transcendental system: non-material systems of logic and belief.

In research, the well-known mechanism is using models to analyse complex situations. A problem that often arises is the use of models from lower system levels in order to analyse higher system levels, with as a consequence loss of information and bad predictability. An example is using simple control models for analysing social systems.
In food quality management, technological problems cannot be typified as cybernetic system problems, as it is the case for most technical products. Food quality problems require modelling and analysis at levels 5 and 6, indeed integrating several technical disciplines. Moreover, quality management problems require modelling and analysis at levels 7 and 8, integrating managerial

disciplines. As a consequence, research in food quality management is complex. It has to combine high but also different system levels, which result in high requirements on modelling and system analysis.
Food quality management research has to cope with another problem. It includes two extremes in research activities, fundamental scientific research and policy-relevant scientific research, that require different methods and models, but are at the same time interdependent. Figure 1.5 typifies the two research activities.

At the left side of the figure fundamental scientific research is characterised as conclusion-oriented research. The process starts with a question, driven by a feeling of curiosity or amazement. Then more and more information is gathered, followed by continuous matching objective facts and creative ideas. The process is a process of cognition resulting in knowledge and is more or less infinite.
At the right side of the figure policy-relevant scientific research is characterised as decision-oriented research. The process starts with a problem and is driven by the necessity of solving the problem. Information is gathered to analyse the situation and for developing alternative solutions in a process of continuous matching goals and ways to reach them. The process is one of selection resulting in a decision at the right moment.
Decision-oriented research is very much supported by conclusion-oriented research. And when research is scientific, conclusion-oriented research is a condition for solid decision-oriented research. Food quality management research very often is decision-oriented. So, in food quality management research, the cognition-selection interaction is very important for coupling both kinds of research.

Both characteristics of food quality management research result in high requirements on modelling and system analysis. At the other hand, models and underlying concepts always will remain a simplification of reality.
This book offers a wide range of technological and managerial models, as well as models and concepts that link them, in order to get more insight and grip on the impressing phenomenon of food quality management.

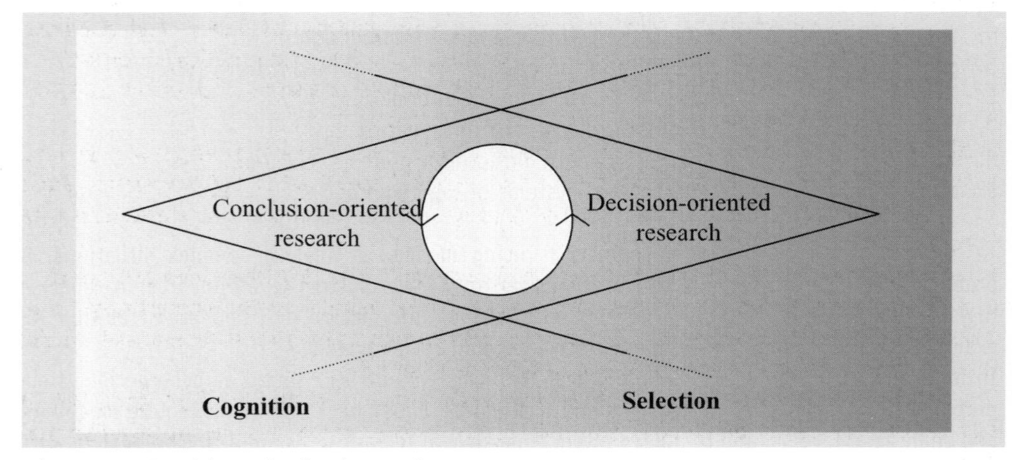

Figure 1.5. Cognition-selection interaction.

2 Food quality

As previously mentioned, quality has become an important issue in agri-business and food industry. Whereas it used to be focused on quantity, in the last decade it shifted towards a quality oriented sector. But what does quality mean for agri-business and food industry? As a matter of fact, there is no unambiguous definition for quality. Moreover, a wide variety of approaches and concepts, about how consumers perceive quality and how it can be influenced by different factors, have been proposed. In section 2.1 a short overview of major definitions and quality concepts is given.

Within the philosophy of the techno-managerial approach, quality is not only considered as physical product quality, but also includes aspects related to production characteristics (*e.g.* environmentally sound, wholesome) and performance of the organisation (*e.g.* minimal costs, service quality). Some of these aspects are influenced mainly by technological factors, whereas others are directly affected by management decisions. A proper understanding of technological factors influencing physical features of agri and food products is assumed to be the basis for the techno-managerial approach. Therefore, sections 2.2 and 2.3 are mainly focused on product quality considered from a more technological viewpoint. General management processes that can influence quality performance are described in Chapter 3.
Last but not least, not only consumers but also international and national legislation put demands on quality. The last section gives a brief overview of international, national and branch regulations, which are relevant for quality and safety of agri and food products and/or production.

2.1 Quality definitions and concepts

Many authors have attempted to define or describe the concept of quality. Some common definitions for quality and a range of quality concepts are described below.

Quality definitions
In the last two decades some people contributed largely to the thinking about quality, *i.e.* the quality experts or guru's. Some of them have attempted to develop a definition for quality.
Juran (1990) defined quality as 'product performance that results in customer satisfaction and freedom from deficiencies, which avoids customer dissatisfaction', in short 'fitness for use'.
Deming (1993), another quality expert, stated that 'a product or a service possesses quality if it helps somebody and enjoys a good and sustainable market'.
Crosby (1979) described quality as 'complying with clear specifications, whereby the management is responsible for establishment of univocal specifications'. More details about these quality philosophies and their role in developing quality management are explained in Chapter 3.
In addition, the Institute of Food Science and Technology (IFST, 1998) has described the term quality, applied for food, as follows. When applied meaningfully to the character of food, 'quality may refer to the degree or standard of excellence, and/or the fitness for purpose, and/or the consistency of attainment of the specified properties of the food'.
The American National Standards Institute and the American Society for Quality Control (ASQC) standardised the definition for quality in 1987 as 'the totality of features and characteristics of a product or services that bears on itself ability to satisfy given needs'.

The International Organisation of Standardisation (ISO, 1998) has defined quality as follows
Within the ISO 9000 standards, the term quality is used in the context of 'achieving sustaine
customer satisfaction through meeting customer needs and expectations within organisationa
environment commitment to continual improvement of efficiency and effectiveness'. Quality, i
this sense, is critical to business success.
Generally the definitions for quality can be summarised as 'quality is meeting or exceedin
customers' expectations'

To meet or exceed customers' expectations it is important to know who are the customers an
what they expect. Customers can be defined as those who receive a product (ranging from ra
materials to finished products) or a service from a supplier within a production chain. Fo
example, the milk factory is the customer from farmers who supply the milk. This type o
customer is called external customer, whereas customers within a company are typified a
internal customers (e.g. the cheese production is customer of the quality control laboratory wit
respect to the quality results).
Customers that are the ultimate purchaser or user of a product or service are referred to a
consumers. **The** consumer doesn't exist and there is no average consumer. According to Jonge
(1998) there is a specific consumer, which in a specific situation and on a certain moment ha
a specific need to which the producer can respond.

Quality concepts

The described quality concepts range from simple illustrations to complex models reflectin
factors that might influence quality expectation and perception by consumers or customer
General concepts for quality as well as specific concepts for quality perception of agri-foo
products have been summarised.

1. Zip model

Van den Berg and Delsing (1999) described quality as the relationship between suppliers o
companies delivering products that comply with specific expectations of the customers o
consumers. This has been portrayed in Figure 2.1.

2. Quality view points

According to Evans and Lindsay (1996), the concept of quality is often confusing becaus
people consider quality with different criteria. They distinguished five criteria, *i.e.* judgmenta
product-based, user-based, value-based, and manufacturing-based criteria:

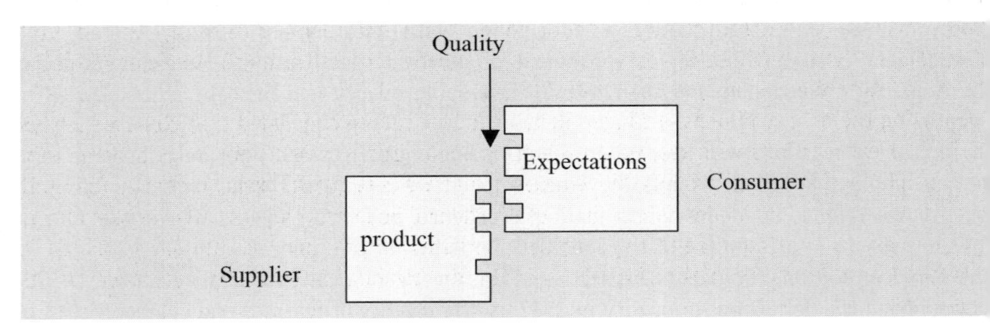

Figure 2.1. Zip model expressing quality (Van den Berg and Delsing, 1999).

From a *judgmental* point, quality can be considered as synonym of excellence or superiority. From this viewpoint quality is loosely related to a comparison of product characteristics, it is (sometimes) more a quality image created by marketing. Typical examples are Rolex watches and Cadillac automobiles, but also Coca-Cola and Nestlé coffee are considered as quality products mainly due to their brand name. Of course these products must comply with consumer demands, but the brand is a kind of guarantee for quality.

From a *product-based* view, quality can be defined as a function of a specific, measurable variable. Differences in quality are thus reflected in a quantitative difference of a certain variable. Quality from this point of view is often associated with price; the higher the price, the better the product.

The *user-based* definition of quality involves the presumption that quality is determined by what a customer wants, in short 'fitness for intended use'.

In the *value-based* criteria, the usefulness or satisfaction can be related to the price of the product. From this point of view a quality product is one that is as useful as competing products and is sold at a lower price or one that offers greater usefulness or satisfaction at a comparable price.

The fifth definition is the *manufacturing-based* one. Here, quality can be described as the desirable outcome of engineering and manufacturing practice, or conformance to specifications. These specifications include targets with tolerances, as specified by the designers of products and services.

Evans and Lindsay (1996) suggested that the criteria used for defining quality depend on one's situation in the production-distribution cycle. Customers generally view quality from the judgmental or product-based perspective. Marketing people are focused on 'consumer needs' and often consider quality from the user-based view. The value-based definition is most useful for product designers who balance between performance and costs to meet marketing objectives. Conformance to product specifications is the major goal in production, and therefore for production personnel the manufacturing-based definition of quality is most practical (Figure 2.2).

3. Quality dimensions

Several authors attempted to define factors, attributes or dimensions, which were assumed to be relevant for the quality perception of a product. Garvin (1984) defined eight principal quality dimensions for (capital) goods. Evans and Lindsay (1996) proposed important dimensions for service quality. Although both authors use different terms, some similarities in meaning can be noticed, e.g. performance and accuracy, conformance and completeness, and serviceability and responsiveness, as shown in Table 2.1. In addition, typical quality dimensions for services are timeliness, accessibility and convenience, whereas aesthetics and perceived quality are typical quality dimensions for manufactured capital goods. The eight quality dimensions for (capital) goods are not just applicable to agri and food products, but an attempt has been made to 'translate' these dimensions as shown in Table 2.1.

Considering the quality dimensions listed in Table 2.1, it can be concluded that consumers have distinctive expectations with respect to (capital) goods, services and agri and food products. For example, agri and food products have aspects that are very typical as compared to (capital) goods and services, like hidden safety risks, shelf life, and healthy aspects. Moreover, consumers have a very critical and emotional attitude towards safety and quality of food. As a consequence, often very high demands are put on product quality and assurance of its production.

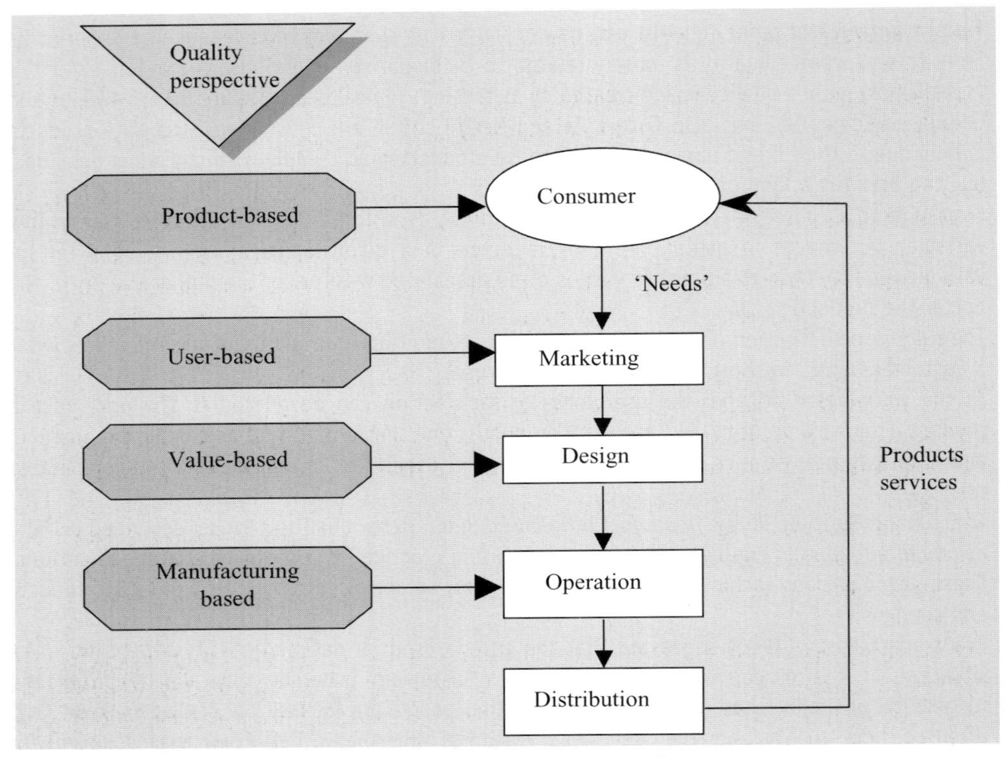

Figure 2.2. Different quality viewpoints (Evans and Lindsay, 1996).

4. Intrinsic and extrinsic quality attributes

Based on the typical aspects of agri and food products we also have proposed a quality concept. This quality concept is focused on technological attributes and factors that can contribute to product quality performance (Figure 2.3). The proposed concept fits within the extended quality triangle (Figure 1.2); it illustrates aspects determining product quality. It is assumed that agri-food products, as such, have no quality. A product has physical features that are turned into quality attributes by the perception of a consumer. With respect to agri-food products, quality perception appeared to be affected by different types of attributes. Relevant attributes for consumers involve safety, nutritional value (health aspect), sensory properties (like taste, flavour, texture, and appearance), shelf life, convenience (e.g. ready-to-eat meal) and product reliability (correct weight, right composition etc.). These attributes can be defined as *intrinsic attributes* and are directly related to the physical product properties. *Extrinsic attributes* refer to production system characteristics and other aspects, like environmental impact or marketing influence. They do not necessarily have a direct influence on physical properties but can affect acceptance of products by consumers. For example, the use of pesticides, of antibiotics to improve animal growth, or the application of biotechnologies to modify product properties, can have a significant effect on food acceptance. These intrinsic and extrinsic attributes and the factors (parameters) in the food production chain that can influence these attributes are reflected in Figure 2.3. The different aspects of this model will be explained in detail in the next sections.

Table 2.1. Quality dimensions for (capital) goods and services in relation to food.

(Capital) goods (Garvin, 1984)	Services (Evans and Lindsay, 1996)	Food
• Performance: primary operating characteristics	• Accuracy: correct performance of the service	• Physical product features: like sensory properties such as taste, odour, texture
• Features: additional properties, the 'bells and whistles'	• Courtesy: friendliness and politeness of employees	• Additional features: e.g. convenience of ready-to-eat meal
• Conformance: degree to which physical and performance characteristics of a product comply with pre-established standards	• Completeness: correctness of the delivery of items • Timeliness: completion of the service in the agreed time	• Product safety: are there no risks for the consumer of foods; consumers implicitly expect a safe product
• Durability: amount of use one gets from a product before it physically deteriorates or needs repair		• Shelf life: agri and food products usually have a restricted storage life
• Reliability: probability of a product to survive over a specified period of time under stated conditions of use	• Expertise and customisation of service personnel	• Reliability: e.g. does the indicated content on the package complies with the actual content?
• Aesthetics: how a product looks, feels, or sounds		• Appearance colour, size, image
• Serviceability: speed, courtesy, and competence of repair	• Responsiveness: quick response of service personnel to unexpected problems	• Complaint service: quick response to defect food products or complete recalls
	• Accessibility and convenience: the easiness of obtaining the service or information • Time: before service is delivered or completed	• Availability: of the product at food service market
• Perceived quality: subjective assessment of quality resulting from image, advertising, or brand names		• Perceived quality: is also relevant for food applications. For example advertisement or brands (marketing) can have a considerable influence on quality perception. • Price of the product

Several other authors also defined intrinsic and extrinsic attributes (Steenkamp, 1989, Van Trijp 1996, Poulsen *et al*, 1996). Intrinsic attributes were described as characteristics that are part of the physical product, whereas marketing efforts mainly determined extrinsic attributes. According to this classification, typical intrinsic attributes included appearance, colour, shape,

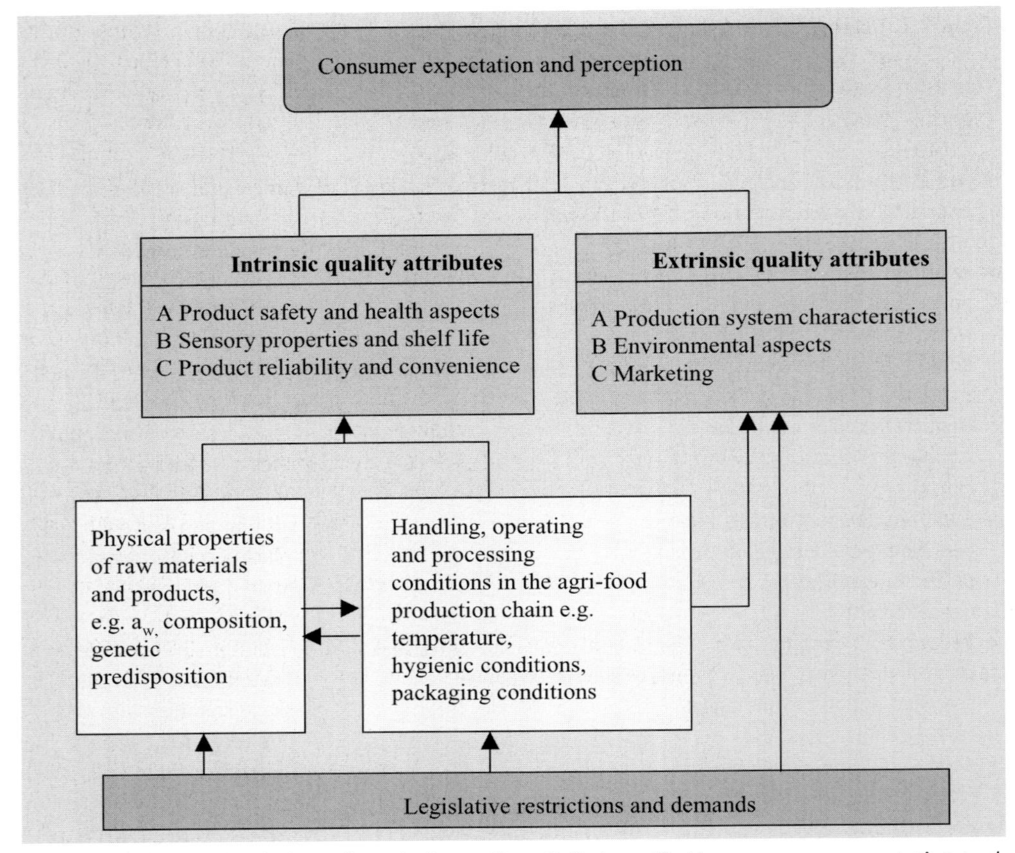

Figure 2.3. General intrinsic and extrinsic quality attributes affecting consumer expectation and perception.

and texture. Typical extrinsic attributes are price, brand name, packaging, labelling, shop, product information etc. According to this classification extrinsic attributes are mainly related to marketing variables, whereas in our model typical food aspects like production characteristics and environmental impact are included.

5. Expectation hierarchy

Van den Berg and Delsing (1999) proposed a hierarchy of expectations. Firstly, a product must be safe and not spoiled, then other attributes, such as taste and convenience become important. However, this hierarchy is not static but dynamic and can be different for different products.

6. Quality guidance model

Van Trijp and Steenkamp (1998) developed a more complex model on quality expectation and experience, which is called quality guidance model. According to their concept, quality perception is formed at two different moments. When the product is purchased, consumers form a quality expectation, whereas upon consumption of the product the actual quality is

experienced. Quality expectation is an important factor in consumer choice behaviour, while quality experience is important for repetitive purchase. In this model, it is assumed that judgement of expected quality could be affected by quality cues. Cues are informational stimuli that can be verified by consumers through their senses prior to consumption. Intrinsic cues are described as part of the physical product, like colour and appearance. Extrinsic attributes are related to the product but are not part of it, like price, brand or store name. These cues have a predictive value for the consumer and thus influence the quality expectation. Steenkamp (1989) found that intrinsic cues are usually weighted more heavily than extrinsic cues.
Steenkamp (1990) also proposed that the actual quality experience is based on the integration of perceptions of quality attributes, whereby a distinction between credence and experience quality attributes was made. Experience quality attributes are those that can be checked by consumers based on actual consumption, like taste, flavour, texture etc. Credence attributes cannot be verified through personal experience, such as the absence of additives, healthiness or environmental friendliness. According to Van Trijp and Steenkamp (1998), these two types of attributes might have a different impact on overall judgement of quality performance.
Jongen (1998) slightly modified the model for food production systems. He added production system characteristics as a factor influencing quality expectation (Figure 2.4). For example, in the last years the numerous problems in the meat chain like mad cow disease (BSE), the dioxin issue and the outbreak of foot and mouth disease, had a large effect on meat consumption. Consumers became more concerned about breeding and meat production conditions. As a consequence, a (temporarily) shift was observed towards consumption of biologically produced meat and meat replacers. Production system characteristics do not necessarily effect physical

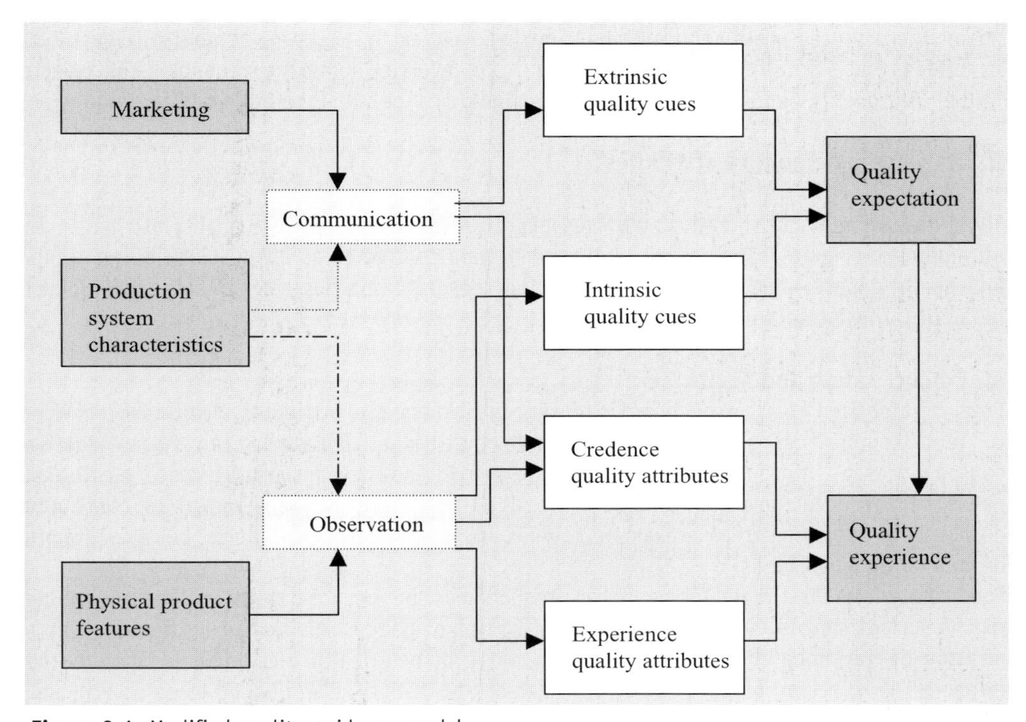

Figure 2.4. Modified quality guidance model.

attributes, *e.g.* use of pesticides will not directly influence product features but might affect quality expectation. However, a change in production system characteristics can influence physical features as well. In a case study in the biological meat production chain, it was shown that the fact that meat was biologically produced positively influenced quality expectation. However, appearance (intrinsic quality attribute) was negatively judged and overruled the positive effect of the production system characteristics. These results support the findings of Steenkamp (1989), where consumers differently weighted intrinsic and extrinsic cues.

Finally, it has to be mentioned that consumer expectations towards quality attributes are often not explicitly expressed. In fact, consumers do not work with accurate specifications (Jongen, 1998). Certain quality attributes, like product safety or packaging integrity are so self-evident that consumers are not aware of their expectations towards these attributes. These expectations are called implicit expectations. Expectations are explicit when consumers consciously choose a product based on specific quality attributes. For example, consumers explicitly choose the delicate taste of hand-made paté or the healthiness of orange juice with additional vitamin C. Nevertheless; implicit expectations can become explicit. For instance, as a consequence of the commotion around the mad cow disease since 1997 and the possible health risks, consumers explicitly chose other 'safe' meat alternatives.

2.2 Quality attributes of food

A clear overview of attributes that influence quality perception of food is an essential part of the techno-managerial approach. Moreover, to control and assure quality a good understanding is required of factors and parameters, which affect these attributes in the agri-food production chain. The proposed quality concept about intrinsic and extrinsic quality attributes (Figure 2.3) is used as basis for the next sections, giving an overview of major technological aspects influencing quality perception; whereas for technological details is referred to literature.

2.2.1 Intrinsic quality attributes

There are different classifications with respect to intrinsic and extrinsic quality attributes. In our approach (Figure 2.3), the following aspects were considered as intrinsic attributes: safety and health aspects of a product, shelf life and sensory properties, convenience and product reliability. These topics are described below.

A. Product safety and health

Product safety and health aspects are important intrinsic quality attributes. Health aspects refer to food composition and diet. For example, nutritional imbalance can have negative consequences on human health. Nowadays, the food industry anticipates these nutritional needs by the development of functional foods. These products are assumed to contribute positively to human health. For example, low-fat and low-cholesterol products, but also vitamin or mineral enriched foods.

Food safety refers to the requirement that products must be 'free' of hazards with an acceptable risk. A hazard can be defined as a potential source of danger. A risk can be described as a measure of the probability and severity of harm to human health. According to Shapiro and Mercier (1994), a food product can be considered safe if its risks are judged to be acceptable.

Different sources can affect food safety, *i.e.* growth of pathogen micro-organisms, presence of toxic compounds, physical agents and occurrence of calamities. Negative health effects can have a different time span. Some can be acute, such as allergic reactions or food poisoning; whereas on the long-term, chronic effects which can occur are cancer, heart and vascular diseases. These chronic effects can be due to, amongst others, unhealthy balances of diet or long term exposure to chemical agents.
An overview of major hazardous agents is presented in Table 2.2. Each group is described in more detail in the next sections.

Pathogenic micro-organisms
Pathogenic micro-organisms include both bacteria and moulds. With respect to these pathogen micro-organisms a distinction must be made between food infection and food poisoning (Table 2.2).

Food infection is caused by the presence of living pathogens in food. The pathogens are transferred trough food to the human (or animal) body. They can penetrate the intestinal mucosa and colonise (*i.e.* multiply) in the gastrointestinal tract or in other tissues. The symptoms consist of nausea, vomiting, abdominal pain, headache and diarrhoea. Major bacteria that are responsible for food infections are members of the genus *Salmonella*, *Shigella* ssp and certain strains of *Escherichia coli* (Jay, 1996), but also *Campylobacter jejuni*. Beef, turkey, chicken, eggs, pork and raw milk products are typical vehicle foods for Salmonellosis outbreaks. Poor personal hygiene is a common factor in foodborne Shigellosis, with shellfish, chicken, salads, fruits and vegetables being prominent among vehicle foods. Traveller's diarrhoea is often caused by specific strains of *Escherichia coli* (Jay, 1996).
Another source of food infection is foodborne listeriosis caused by *Listeria monocytogenes*. Listeria species are widely distributed in soils, animal faeces, sewage, silage and water. Listeria infections also occur in foods, especially raw milk, soft cheeses and other dairy products. It is suggested that Listeria infections in some foods might be due to zoonotic transmissions, *i.e.* disease transmission from animal to human (Hird, 1987). Listeriosis forms mainly a risk for pregnant women, young children and individuals with a suppressed immune system (for example by AIDS, diabetes etc.).
Campylobacter jejuni is assumed to be the most common cause of acute bacterial diarrhoea in humans. It is not an environmental organism but the bacteria are carried through the animal intestinal tract. A large percentage of all major meat animals contain *C. jejuni* in the faeces, especially poultry (Jay, 1996).

Food poisoning is caused by toxic compounds produced by pathogenic bacteria (enterotoxins) and moulds (mycotoxins) in the food. The compounds can be released in raw materials or processed food products. Consumption of the poisoned food can result in symptoms that range from acute abdominal pain and diarrhoea to long term diseases such as cancer or histological changes in the liver.
Clostridium botulinum, Staphylococcus aureus and Campylobacter jejuni are well known bacteria responsible for production of enterotoxins in food. *Clostridium botulinum* is an anaerobic spore-forming bacteria widely spread in soil and water. Especially food packed under low oxygen concentrations containing low-acid foods may function as vehicle food for *Clostridium* (*e.g.* canned foods, 'sous-vide' processed or modified atmosphere packed foods). Staphylococci have been detected in a large number of vehicle foods (like meat, chicken and bakery products).

Table 2.2. Overview of factors that can affect product safety, including examples.

Pathogenic micro-organisms		
Pathogenic bacteria	**Toxegenic moulds**	
FOOD INFECTION *E.G.*: • *Salmonella spp* • *Shigella spp* • *Escherichia coli* • *Listeria monocytogenes* • *Campylobacter jejuni* FOOD POISONING BY PRODUCED TOXINS *E.G.*: • *Clostridium botulinum* can produce highly toxic neurotoxins • *Staphylococcus aureus* may produce different enterotoxins	FOOD POISONING BY MYCOTOXINS *E.G.*: • *Aspergillus flavus* and *A. parasiticus* can produce aflatoxins • *Penicillium citrinum* may produce citrinin • *Penicillium patulum* and other penicillia may produce patulin • *Fusarium spp* can produce zearalenone and trichothecens, such as DON (deoxynivalenol)	
Chemical toxic compounds		
Natural toxins	Toxins formed during processing, storage, handling	Contaminants
• Protease inhibitors in beans, peas, potatoes and cereals • Glucosinolates in cabbage and related species • Cyanogenes in peas, beans, cassava, bitter almonds • Pyrrolizidine alkaloids in potatoes, tomatoes • Hydrazine in mushrooms • Hemagglutaninins in beans and peas	STORAGE AND HANDLING • Seafood poisoning PROCESSING *E.G.*, • Heterocyclic amines • Maillard reaction products • Formation of nitrite from nitrate at cooking	ENVIRONMENTAL CONTAMINANTS *E.G.*, • PCB's • Nitrate RESIDUES *E.G.*, • Pesticides, like DDT • Veterinary drugs, *i.e.* antibiotics and hormones • Disinfectants and detergents • Packaging compounds, such as terephtalic acid, BADGE
Foreign objects		
Glass, wood, metal pieces, stones, pests and insects etc.		

Moulds can also cause food poisoning; A well-known mycotoxin is aflatoxin, which can be produced by *Aspergillus flavus* and A. *parasiticus* (Table 2.2). Major vehicle products are peas and cereals such as rice, corn and wheat stored under warm and humid conditions.

The factors that often contribute to outbreaks of pathogen micro-organisms include (Jay, 1996):

- inadequate refrigeration;
- preparing foods far in advance of planned service;
- infected persons practising poor personal hygiene;
- inadequate cooking or heat processing;
- holding food in warming devices at bacterial growth temperatures.

Therefore, these factors should be considered when controlling the safety during production, storage, transport and preparation of food.

Toxic compounds

Toxic compounds can originate from different sources in the agri-food production chain (Table 2.2). Toxins can occur as natural compounds in raw materials (*i.e.* natural toxins), but they can also be formed during storage and processing. Other sources of toxic compounds are contaminants from the environment (like polychlorinated biphenyls -PCB's- and nitrate), residues from pesticides, veterinary drugs and disinfectants. In addition, consumers often consider food and colour additives as unsafe toxic compounds as well. However, the use of these additives is strictly regulated and the scientific dossiers are permanently reconsidered to guarantee humans safety.

To estimate the risk of toxicants for food safety, the following points have to be considered:

- The **origin** of the compound? For example, fresh raw mushrooms contain hydrazine, it is a natural toxicant. Other toxins are formed due to storage and transport conditions, such as aflatoxin, which formation is favoured by humid and warm storage. Toxic substances can also be introduced during processing, such as the development of heterocyclic amines in broiled, smoked or deep-fried fish or meat. The heterocyclic compounds are due to the reaction between creatinine with amino acids and sugars (Wolf, 1992; Shapiro and Mercier, 1994).
- What are the **properties** of the toxic compounds?
  - The *fat solubility* of the toxic compound is an important aspect. Fat-soluble toxic compounds can accumulate in the food cycle production. For example, small amounts of polychlorinated biphenyls (PCBs) accumulate in the adipose tissue of fatty fish species; by far the biggest source of PCBs in the human diet is fatty fish (Shapiro and Mercier, 1994).
  - Another important property is the *degradability or inactivation*. Some toxic compounds are very stable and therefore remain in the food production cycle for a long period. DDT, a very stable pesticide that has been used in the 40-50's, has accumulated in the food production cycle via milk fat, meat and breast milk. For many years, human breast milk contained higher amounts of DDT compared to cow milk.
  - The toxicity of the toxic compound itself and the toxicity of its degradation products also determine the risk. Compounds can be toxic at very low levels and can have a direct or indirect toxic effect. For example, the enterotoxin produced by *Clostridium botulinum* is highly toxic and the fatality rate is much higher than the enterotoxin produced by

Clostridium perfringens (Jay, 1996). Oxalates and alkaloids have an indirect effect; they can bind bivalent metals and thus effect their bioavailability. Sweet peas can contain lathyrogens (an unusual amino acid) that can cause directly skeletal and aortic abnormalities. The toxicity is assessed by approved tests, including tests to study the effects of toxic compounds on reproduction and teratogenicity, tests to assess mutagenicity and carcinogenicity, tests for allergic reactions and tests for immunity function. Based on these toxicity tests the acceptable daily intake (ADI) is assessed. The ADI is the daily amount of the substance that can be ingested during the whole life span without a negative effect on health.

- How can you **affect** the toxic compounds? Can the toxicant be *prevented*, *controlled* or *reduced*. Genetic engineering or plant breeding can be used to eliminate some undesirable substances from plants (prevention). Cultivation and storage conditions can also prevent or sometimes reduce the toxic compound. For example, the amount of glycoalkaloids in potatoes can be kept low by controlling cultivation, harvesting (proper ripeness stage and bulb size) and storage conditions (light, temperature and humidity). In some cases, the toxic compound can be reduced or eliminated by processing; for example, heating eliminates trypsin inhibitor.
- What is the **damage**? The spectrum of potential risks associated with toxic compounds in food: range from acute poisoning to the undefined risk of lifetime exposure to potential toxic substances. The extent to which these latter compounds contribute to the disease burden of humans cannot be accurately assessed (Wogan and Marletta, 1985).

Foreign objects

Foreign objects belong to the third type of safety influencing factors that can be distinguished. Foreign compounds includes physical objects like pieces of glass, wood, stones but also pests and insects.

Calamities are all other (unexpected) phenomenons that negatively affect food safety. A typical example, is the disaster with the nuclear plant in Tsjernobyl which was unexpected, but had large consequences on food safety.

B. Sensory properties and shelf life

Sensory properties and shelf life are also intrinsic quality attributes (Figure 2.3). The sensory perception of food is determined by the overall sensation of taste, odour, colour, appearance, texture and sound (*e.g.* the sound of crispy chips). The physical features and chemical composition of a product determine these sensory properties. In general, agri-food products are perishable by nature. After the harvesting of fresh produce or the processing of foods, the deterioration processes starts, which negatively affects the sensory properties. Processing and/or packaging are aimed at delaying, inhibiting or reducing the deterioration processes in order to extend the shelf life period. For example, freshly harvested peas are spoiled within 12 hours, whereas canned peas can be kept for 2 years at room temperature.

The shelf life of a product can be defined as the time between harvesting or processing and packaging of the product and the point at which it becomes unacceptable for consumption. The unacceptability is usually reflected in decreased sensory properties, for example formation of rotten odour or sour taste by bacteria spoilage. The actual shelf life of a product depends on the rate of the deterioration processes. Often one type of deterioration process is limiting for the shelf life; for example, cured ham can colour grey very quickly upon exposure to oxygen. Although the product is still safe, because bacteria did not spoil it, it will become unacceptable because of the grey colour (*i.e.* the shelf life limiting reaction). Below are described some major shelf life limiting deteriorating processes typical for food products.

Shelf life can be restricted by microbiological, and/or (bio) chemical and/or physiological and/or physical processes (Fennema and Tannenbaum, 1985; Singh, 1994; Jay, 1996), as shown in Table 2.3.
Microbiological processes in foods can result in food spoilage with the development of undesirable sensory characteristics, including loss of texture, development of off-flavours and off-colours. In some situations the food contains pathogens; the food can become unsafe prior to detecting any changes in sensory characteristics.
Typical **chemical reactions** that can limit product shelf life are non-enzymatic browning and oxidative reactions. Non-enzymatic browning (or Maillard reaction) causes mainly changes in appearance and lowering of nutritional value by loss of essential amino acids. Oxidative reactions, especially autoxidation of lipids can alter the flavour. Bleaching of plant pigments (*e.g.* carotenoids) can be due to autoxidation as well. Generally, chemical changes occur during processing and storage of agri-food products. Finally, the non-enzymatic browning reaction also leads to desirable quality characteristics, such as browning of bread crust and the brown colour of fried meat.
Biochemical reactions involve enzymes. Generally enzymes are compartmentalised, which means that the enzymes are separated from their substrate on subcellular or tissue level. Disruption of the integrity of the plant or animal tissue leads to enzymatic reactions. For example, cutting of fresh vegetables initiates several enzymatic reactions, such as browning by phenolase and formation of off-flavours by lipoxygenase. In some instances biochemical reactions are controlled and used to produce better digestible food, *e.g.* fermentation of cabbage. A typical example is accelerated ageing of meat by increased hydrolase activity. This process is accomplished by exposing the meat to higher than normal temperatures while surface growth of bacteria is controlled by ultra violet light (Richardson and Hyslop, 1985).
Physical changes are often due to mishandling of agri-food products during harvesting, processing and distribution. For example, bruising of fruits during harvesting and post-harvest storage handling favours the development of rot. Mishandling of manufactured products during distribution might result in high numbers of broken or crushed products. During storage and distribution, fluctuating temperature and humidity conditions can, depending on the conditions, result in desiccation of humid products, swelling of dried products or phase changes (Singh, 1994). The breaking of emulsions and phase separation are also typical physical processes resulting in negative product features.
Physiological reactions commonly occur during post-harvest storage of fruits and vegetables and strongly depend on storage conditions. The products still have a respiration rate and often ethylene is produced, which can have a considerable effect on typical post harvest defects (see also section on influencing factors).

The shelf life of a product is often limited by one major reaction, but sometimes a typical quality defect may be due to different mechanisms. For example, rancid off-flavours can be due to lipase activity producing short fatty acids chains or by oxidation of fatty acids. It is therefore necessary, to be able to control this quality defect, to first identify the responsible mechanism. Often the fastest reaction is responsible for the shelf life limitation. For example, undesirable changes in texture of bread usually occur prior to growth of moulds. In this case physical degradation is faster than microbiological changes; degradation is thus the shelf life limiting

Table 2.3. An overview of major shelf life limiting reactions in agri-food products.

Shelf life limiting (changes)	Causes or type of reaction	Undesirable effects
Microbiological changes	Growth of spoilage micro-organisms; typical spoilage bacteria, moulds and yeast's:	Common effects: loss of texture, development of off-flavours, off-taste and off-colours, formation of slime, rotting
	• Acinetobacter (fresh meat, poultry)	
	• Aeromonas (fresh meat)	
	• Erwinia (fruit and vegetables)	
	• Pseudomonas (fresh meat , poultry)	
	• Cladosporium (fresh/frozen meat)	
	• Rhizopus (meat, vegetables)	
	• Candida (fresh/frozen meat)	
	• Lactic acid bacteria (various food)	
Chemical reactions	Non-enzymatic browning (Maillard reaction)	• Browning, *e.g.* dried milk powder • Formation of toxic compounds
	Oxidation reactions	• Formation of rancid off-flavours bv lipid autoxidation • Bleaching of carotenoids by autoxidation
Biochemical reactions	Enzymatic reactions; typical enzymes:	
	• Phenolase	• Browning on cut surfaces of light coloured fruits and vegetables
	• Milk proteinase	• Gelation in milk subjected to ultra-high short-time (UHST) temperature processing
	• Lipases	• Short-chain fatty acids in milk lead to undesirable hydrolytic rancidity; the fatty acids formed can also lead to other undesirable flavours
	• Lipoxygenase	• Degradation products of the polyunsaturated fatty acids can have undesirable effects on flavour, colour, texture and nutritional value
	• Chlorophyllase	• Released fatty acids are substrates for lipoxygenase leading to off-flavours in affected plant foods or undesirable textural changes in frozen fish
	• Phospholipase	• Degradation of green chlorophyll pigments during post harvest storage
Physical changes	Mishandling of agri-food products	
	• Bruising or crushing	• Development of rot during post harvest storage of fruits and vegetables
	• Temperature fluctuations	• Phase changes *e.g.* due to thawing and re-freezing of foods
	• Humidity conditions	• Undesirable desiccation or moisture pick-up
	Other typical physical reactions	
	• Retrogradation of starch	• Undesirable texture changes of starch containing bakery products
Physiological reactions	Processes in respiring agri-foods	
	• Respiration process	• Accelerated ripening by ethylene or by incorrect storage conditions
	• Post harvest defects by ethylene	• Typical effects like bittering of sprouts, browning of leave nerves
	• Chilling injury	• Some fruits and vegetables get brown spots upon storage at too low temperatures

factor. Inhibition, reduction or prevention of the main shelf life limiting factor often results in an extended shelf life for that specific factor. However, during the extended storage slower degradation processes may become prominent. For example, freezing of foods extends the microbial-free shelf life but after 1 to 1 ½ year colour and texture changes occur by chemical and physical reactions (Ellis, 1994).

To control the technological product quality it is important to understand the different processes that limit product shelf life and affect sensory properties.

C. Product reliability and convenience

Two other intrinsic quality attributes, shown in Figure 2.3, are product reliability and convenience.

Product reliability refers to the compliance of actual product composition with product description. For example, the weight of the product must be correct within specified tolerances. But also claims, such as enriched with vitamin C, must be in agreement with actual concentration in the product after processing, packaging and storage. Deliberate modification of the product composition will cause damage to the product reliability, *i.e.* product falsification. An example is when (cheaper) alternative raw materials are used and not mentioned on the label. Product reliability is generally an implicit expectation, consumers just expect that a product is in compliance with the information mentioned on the packaging.

Convenience relates to the ease of use or consumption of the product for the consumer, and thus contributes to product quality. Product convenience can be accomplished by preparation, composition and packaging aspects. Convenience food has been defined as food offered to the consumer in such a manner that purchase, preparation and consumption of a meal costs less physical and mental effort and/or money than when the original and/or separate components are used.

The increased consumption of convenience food is assumed to be due to decreasing volume of households, decreasing appreciation for housekeeping, increasing labour participation of women and increasing welfare (Van Dam *et al*, 1994). Convenience foods range from sliced and washed vegetables to complete ready-to-eat meals that only have to be warmed in the microwave or oven. Much attention is paid in the food industry to development of ready-to-eat meals that can be easily and quickly prepared while having good sensory and nutritional properties. Also packaging concepts are more and more designed to fulfil the consumer's need for convenience; typical examples are easy-to-open and close, good pouring properties and extra light packaging.

2.2.2 Extrinsic quality attributes

Production system characteristics, environmental implications of food products and their production and marketing aspects can be considered as extrinsic quality attributes (Figure 2.3). Extrinsic quality attributes do not necessarily have a direct influence on physical product properties, but can influence consumers' quality perception. For example, marketing activities can influence consumer expectations but have no relationship with any physical property.

A. Production systems characteristics

Production characteristics refer to the way a food product is manufactured. It includes factors such as the use of pesticides while growing fruit and vegetables, animal welfare during cattle breeding, use of genetic engineering to modify product properties or use of specific food preservation techniques. The influence of production systems characteristics on product acceptance is very complex. For example, there has been much concern about public acceptance of new genetically modified food products and genetic engineering in general. Nevertheless, Frewer and co-authors (1997) indicated in their study on the consumer's attitude towards different food technologies used in cheese, that product benefits (such as product quality, animal welfare, environment) was a more important factor in their decision than process considerations. However, they emphasised that the results could not be merely translated to other applications of genetic engineering in food production; each application should be considered on a case-by-case basis.

B. Environmental aspects

Environmental implications of agri-food products refer mainly to the use of packaging and food waste management. Wandel and Bugge (1997) proposed that intrinsic quality properties such as taste or nutritional value are related to personal interests, whereas environmental properties of food may be related to wider community oriented interest. Consumers may express interest in buying foods from environmentally sound production, either because of concern for their own health or because of concern for the external environment. With respect to the environmental consequences of packaging waste, European directives have been enacted to reduce this environmental impact. Since 1997, the food packaging industry is legally liable to improve material recycling and to reduce the amount of packaging materials, so reducing packaging waste. In response to this legislation, one of the largest retailers in the Netherlands stimulates their packaging suppliers to produce according to this Packaging Convenant. This was not only because of legal requirements but also because consumers become increasingly environmentally concerned.

With respect to waste management, inefficient processing is mainly a cost problem for processors and not yet a major quality concern of consumers. For example, in fresh produce processing, recoveries vary from 47 to 52% of finished product yields; the product loss can be found on the processing floor. Hurst and Schuler (1992) suggested that more emphasis should be put on waste management, to improve waste reduction methods and thus decrease the environmental wasteload.

Though citizens may be concerned about environmental issues, it can not be assumed that behaviour has changed accordingly. This is particularly the case when the product represents a conflict between environmental soundness and other perceived benefits such as convenience, performance, various other quality attributes and price (Alwitt and Berger, 1993; Wandel and Brugge, 1997). Moreover, although consumers pretend to be interested in foods produced in an ecologically sound manner and with less environmental impact, only a small group is willing to pay the present high prices for these products (Wandel and Brugge, 1997). In fact as citizens, people are concerned with social and environmental issues, however as customers the economic aspects are often prevalent. However, currently in the Netherlands the major retailers are discussing the purchase, on large scale, of environmentally sound produced meat. The large-scale purchase may lower the price and thus stimulate the consumers to purchase.

C. Marketing (communication)

The effect of marketing on product quality is complex. According to Van Trijp and Steenkamp (1998) consumers form an impression about the product's expected fitness for use at purchase;

this is the consumer's judgement of 'quality expectation'. In their quality guidance model they proposed that marketing efforts (*e.g.* communication via branding, pricing and labelling) determine extrinsic quality attributes, affecting quality expectation. However, marketing can also affect credence attributes (which can not be checked by consumers themselves), influencing quality experience. This model has been explained in more detail in section 2.1 (Figure 2.3).

2.3 Factors affecting physical product features in the agri-food chain

The manufacturing of agri-food products aims at extending shelf life by controlling the typical shelf life restricting factors and so assuring food quality. With respect to control from a technological perspective, it is of vital importance to understand where in the agri-food chain and how intrinsic and extrinsic quality attributes are affected. In which stages of production can potential hazards occur? Is it possible to remove these hazards, or are preventive measures needed to obtain a safe food product? For example, processing can eliminate many pathogens and spoilage micro-organisms, whereas heat-stable toxins, many environmental contaminants and various residues cannot be removed by processing. Their occurrence should be kept low by preventive measures as early in the food production chain as possible. Figure 2.5 gives an overview of the characteristic food chain. Different conditions in the sub-chains that can affect quality attributes are shortly described below.

2.3.1 Animal production conditions

Animal production can be divided in meat production, *i.e.* pork, beef, poultry, sheep, fish and shellfish, and animal products like eggs and milk. Production conditions can have a direct or indirect effect (*e.g.* via milk) on intrinsic quality attributes, such as food safety and sensory properties. Moreover, the conditions applied determine the production system characteristics and thus contribute to extrinsic quality. Major aspects involved in animal production are choice of breed, feeding, living conditions and animal health (Figure 2.5).

Choice of breed

Most breeding programmes are more focused on yield increase than improving product quality. For example, the breed choice of cows has a profound effect on milk yield but has less effect on milk composition (*i.e.* nutritional quality); the heavier breeds tend to produce more milk (Nickerson, 1995). However, some pig breeds have a genetic predisposition for typical quality parameters of meat. Like, the Duroc breed that is associated with darker, redder muscle colour, increased fat firmness and increased tenderness, although carcass conformation may be poorer (Edwards and Casabianca, 1997). These breeds are often crossed with other breeds to obtain optimal characteristics including meat quality. However, attention is shifting towards varieties with high quality meat characteristics. A typical example is the Wagyu breed, which gives meat with a marbled fat structure and has a delicate taste. This type of meat is a delicatessen in Japan, but is now gaining attention in Europe as well.

Animal feeding

Animal feeding can affect food quality in different ways, *i.e.* directly or indirectly. It can directly influence nutritional value by affecting composition of the product. For example, feed composition is a key factor influencing milk yield and composition (*i.e.* mainly lipid content).

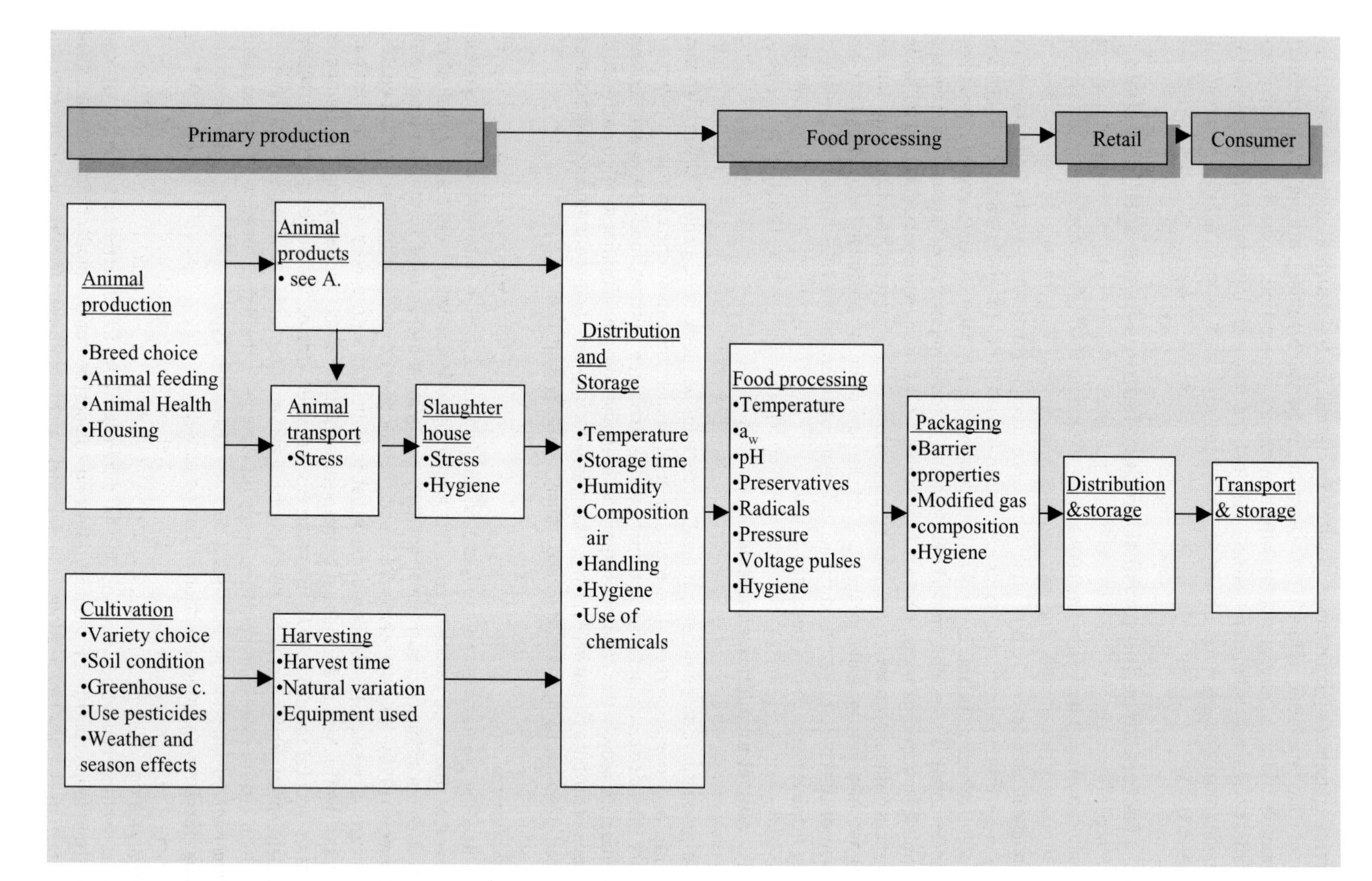

Figure 2.5. Quality affecting factors in the agri-food chain.

Precursors required by the mammary gland to synthesise fat are generated during fermentation of feedstuffs in the rumen. Therefore, diets that change fermentation can affect fat content of the milk. Starch is also needed to maintain microbial digestion and subsequent protein synthesis; both are positively correlated with yield and percentage of milk content (Nickerson, 1995).
The quality of the animal feeding itself, *i.e.* presence or absence of Mycotoxins or environmental contaminants, can have an indirect effect on safety of the final food product. For instance, aflatoxin-containing forage fed to dairy cows, can be hydroxylated to a metabolite of aflatoxin. About one-fiftieth of the ingested aflatoxin is carried over to milk as the hydroxylated metabolite (Heeschen and Harding, 1995).

Housing conditions
The *housing conditions* of animals determine to a certain extent the bacterial load on their exterior surfaces. In general, the cleaner the housing conditions the lower the load. However, to achieve a high bacteriological quality at farm level it is important for farmers to know the sources of contamination and to understand how to control them (Harding, 1995). Large numbers and a great variety of micro-organisms are found on the exterior surface (*e.g.* skin or teats) and in the intestinal tract of cattle, sheep and pigs; pigs generally have higher counts of micro-organisms than cattle (Forsythe and Hayes, 1998).
For meat production, both the exterior and interior load are important factors for food safety. Although the underlying tissue of the slaughter animals is assumed to be sterile, high bacterial loads on the interior and/or exterior surfaces can lead to contamination of other animals during transport or contamination of the sterile meat during slaughter.
For milk production, hygienic preventive measures must be taken like cleaning of teats, cleaning and sterilising of milk equipment and exclusion of milk from mastitic cows.
For fish, the bacterial flora on the skin and gill surfaces are affected by variations in marine environments. Psychrotrophic gram negative bacteria dominate in cooler seas, whereas in warmer seas mesophiles and micrococci are the major spoilage bacteria (Forsythe and Haynes, 1998).

Another aspect of housing is the *stocking ratio*, *i.e.* intensive versus extensive (free-range) housing. Although differences in product quality were expected, only a few studies reported improved meat tenderness from outdoor reared animals (Dufey, 1995). In a sensory evaluation study it was shown that when panellists were aware of the origin of the animals, they rated free-range pork as juicier, less bland, more tender and more pleasant. However, when the panellists were unaware of the origin, no significant differences were observed in sensory attributes (Oude Ophuis, 1994). Peoples' perception of the merits of a production system seems to influence their perception of the quality of the product produced by that system (Edwards and Casabianca, 1997).

Animal health
Animal health and use of veterinary drugs can also influence product quality. For example, occurrence of mastitis (*i.e.* inflammation of the mammary gland and disturbance of milk secretion) leads to alterations in the composition and chemical-physical properties of the milk. The number of somatic cells is an indicator of the qualitative and hygienic properties of milk and reflects the mastitic situation of the given animal (Heeschen, 1998). Animal diseases are often treated with antibiotics and their residues in animal products (milk, meat) are assumed

to be hazardous for human health. For protection of consumers, the European Union (1990) and the Codex Alimentarius Commission (1995) have set maximum residue limits of veterinary drugs in foods of animal origin.
Additionally, there is concern about the addition of antibiotics to feed to accelerate growth of the animals. Animal pathogens may become resistant to the antibiotics and resistant pathogens might be transferred from animals to human. There is a potential risk that these resistant pathogens cannot be treated with human antibiotics and this might have large consequences for human health.
Whereas antibiotic residues can have a direct effect on food safety, the use of veterinary drugs in itself can also be an important extrinsic quality attribute, since the production technique can affect the consumers' acceptation.

2.3.2 Animal transport and slaughter conditions

Transport and slaughter conditions can affect intrinsic attributes like sensory properties (*e.g.* texture), food safety (contamination by pathogens) and microbial shelf life (initial contamination).

Stress
Stress factors such as exercise, fear, hot and cold temperatures can negatively affect meat quality during transport and handling of slaughter animals (Figure 2.5). Stress can result in different quality defects depending on the type of animal. For example, meat of pigs exposed to stress has a soft texture, poor water-holding capacity and a pale colour, also called pale-soft-exudative meat (PSE). Due to stress, glycolysis is stimulated, which results in too rapid pH decrease while cooling is not yet completed. As a consequence, the sarcoplasmic proteins degrade to contractile proteins and thus modify physical properties. In cattle, too high post mortem pH due to antemortem stress can result in meat with dark, firm and dry characteristics (DFD meat). Moreover, the more neutral PSE and DFD meat (*i.e.* pH not low enough) is also more susceptible to microbial spoilage (Hultin, 1985).
Measures to prevent or reduce stress during transport and handling include:

- proper **loading density**; a too high density results in PSE meat for pigs, DFD meat for cattle and serious transport-induced hematomas in musculature (Tarrant and Grandin, 1993; Troeger, 1996)
- **loading** and **unloading facilities**, *i.e.* angle of inclination of ramps. Steep slopes result in high heart rates thus stress. Moreover, pitchforks are often used as driving agents, which leads to carcass rind or even to bleeding in the back fat, which eventually has a negative effect on meat quality (Troeger, 1996)
- **duration of transport** can also negatively effect meat quality. Too short transportation times of pigs can increase the number of carcasses showing PSE characteristics. Longer transportation times probably lead to habituation and calming, normalising the metabolic disturbances due to loading (Fortin, 1989)
- **mixing of animals** unknown to each other at the slaughterhouse has been shown to be particularly stressful and results in DFD or PSE meat. It has proved worthwhile in practice to get the animals into individual passageways with limited inclination (Warriss *et al*, 1984).

Slaughter conditions
The slaughter procedure includes many steps like killing, bleeding, scalding, skinning and evisceration, during which the sterile underlying muscle tissue can be contaminated by *e.g.* the intestinal tract, the exterior surface, hands, knives and other utensils used. Total bacterial counts for freshly cut meat surfaces are likely to vary between 10^3 and to 10^5 organisms per cm^2. Reduction of the microbial load of freshly killed animals can be achieved to a certain extent by spraying the carcasses with hot water containing chlorine, lactic acid or other chemicals (Smulders, 1988; Forsythe and Hayes, 1998). Moreover, strict hygienic conditions in the abattoir such as proper cleaning of walls, floors, knives and other utensils reduces the risk for contamination.

2.3.3 Cultivation and harvest conditions of vegetable products

Cultivation and harvest conditions greatly influence properties of fresh produce and processed products, including nutritional composition, sensory properties (taste, odour, texture and colour), and contents of natural toxins, anti-microbial agents and anti-oxidants.

Important quality affecting factors during *cultivation*, as shown in Figure 2.5, include:
- selection of suitable varieties by plant breeding and/or genetic engineering
- cultural practices including sowing date, nutrient supply (*e.g.* use of nitrogenous fertilisers) irrigation and plant protection (*i.e.* use of pesticides and herbicides)
- environmental effects like temperature, day length and amount of rainfall (Feil and Stamp, 1993)

Quality requirements are strongly crop dependent. For example, the most important quality criterion for wheat is the grain protein content that correlates positively with baking quality and contents of essential amino acids. Weather conditions and a well-balanced nitrogen supply (Sanders *et al,* ?) mainly influence these quality parameters. For rape seed, for example, nutritional and technological quality is defined by the fatty acid composition of the oil. Since the fatty acid composition is genetically fixed, the plant breeders, and not the farmers, control to a large extent the quality of the oil (Feil and Stamp, 1993).

With respect to *harvest conditions*, time of harvesting and occurrence of mechanical injury during harvesting are factors that influence product quality. During growth and ripening of fruits and vegetables many biochemical changes take place, including:
- changes in cell wall constituents, *e.g.* softening of fruits
- starch-sugar transformations. For example, degradation of starch to simple sugars results in a sweet taste in bananas. In other commodities like potatoes, corn, and peas the synthesis of starch predominates
- pigment metabolism. The green chlorophyll pigments often disappear upon ripening while other pigments like carotenoids and flavonoids are synthesised
- formation of aroma compounds and their precursors during ripening (Haard, 1985; Salunke *et al*, 1991)

Most of these processes continue after harvesting. However, the time of harvesting can affect the further course of the processes. For examples, if green bell peppers are harvested too early then they will not turn to red upon storage and thus will not obtain the final quality characteristics.

Mechanical injury can occur upon harvesting and transport. Damaging of plant organs can result in:

- **wound healing**, *i.e.* formation of physical barriers such as a waxy barrier or lignin deposition in asparagus
- formation of **stress metabolites**. These metabolites appear to have a protective function for the plant but can have negative effects on food quality. For example, the stress metabolites, coumarins in carrots and glycoalkaloids in white potato tubers have a bitter taste. Moreover, some stress metabolites exhibit toxic properties or are precursors for undesired enzymatic reactions
- **enzymatic-catalysed browning** occurs upon mechanical injury and leads to undesired brown coloration of fresh produce
- **ethylene production** upon wound healing stimulates respiration and accelerates the ripening and senescence process leading to a shorter shelf life (Haard, 1985; Salunkhe *et al*, 1991)

The processes described above have become of increasing concern with the progress in mechanical harvesting techniques.

2.3.4 Food processing conditions

Physical properties of manufactured foods are determined by compositional characteristics of individual ingredients and/or raw materials (*e.g.* pH, initial contamination, presence of natural anti-oxidants), composite (*e.g.* addition of preservatives) and by processing conditions (*e.g.* temperature, pressure). Effects of different processes and handling on quality attributes from a technological perspective are presented in Table 2.4.

When considering current food preservation techniques, theseare limited to a relatively small set of parameters, including time and temperature (t-T), pH (*i.e.* acidity), a_w (*i.e.* water activity), use of preservatives, modification of gas composition or combinations (Figure 2.5). The use of a combination of preservation techniques is also called the hurdle technology. The newer food processing techniques use additional parameters such as high hydrostatic pressure, high voltage electric discharge, γ-irradiation and manothermosonication *i.e.* ultrasonication combined with increased temperature and raised pressure (Ohlsson, 1994;Gould, 1996; Van Boekel, 1998). Most of these newer techniques are still under development and are not yet commercially feasible. Major factors that influence physical product properties during food processing are described below in more detail.

Temperature-time

Elevated temperatures are applied to *reduce* number of micro-organisms, to *inactivate* enzyme activity and to *increase* (bio) chemical reactions.

Low temperatures are used to *inhibit* growth of micro-organisms or *delay* (bio) chemical and physiological reactions. For all these processes not only the level of temperature (T), but also the duration (t), determines the degree at which the process occurs. The temperature-effect profiles differ for each specific reaction. For example, bacteria, yeast and moulds species each have an optimum growth temperature; At high temperatures they are inactivated, whereas at low temperatures their growth is delayed. Likewise, enzymes have an optimum and an inactivation temperature profile. Chemical reactions, however, are accelerated at increasing temperatures; the higher the temperature the faster the reactions. For example, temperature

Table 2.4. Overview of effects of food processing conditions and handling on quality attributes.

Processing and handling	**Influencing factor**	**Quality effects**
Current food processing systems		
Heating (*e.g.* pasteurisation, sterilisation)	Time (t) - temperature [T] condition	• Shelf life extension and improvement of food safety by *inactivation* of micro-organisms • Modification of texture, flavour, colour by chemical and physical reactions
Freezing and cooling	Time (t) - temperature [T] condition	• Shelf life extension by *stabilisation*, *reduction* or *inhibition* of growth of micro-organisms and by delaying chemical, physiological and biochemical reactions
Curing and drying	Lowering a_w	• Shelf life extension and improvement of food safety by *stabilisation*, *reduction* or *inhibition* of growth of micro-organisms • Modification texture, flavour, colour by chemical and physical reactions
Fermenting	Decreasing pH Alcohol production	• Shelf life extension and improvement of food safety by *stabilisation*, *reduction* or *inhibition* of growth of micro-organisms • Modification of texture, flavour, colour by chemical and biochemical reactions
Irradiating	Radicals	• Shelf life extension and improvement of food safety by *inactivation* of micro-organisms • Initiation of chemical reactions (*e.g.* lipid oxidation) • Extrinsic quality factor is the poor acceptance by consumers of irradiation as processing technique
Extrusion	Time (t) - temperature (T) and pressure (p) conditions	• Shaping of products • Texture, colour and taste formation
Separation techniques (membrane separation)	Size of membrane holes	• Removal of cell particles • Removal of fat particles
Diverse manual handling (*e.g.* filling, closing, sorting, peeling etc.)	Hygienic conditions	• Hygienic conditions determine microbial contamination
Mechanical handling (*e.g.* washing, slicing, deboning, cutting)	Reduction of preparation time at home [t] Hygienic conditions	• Improving product convenience • Hygienic conditions determine microbial contamination

Table 2.4. Continued

Processing and handling	Influencing factor	Quality effects
Packaging (*e.g.* vacuum, modified atmosphere, active packaging)	Modified gas conditions, *e.g.* low oxygen, or extreme high oxygen conditions Oxygen barrier Moisture barrier	• Maintenance of processed foods or fresh produce • Shelf life extension by providing a barrier against physical and chemical contaminants, by lowering respiration velocity of fresh produce (physiological) and by delaying chemical reactions (*e.g.* lipid oxidation, decolouration)
New developments (Van Boekel, 1998)		
High pressure treatment	High hydrostatic pressure	• Shelf life extension and improvement of food safety by *inactivation* of micro-organisms and inactivation of enzymes (biochemical reactions) • Enhanced chemical reactions like lipid oxidation • Improvement of sensory properties by milder preservation conditions
High electric field pulses	High voltage pulses	• Shelf life extension and improvement of food safety by *inactivation* of micro-organisms; no spores inactivation

raise can result in loss of nutritional value (*e.g.* destruction of vitamins), formation of undesired (*e.g.* off-flavours) or desired (*e.g.* brown coloration of bread crust) sensory changes. The rate of temperature change (dT/dt) in a process determines the relative rates of competitive chemical reactions and the rate at which micro-organisms are destroyed. In food manufacturing there is a sustained effort in optimising time-temperature combinations to produce safe and stable foods with the least heat damage as possible (Fennema and Tannenbaum, 1985; Van Boekel, 1998). A schematic overview of time-temperature dependence of different reaction types is given in Figure 2.6.

Water activity (a_w)

Water activity is also an important factor in regulating growth of micro-organisms, enzyme activity and occurrence of chemical reactions. Fennema and Tannenbaum (1985) provided typical reaction rate-a_w relationships for several important reactions in the temperature range of 25-45°C (Figure 2.7). It is obvious that no microbial growth occurs at a water activity below 0.6, whereas lipid oxidation rate increases both at very low and at higher a_w values. It is common knowledge that foods protected against microbial spoilage by low water content are still susceptible to enzymatic spoilage (Richardson and Hyslop, 1985). The exact reaction rate, position and shape of the curve, can change depending on composition, physical state and food structure.Gas composition (especially oxygen) and temperature can also influence the a_w-profiles (Fennema and Tannenbaum, 1985).

Acidity (pH)

The pH is another factor in controlling bacteriological, enzymatic and chemical reactions.
Most micro-organisms grow best in the pH range 6.6-7.5 and a few grow below 4.0. The pH of food products range from 1.8 (limes) to 7.3 (corn). Bacteria generally have a smaller pH range

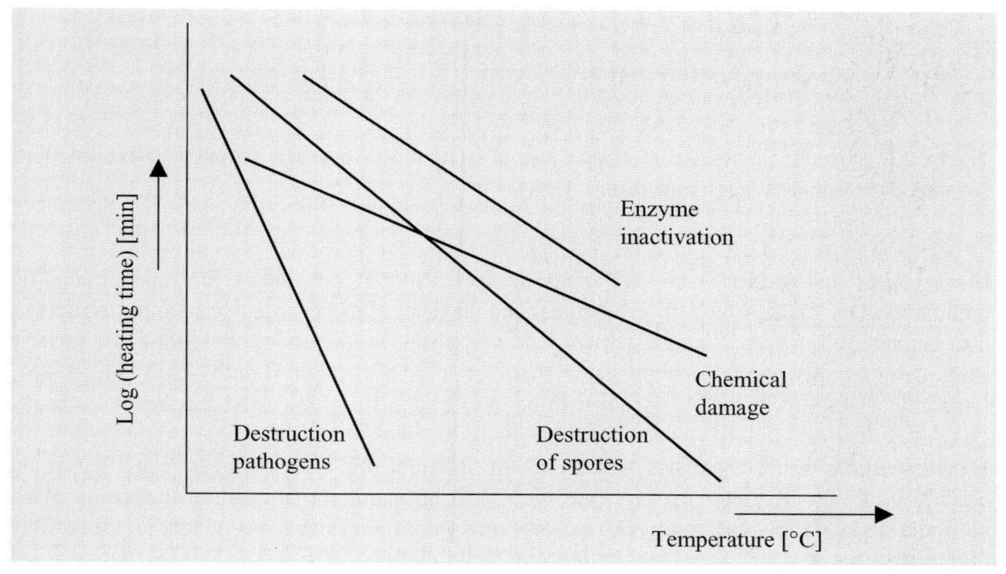

Figure 2.6. Schematic overview of destruction/damage lines for different reactions (microbial and (bio)chemical) at specific time-temperature combinations (adapted from Van Boekel, 1998).

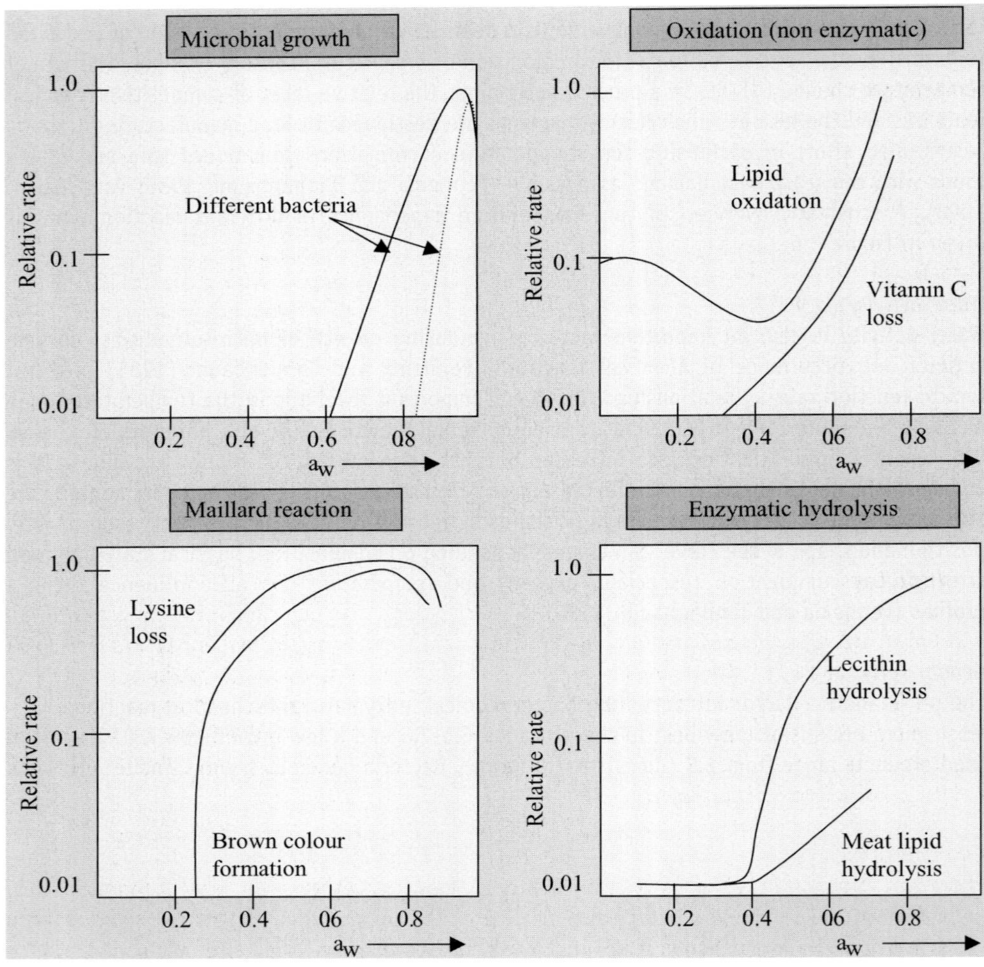

Figure 2.7. Schematic presentation of examples of relationships between a_w and different reactions (adapted from Fennema and Tannenbaum, 1985).

than moulds and yeasts. Table 2.5 gives an indication of pH ranges for some foodborne organisms (Jay, 1996). However, pH minimal and maximal of micro-organisms are not strict and also depend on the type of acid used. For instance, citric, hydrochloric, phosphoric and tartaric acids permit growth at a lower pH value than acetic acid or lactic (Gould, 1996; Jay, 1996). Generally, a pH< 4.5 in food is assumed to be low enough to inhibit most foodborne bacteria.

Enzymes usually exhibit maximal activity at an optimum pH. Most enzymes reveal maximum activities in the pH range between 4.5- 8.0. The optimum for a specific enzyme is often confined to a narrow pH range. At extreme pH values, enzymes are generally irreversibly inactivated because of a change in protein conformation, whereas at less extreme values enzymes can be reversibly inactivated. The pH can be controlled to maximise, to prevent or to inhibit an enzymatic reaction (Richardson and Hyslop, 1985).

Table 2.5. Indication of pH ranges for some food borne organisms (Jay, 1996).

Micro-organisms	pH range
Moulds	0 - 11
Yeasts	1.5 - 8.5
Lactic acid bacteria	3.2 - 10.5
Staphylococcus areus	4.0 - 9.8
Salmonella spp.	4.2 - 9
Escherichia coli	4.5 - 9
Clostridium botulinum	4.8 - 8.2
Bacillus cereus	4.7 - 9.3
Clostridium perfringens	5.4 - 8.6
Campylobacter spp.	5.9 - 9

Rates of chemical reactions can also be influenced by pH. For example, extreme pH's can accelerate acid- or base catalysed reactions.

Food additives

A wide variety of compounds are added to food for functional purposes, of which several are naturally present in foods, such as natural anti-oxidants and anti-microbial factors. Food additives are allowed if they provide a useful and acceptable function (or attribute) to justify their use. Acceptable functions include improved shelf life, enhanced nutritional value and improvement of processing facilitation. Food additives are basically applied to improve physical product properties. However, the trend toward foods with fewer additives forces the food industry to find other solutions.

Gas composition

The gas composition of headspaces in packed foods together with barrier properties of packaging materials greatly influences shelf life and food safety. Especially lower oxygen concentrations are applied to delay oxidative reactions (*e.g.* lipid oxidation, discolouration), to inhibit growth of aerobic micro-organisms and to decrease respiration rates, thus extending the shelf life of food products (Phillips, 1996). Several packaging concepts such as vacuum, modified atmosphere and active oxygen scavenging are used to obtain these conditions. Since low oxygen levels may favour the growth of some anaerobic foodborne bacteria, preventive measures must be taken, such as correct heat treatment, and/or low pH, and/or low a_w, and/or hygienic handling. A new development with respect to gas composition is the use of extreme high oxygen conditions to extend shelf life.

Combination of factors, hurdle technology

Nowadays there is a shift from individual preservation factors to a combination of factors to guarantee food safety, while maintaining sensory properties and nutritional value, also called the hurdle technology (Gould, 1996). Some typical combination treatments include:

- enhancement of the effectiveness of an anti-microbial acid by lowering the pH
- anti-microbial efficacy of carbon dioxide in modified atmosphere is greatly enhanced at reduced temperature

- foods with low water activity and/or reduced pH require a milder heating treatment because of the synergy between a_w, pH and temperature
- the combination of mild heating of vacuum-packaged food with well-controlled chill storage also appeared to be successful, *i.e.* "sous-vide" production

New preservation techniques are as well often based on combinations *e.g.* the strong synergism of pressure with heat is applied in the high hydrostatic pressure technique.

Initial contamination and hygienic production

Last but not least, initial and cross contamination at food manufacturing influence shelf life and food safety. The factors during harvesting (*e.g.* soil residues, bruising) and slaughter (*e.g.* hygienic conditions, natural differences between animals) affect the initial contamination of raw materials. Contamination during processing can origin from improper personal hygiene, unfiltered air, cross-contamination between products (*e.g.* contamination of pasteurised product with raw product) and insufficient cleaning of equipment and machines

2.3.5 Storage and distribution conditions

To consider effects of storage and distribution on physical product feature, a distinction must be made between fresh produces and manufactured foods. An overview of relevant factors for both groups is given in Table 2.6.

Fresh produce such as fruits and vegetables are characterised by their respiratory activity after harvesting. Most fleshy fruits show an increased respiration rate accompanied with ripening (*i.e.* colour, flavour and texture changes). Temperatureand composition of the storage atmosphere markedly influence the respiration process in plant tissues. Normally the rate of respiration decreases as temperature declines, within the physiological temperature range for a particular commodity. Exposure outside the recommended range for more than short periods

Table 2.6. Factors affecting product features during storage and distribution.

Relevant factors	Fresh produce (bulk storage)	Manufactured food
Initial bacterial load	High bacterial load reduces shelf life	Hygienic conditions during manufacturing
Temperature-time [T-t]	Low temperature decreases respiration Too low temperature → chilling injury	Low temperature delays (bio)chemical and microbial reactions
Relative humidity [RH]	Too high RH favours growth of moulds	RH in combination with packaging barrier properties determine effects
Controlled atmosphere	Oxygen reduction combined with increased carbon dioxide tension slows respiration and subsequent decay	Modified atmosphere packaging
Germination inhibitors	Delay *e.g.* sprouting of potatoes	
Infestation control	Prevention of product loss by insects and mites; use of pesticides	Prevention of product loss by insects and mites; proper packaging
Handling	Mechanical injury (see harvesting)	Broken packages Cross contamination due to mixed product storage

causes injury and decrease in quality and shelf life, *i.e.* chilling injury. Visible effects of chilling injury include tissue necrosis (*e.g.* apples), failure to ripen (*e.g.* bananas), black spots (bell peppers, aubergines) and wooliness in texture (*e.g.* peaches) (Haard, 1985). Furthermore, manipulation of oxygen and carbon dioxide concentration of the storage atmosphere can considerably delay ripening and subsequent decay process.
Another important factor for fresh produce is the relative humidity (RH). A proper RH regulation is required to prevent both desiccation (RH too low) and mould growth (RH too high). Furthermore, chemicals can be applied during storage to prevent *e.g.* sprouting or damage by insects.

The major factors for *packaged manufactured foods* are storage temperature and duration, and packaging materials with good barrier properties to prevent contamination and/or diffusion of moisture and/or oxygen.

2.4 Legislative demands on quality of agri-food products

Besides consumers and customers, legislation puts as well high demands on quality of agri-food products and their production. As a matter of fact, a whole range of laws, acts, regulations, provisions, directives and norms related to food production are intended to protect human health, to minimise environmental implications and to prevent unfair competition (Gardner, 1995). The legislative measures can refer to different aspects of agri-food production, ranging from animal feed preparation, use of pesticides, residue levels, food hygiene, product-related demands and control systems, up to labelling requirements. The various legislative systems act on different levels namely:

- World-wide, *i.e.* Codex Alimentarius
- European level, *i.e.* European food legislation
- National level, *e.g.* Food and Commodity Act of the Netherlands
- Branch level, *i.e.* regulations drawn up by commodity boards for e.g. dairy products

Each law system has its specific structure (or principle instruments) and control and enforcement authorities (Table 2.7). The authorities are responsible for monitoring if agri-business and food industry comply with the relevant regulations. However, legislation with respect to food production is constantly on the move because of continuous increasing international trade and increasing public concern with respect to food safety. Therefore major legislative systems are shortly described below.

Codex Alimentarius

The Food and Agricultural Organisation (FAO) and the World Health Organisation (WHO) formed the Codex Alimentarius in 1962. They elaborated food standards ranging from specific raw and processed materials (*i.e.* commodity subjects) to food hygiene, pesticide residues, contaminants and labelling, up to analysis and sampling methods (*i.e.* general subjects), as shown in Table 2.7. In addition, the Codex Alimentarius included provisions of an advisory nature in the form of codes of practices, guidelines and other recommended measures. The Committees of Governments Experts were hosted in different countries *e.g.* the Committee for Food Additives and Contaminants is situated in the Netherlands.
Codex standards used to have a non-obligatory character. However, the relevance of the Codex Alimentarius standards increased since the foundation of the World Trade Organisation (WTO) in

Table 2.7. Overview of different legislative systems.

World-wide legislation: Codex Alimentarius		
General subjects (host government of committee)	*Product-related subjects (host government of committee)*	*Control*
• Residues of veterinary drugs (USA) • Pesticide residues (Netherlands) • Food additives and contaminants (Netherlands) • Food labelling (Canada) • Food Hygiene (USA) • Analysis and sampling (Hungary)	• Processed fruits and vegetables; cereals, pulses and legumes (USA) • Sugars; fats and oils (UK) • Cocoa products and chocolate; soups and broth (Switzerland) • Edible ices (Sweden) • Fish and fishery (Norway) • Foods for dietary use (Germany) • Processed meat and poultry (Denmark) • Vegetable proteins (Canada) • Meat hygiene (New Zealand) • Tropical fresh fruits/vegetables (Mexico)	• Control and enforcement rests with the Member States

EU-legislation			
Horizontal rules	*Vertical rules*	*Miscellaneous*	*Control*
• Labelling • Pre-packaging • Food contact materials and articles • Additives • Processing aids • Water for human consumption • Pesticides residues • Contaminants	• Products including: • Sugar • Natural mineral water • Quick frozen foods • Novel foods • Novel food ingredients	• Sampling and analyses methods • Official control of foodstuffs • Free movement of foodstuffs • Geographical indication and designation of origin • Foodstuffs hygiene	• Control and enforcement rests with the Member States • General principles for inspection • sampling and control of food is embodied in EU directive • Veterinary Inspectorate of EU controls establishments in 3rd countries producing animal products for export

Table 2.7. Continued.

National legislation (Netherlands)			
e.g. Food and Commodity Act	*Control*	*e.g. Agricultural Quality Act*	*Control*
• Framework filled in by general royal decisions • Horizontal provisions including additives, packaging and utensils, food hygiene etc. • Vertical provisions are product-group related And include both food and non-food (*e.g.* asbestos, cosmetics) commodities	• Supervision by Inspectorate for Health Protection and Veterinary Public Health	• Framework filled in by general decisions • Product-aspect topics *e.g.* origin, quality, grade, care, weight, etc. • Miscellaneous topics *e.g.* design and use of buildings, transport, equipment, and quality payment system	• General inspection service or • Private law authorities *e.g.* Central Organ for Dairy Control (in Dutch COKZ)
Branch legislation (Netherlands)			
Commodity boards			*Control*
e.g. Commodity Board for Cattle and Meat e.g. Commodity Board for Fruits and Vegetables e.g. Commodity Board for Margarine, Fats, and Oils			• General inspection service or • Private law authorities or • Inspectorate for Health Protection and Veterinary Public Health

1995, where the role and responsibilities of the Codex Alimentarius were officially established. Since that time food standards have a more compelling character (Damman, 1998; Micossi, 1998). The role and responsibilities were recorded in the Agreement on Sanitary and Phytosanitary Measures (SPS-agreement) in addition to the already existing Technical Barriers of Trade (TBT-agreement). These agreements include all kinds of measures and are aimed at protecting human health while keeping fair trade.
The SPS-agreement covers all measures concerned with the protection of

- human or animal health from food-borne risks
- human health from animal- or plant-carried diseases
- animals and plants from pests or diseases (SPS agreement, 1996)

The TBT-agreement includes all technical regulations and covers any subject from car safety to shape of food cartons. In terms of food, included relevant topics in the TBS agreement are labelling requirements, nutrition claims and concerns, and packaging regulations (SPS agreement, 1996).
Both agreements encourage governments to base their national measures on international standards, guidelines and recommendations developed by the WTO members (including Codex Alimentarius). More stringent requirements than the international standards are allowed if they are based on scientific arguments from an assessment of potential health risk. Moreover, the higher standards should not lead to trade barriers.

EU regulation
During the last years, food safety has attracted increasing public concern in the European Union. Governments have found themselves compelled to take more responsibility for the protection of public health and safety as affected by food. Moreover, national regulations have not always been consistent, which resulted in unnecessary trade barriers. The objective of the European food policy is to obtain a single market in foodstuffs (without trade barriers) while maintaining a high level of food safety, public health and consumer protection. Therefore, the EU has adopted a legislation for common agricultural policy and harmonisation legislation for the internal market (Micossi, 1998).
The EU regulations are very complex, but the most commonly used instrument for harmonisation are the Directives. A Directive is a measure addressed to the Member states, which must be incorporated in the national legislation within a set time-schedule (de Sitter, 1998). For example, the Council Directive 93/43/EC on Food Hygiene has been implemented in the Dutch law via the Commodity Act regulation 'Hygiene for Food'. In this way all EU directives are being implemented in the national laws of the Member States.
The European food legislation can be divided in three main families (Micossi, 1998):

- Legislation related to **product safety**, which covers areas such as food hygiene, additives, food contact materials, novel foods and the control system. Most of these measures have a horizontal nature, which means that the requirements are across a wide range of products and processes. Besides, there are several product-specific (vertical) food safety measures.
- The second family relates to **consumer information** that is mainly provided through labelling.
- The third group, **quality requirements**, are aimed at protecting quality and include vertical directives, *e.g.* milk products, dietetic foods and special regionally produced foods.

The EU legislation respects its obligations under WTO and is in line with the Codex Alimentarius where relevant (Micossi, 1998).

With respect to control and enforcement of EU-Directives, each Member State is responsible for it themselves. The general principles for inspection, sampling and control of foodstuffs have been established in a framework Directive as well. The Member States must only inform the EU Commission about their control activities. The European food policy tends to move responsibility for control of foods towards the industry. Preventive food safety systems have been developed and are being implemented, to identify and control potential hazards. This system aims at safety control prevention instead of inspection during food production. As a consequence, the function of national authorities is merely to inspect if these systems are well established and controlled.

However, due to the large number of problems in the agri-food chain in the last five years (BSE, dioxin issue, foot and mouth disease, increased number of food poisoning etc.) a number of other initiatives have been taken to support food safety besides legislative measures.
Since the year 2000, initiatives have been taken on European level to found the *European Food Safety Authority*. Major tasks of this authority will be monitoring (*quality assurance*), risk assessment, scientific advice (*scientific research*) and dissemination of food safety information (*communication*). The actual responsibility for legislation, control and enforcement will be kept at the European parliament and Council level. Nevertheless, the European Food Safety Authority will be the supplier of scientific and technical information for the European Commission to develop policy and legislation for food safety issues in the agri-food chain.
Two other initiatives on European level are the European Food Safety Network (EFSN) and Safe Food in Europe (SAFE). The EFSN consists of 16 governmental research institutes whose objective is to identify and collect relevant knowledge and expertise with respect to food safety issues in order to handle calamities and crisis more efficiently. SAFE is a collaboration between different contract research institutes and Wageningen University. The objectives of SAFE are to collect and analyse relevant scientific information, develop food safety research programmes and communicate to the public about food safety issues. Both EFSN and SAFE may support the European Food Safety Authority in fulfilling the scientific basis and communication role.

National regulations

Food and Commodity Act

In the Netherlands, the major law to protect public health and safety in relation to foodstuffs is the Food and Commodity Act, which was enacted in 1919. The act not only covers food, but also regulates several non-food commodities such as packaging and utensils, children's toys and beds, with respect to safety of their design and function (de Sitter, 1998). The Food and Commodity Act is a framework in which detailed executive regulations are embedded. Horizontal provisions cover topics which are not specific to a product, *e.g.* labelling, packaging, utensils and colorants. The vertical provisions cover a wide range of specific foods and non-food commodities. The current process with respect to European food legislation will lead to a further incorporation of relevant EC-Directives in the Food and Commodity Act, resulting in an increasing harmonisation with other EC members.

Agricultural Quality Act

Another important national law in the Netherlands is the Agricultural Quality Act enacted in 1971. The Act has been developed to stimulate Dutch export and to guarantee the quality of the export products. The regulations can refer to *product-aspects* like origin, quality, grade, care, packaging, shape, indication, measure and weight of the products.

Other topics covered are *design and use* with respect to products, buildings, means of transport, machines and equipment, quality and use of packaging materials.
In addition, a restricted number of *product-oriented measures* are embedded in the Act, such as Agricultural Quality Measure for Raw Milk and Dairy Preparation or for Cheese products.
Finally, payment-for-quality systems can also be regulated in the Agricultural Quality Act, *e.g.* 'Regulation 1994 payment system for farm-milk quality' (Regulation Milk Quality Payment, 1996). The regulation refers to:
Sampling, transport and treatment of the milk samples. A professional collector of raw milk is responsible for the collection of representative milk samples. The collector must record the sampling procedure and data in a handbook approved by the Central Organ for Dairy Control (in Dutch COKZ).
Quality investigation aimed at the evaluation and assessment of milk quality. The quality of the milk is graded upon the results of the following investigations:

- control on the presence of antibiotics or inhibitory substances
- determination of total bacteria count
- examination of contaminants
- analyses of somatic cell count
- determination of butyric spores
- analyses of the lipolysis index
- examination of freezing point of the milk to assess if no water is added

The analyses are performed on regular basis as approved by COKZ. The collector will give written feed back information if the milk quality is below the norm.
Withholding of the price reduction or *payment of the quality bonus*. The collectors are obliged to pay a bonus over a 12-weeks period if the milk composition in that period complied with the quality norms. The bonus is calculated over the total amount of milk produced during that period. The amount of price reduction is determined at set intervals by the collector, based on the number of reduction points. The reduction price is calculated on the total amount of milk collected in that period.
With respect to administration, the collector must administrate the following topics: weight of milk, the results of the quality investigations and the calculation of the adjustment for price reduction or bonus payment.
Many countries have a similar payment system for farm-milk. Although the differences in payment systems are large, the most important criteria were antibiotics or inhibitory substances and added milk. The base price for milk was determined mainly by the dairy responsible for collecting and/or processing the milk and in some instances by professional organisations or public authorities (Heeschen, 1998).

Commodity boards

The law on industrial organisations has established several vertical commodity boards (*e.g.* fruits and vegetables, dairy produce) and horizontal trading organisations (*e.g.* retail, industry). Each (vertical) Commodity board has its own policy towards regulations with respect to the specific product group covered by the board. Some boards established specific measures, whereas legislation for other product groups is incorporated in the Food and Commodity Act and/or the Agricultural Quality Act.
The General Inspection Service or private authorities (like COKZ) control the compliance of Commodity board regulations (Table 2.7).

3. Quality management

Besides technological factors, management activities are essential for the actual performance of product and production quality. To understand the influence of management activities it is necessary to have a proper knowledge of the basic principles. This chapter gives a brief introduction to general principles of quality management.
The four basic management functions are planning, organising, leading and controlling. These functions provide the main structure of this chapter. The key element in all these management functions is decision-making. Therefore, the chapter starts with a concise description of the theory of decision-making. The effect of organisational conditions on decision-making is reflected in the four administrative concepts.
In the next section an overview is given of the historical development of quality management. Moreover, the contribution of the major quality guru's to the evolution of quality thinking is described.
The subsequent sections focus on translating the basic management functions planning, control, leading and organising to quality management. With respect to planning and control, two additional functions are important for quality management. The additional functions, *i.e.* quality improvement and quality assurance, are explained in detail. Besides the general principles of the management function leadership, attention is paid to quality behaviour and empowerment. With respect to organising and quality management, the position of the quality assurance department is described in more detail.
The chapter is closed with the topic chain management, including aspects such as customer-supplier relationships, supply chain management and partnerships.
Whereas this chapter includes general principles, the following chapters contain a more detailed analysis of specific aspects of quality management activities with respect to agri-food production.

3.1 Management functions and decision-making

The functions of management are planning, controlling, leading and organising within a system (*i.e.* the organisation). Subsequently, the function of an organisation, the principles of decision-making, factors influencing these decision-making processes, decision-making models and administrative concepts are explained.

3.1.1 The organisation

An organisation can be defined as a collection of people working together to achieve a common purpose. Within an organisation, people are able to accomplish tasks, which were not possible when acting alone. The basic purpose of an organisation is to produce goods and/or services that satisfy the needs of customers. A clear focus on "quality products" and "customer satisfaction" is increasingly viewed as a source of strength and performance advantage for the organisation (Schermerhorn, 1999).

Organisations are generally considered as open systems that interact with their environments in the continuous process of transforming resource inputs into product outputs in the form of finished goods and/or services (Figure 3.1). Hereby the environment plays a vital role, on the one hand as an input of resources and on the other hand as a user of products and services. The organisation has to justify its existence by realising added value in the eyes of customers and other stakeholders.

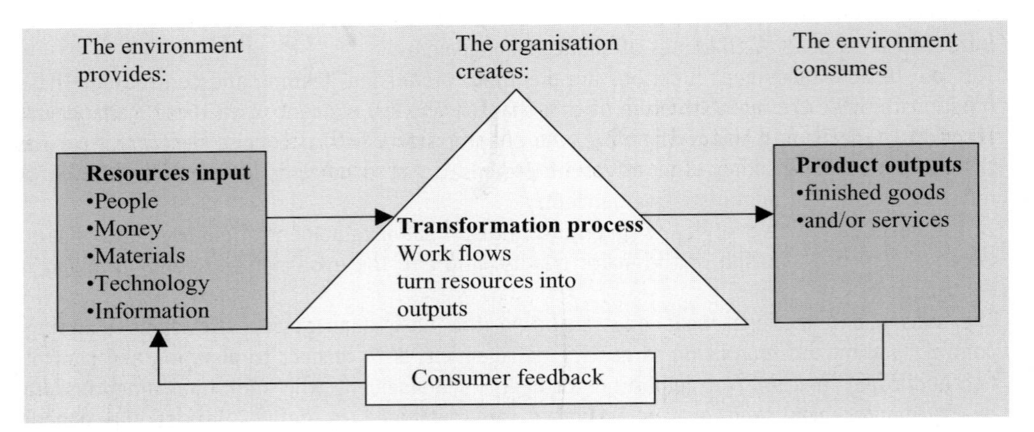

Figure 3.1. Organisation as open system.

Organisational performance can be measured in terms of effectiveness and efficiency as shown in Figure 3.2. Effectiveness is a measure for the output. Using the extended quality triangle (Figure 2.3) effectiveness reflects to which extent quality output meets customer requirements. Efficiency is a measure of resource costs, which are required to accomplish goals.

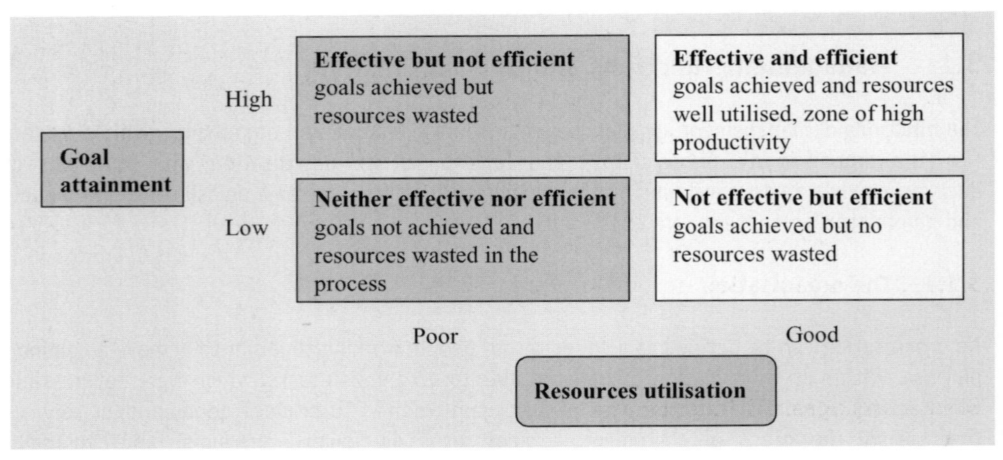

Figure 3.2. Productivity and organisational performance.

Having success in both effectiveness and efficiency leads to high productivity, thus relating output and input. In this case customer requirements are accomplished while resources are optimally used.

Total Quality Management is widely considered as a management concept that is able to ensure high productivity. The major elements of this concept include customer-orientation, organisation-wide quality behaviour and continuous improvement. In comparison, the traditional organisation hierarchy is a poor condition for quality management. Therefore, organisations need to be changed for implementing the TQM concept. The TQM concept will be discussed in detail in Chapter 8.

Figure 3.3 shows the "upside-down" view of organisations. It illustrates the new organisational thinking that is linked to total quality management. Operating workers serve clients, whereas managers are seen as coaches that support workers in their customer-oriented activities.

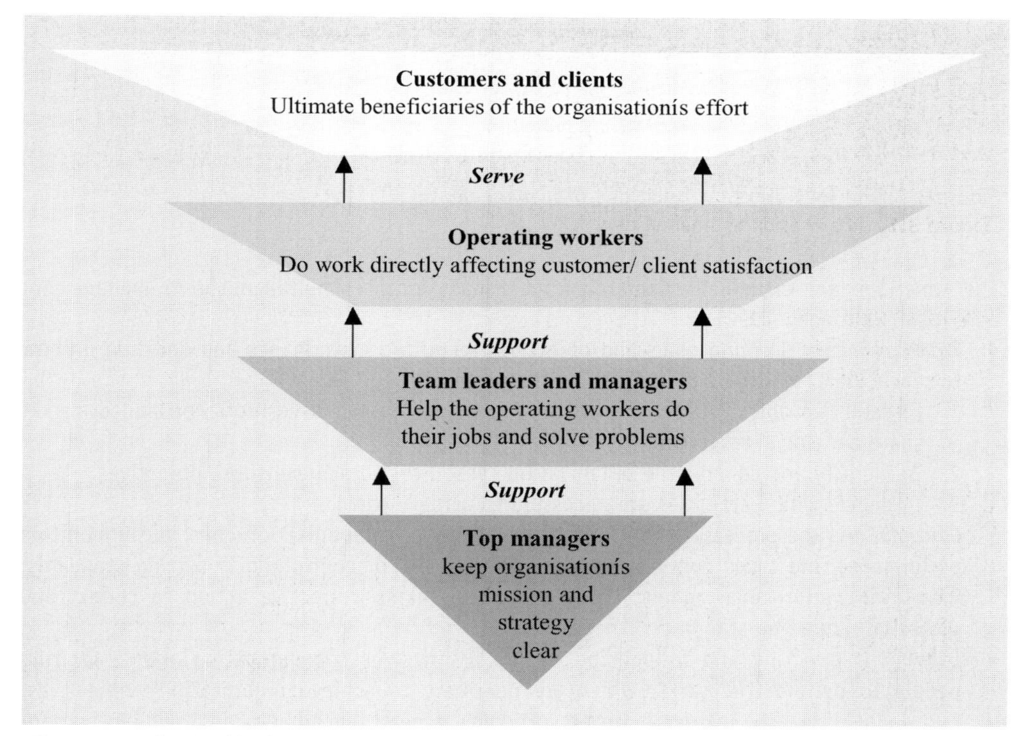

Figure 3.3. The 'upside-down-pyramid' view of organisations.

3.1.2 Management functions

As previously mentioned, all management activities can be classified into four major functions, *i.e.* planning, controlling, leading, and organising (Figure 3.4).

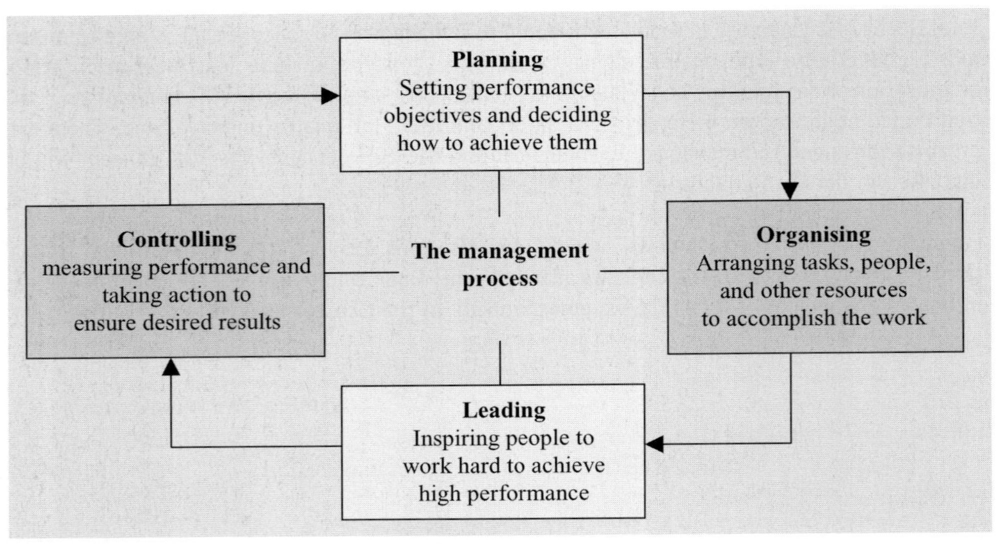

Figure 3.4. The four different functions of management.

The functions are explained separately, but usually occur simultaneously in management activities (Hellriegel, 1992).

- *Planning* means defining goals and objectives for future performance and deciding on how to reach them. Managers plan for three reasons:
  1. To establish objectives for the organisation, such as increased profit, expanded market share and social responsibility.
  2. To identify and commit the organisation's resources to achieve those objectives.
  3. To decide which activities are necessary to achieve them.
- *Controlling* is the process by which a person or group or organisation consciously monitors performance and takes corrective action. The process of controlling is setting standards, measuring performance against these standards, taking corrective action to correct any deviation, adjusting standards if necessary.
- *Leading* is directing and influencing people. It involves communicating and motivating people to perform the tasks, which are necessary to achieve organisation's objectives. Knowledge of leading is directly linked with knowledge of human behaviour and motivation.
- *Organising* is the process of creating a structure of relationships among employees. This means creating an organisational structure (departments, jobs, responsibilities and authority) and a set of rules and procedures, so co-ordinating human behaviour. Organising will enable people to carry out management's plans and meet overall objectives. Effective organising will enable managers to obtain better co-ordination of human and material resources.

Management processes take place within organisations. While executing these processes, the organisation itself is changing. In fact, organising changes the organisational structure, whereas leading changes people.

Kampfraath and Marcelis (1984) proposed that an organisation should be considered as a condition for management processes (Figure 3.5). It is a mid-term or long-term administrative

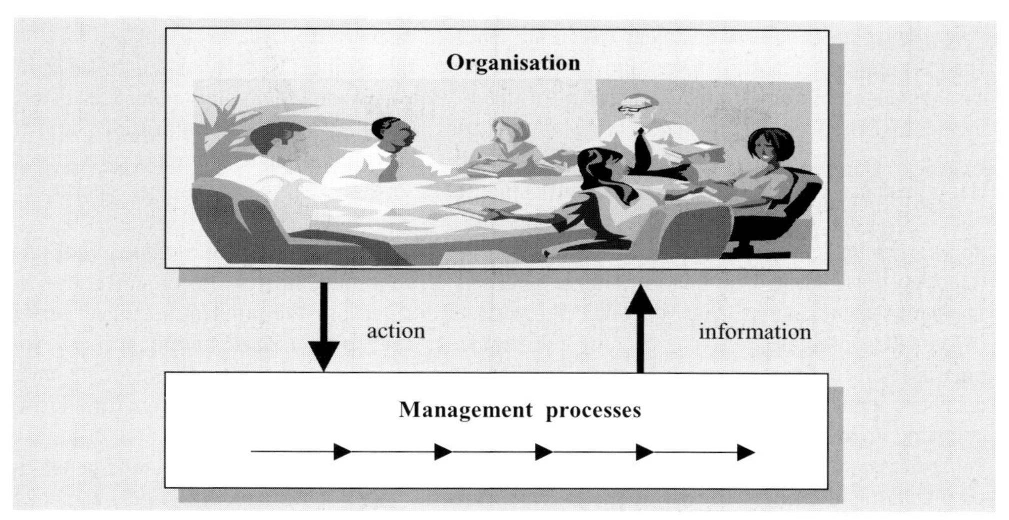

Figure 3.5. The organisation versus the management processes.

infrastructure having its own dynamics and life cycle. Under various organisational conditions, management processes and therefore management decisions will be different. For instance a quality department, reporting to the production managers, hardly will be able to influence quality decision-making in the purchasing or marketing department. However, a quality department, reporting to the chief executive officer and being a member of the management team, will not meet this problem.

Compare management processes as the flow of the river between its banks. The banks are the organisation, which determine how the river flows. At the same time the flow (management) is changing its banks so creating new conditions for itself. Therefore, effective managers not only make decisions themselves but they also influence decision-making of people in the organisation by shaping good organisational conditions.

Finally, it should be noticed that management functions are not exclusively the task of managers alone. In fact, everyone in the organisation is making decisions and contributes to one or more management functions. For example, in self-managing teams, operators are able to do much of the managerial work themselves.

3.1.3 Decision-making

For all four management functions (planning, controlling, leading and organising) decision-making is the basic activity. Decision-making is the process of defining problems and selecting a course of action from the generated alternatives. Decisions are made upon collecting and using information. The decision-making process is complicated by the complexity of collected information, the uncertainties in the decision process and the methods for implementation. For example, a correct decision can be useless if it is not implemented properly. Decisions can be divided into either programmed or non-programmed decisions.

Programmed decisions are decisions which are made in response to situations that are routine. They are more or less structured and fairly repetitive. These kind of decisions are suitable for establishment in some type of procedure (Simon, 1960). Examples of programmed decisions include deciding to re-order inventory when qualities fall below a certain level, or deciding to reduce temperature when temperature has exceeded a certain level in a production process. Many business decisions on routine activities and basic operations are programmed decisions.

Non-programmed decisions are made in response to situations that are unique. They are relatively unstructured, undefined and/or have major consequences for the organisation. No standardised procedures exist to resolve these situations because the problem has never arisen before. The problems are complex or uncertain, or they are of such significance to the organisation that it requires tailor-made decisions (Simon, 1960). Examples of non-programmed decisions include deciding to build a new plant, to enter into a strategic alliance with another company or to develop a new product.
Regardless whether a decision is programmed or non-programmed, decision-making is generally considered as a process consisting of the next essential steps (Schermerhorn, 1999) as shown in Figure 3.6.

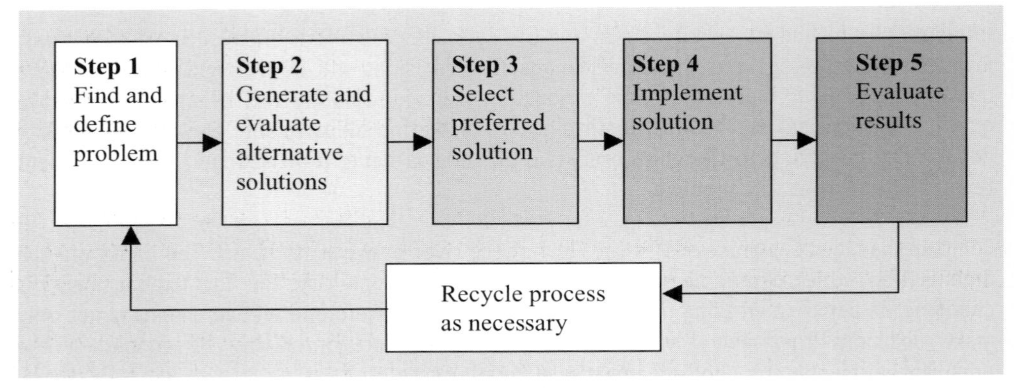

Figure 3.6. Steps in decision-making and problem-solving.

1. *Identify and define the problem*
Identification and defining the problem starts with collecting, processing and reflection of information. It often begins with the appearance of problem symptoms. A problem is a difference between an actual situation and a desired situation. Problem analysis is a critical part of decision-making, because its determines the direction of both the decision-making process and ultimately the decision that is made. Special care must be taken not just to address a symptom while ignoring the true problem.

2. *Generate and evaluate possible solutions*
This step involves identifying courses of action that can lead to solution of the problem. At this stage more information is gathered, data are analysed and the advantages and disadvantages of possible solutions are identified. The better the pool of alternatives, the more likely a good solution will be achieved. Common errors at this stage include selecting a particular solution

too quickly (*i.e.* jumping to a conclusion) and choosing an alternative that, although convenient, has too much side effects.

3. *Choose a solution*
At this point a decision is made to select a particular course of action. How this decision is made and by whom must be properly resolved for each problem situation. In some cases, the best alternative may be selected based upon cost-benefit criteria, whereas in other situations additional criteria maybe relevant, such as stakeholders interests or ethics. However, once alternatives are generated and evaluated, a final choice must be made. This is the point of ultimate decision-making.

4. *Implement the solution*
Based upon the preferred solution, appropriate action plans must be established and fully implemented. This is the stage at which directions are finally set and problem-solving actions are initiated. People not only use the strength of mind and creativity to arrive at a decision. They also need the ability and willingness of employees to implement the decision. Difficulties at this stage are often related to lack of participation or failure to involve the right persons.

5. *Evaluate results*
The decision-making process is not complete until results are evaluated. If desired results have not been achieved, then the process must be restarted to allow for corrective actions. Moreover, in any evaluation both the positive and negative consequences of the chosen course of action should be examined. If the original solution appeared inadequate, a return to earlier steps in problem solving may be required to generate a modified or new solution.

The concept of organisational and individual learning is directly linked to decision-making. Senge (1990) described a learning organisation as "an organisation that by advantage of people, values and systems is able to continuously, change and improve its performance based upon the lesson of experience".
Systematic evaluation and analysis of the outcome of decision-making processes, together with personal characteristics, are the basis for individual learning. Senge (1990) emphasised the importance of having the ability to learn and make that learning continuously available to all members of the organisation.

3.1.4 Factors influencing decision-making

Management can be defined as goal-oriented decision-making. Decisions have to be made in order to stimulate actions towards more or less specified objectives. There are two main factors having great influence on decision-making, *i.e.* the availability of information and the existence of interests (Figure 3.7).

Starting with a theoretically unending number of alternatives, lack of information results in uncertainty and existing interest result in constraints. Both uncertainty and interest reduce the room of decision, because some alternatives must be excluded, while others may become preferable.
Decision-making processes could be characterised on a scale ranging from the extreme 'fully depending on information availability and information analysis' to the other extreme 'fully

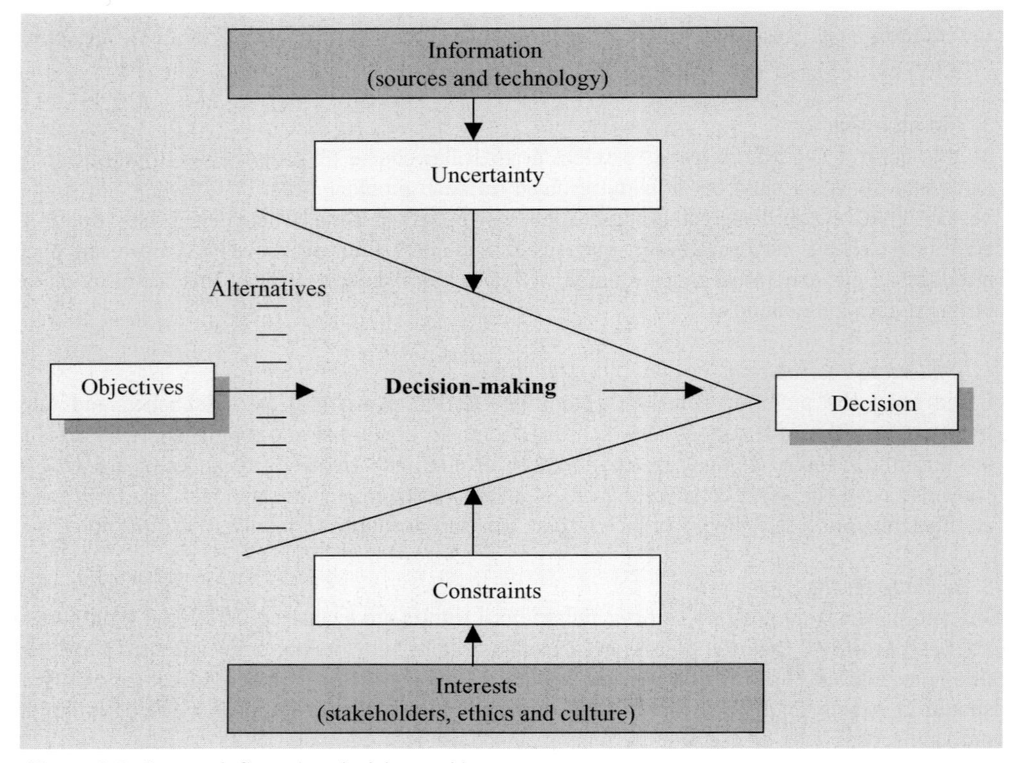

Figure 3.7. Factors influencing decision-making.

depending on interests and the way conflicts are managed'. Most decision-making processes show a mix of both elements, information and interests. The two major affecting conditions, information and interest, are discussed in more detail below.

1. Information

The availability of the information is an important condition affecting decision-making. In an ideal situation one would have all the information needed to make decisions with certainty. However, most business situations are characterised by incomplete or ambiguous information, which affects the level of certainty and so reduces the room for decision. With respect to the availability of information, three major situations can be distinguished *i.e.* certainty, risk and uncertainty (Radford, 1975; Hellriegel 1992) as reflected in Figure 3.8.

- Certainty is the situation that exists when decisions are made in a context of being fully informed about a problem, its alternative solutions and their respective outcomes. Under this condition one can anticipate and even exercise some control over events and their outcome. However this situation almost never occurs.
- Uncertainty is the situation that exists when decisions are made in a context of being incompletely and improperly informed about a problem, its alternative solutions and their respective outcomes.
- Risk is the situation that exists when decisions are made in the context of incomplete, yet reliable information. Under a state of risk one does not know with certainty the future

outcomes associated with alternative courses of action. The results are subject to chance. However, the decision-maker has enough information to determine the probabilities associated with each alternative. On the certainty continuum the condition of risk means that a problem, its alternative solutions and their respective outcomes lie between the extremes of being known and well defined (certainty) and being unknown and ambiguous (uncertainty).

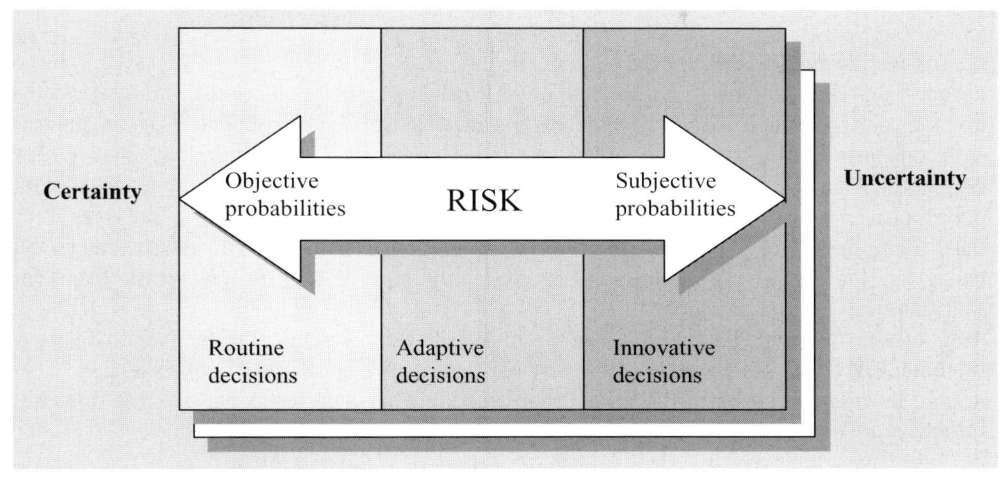

Figure 3.8. State of nature and classes of decision.

To gain access to accurate, timely, complete and relevant information, many organisations have turned to management information systems to manage the sometimes overwhelming amount of information. A management information system (MIS) organises past, present and projected data from both internal and external sources and processes them into usable information. This information is then available for people at all organisational levels. Because people have different needs, information systems must be able to organise data into usable and accessible formats (Gatewood, 1995).

A decision support system (DSS) uses special software to allow users to interact directly with a computer to help in making decisions for solving complex problems (Schermerhorn, 1999). With a decision support system (DSS), managers can anticipate the possible outcomes of alternative actions and so cope better with uncertainty. Decision support systems may incorporate artificial intelligence, which enable computers to "think" as much as possible like the human mind. This resulted in the development of expert systems, which mimic human decision-making processes by using a collection of thousands of "if-then" rules to solve complex problems (Gatewood, 1995).

2. Interests

Interests are a second condition affecting decision-making and reducing the room for decision. By making a decision, interests of persons or groups will be served, while interests of others will be neglected. Conflicting interests may result in stress situations for the decision-maker.

In-group decision-making conflicting interests can lead to differences in opinion and struggle between participants.

An interest may originate from three resources, *i.e.*

a. From society ethics, this provides a set of rules that define right or wrong behaviour
b. From the firm environment, like stakeholders which are more or less involved in the firm's behaviour
c. From inside the organisation, the organisation's culture can influence decision-making behaviour.

a. Interest originating from society ethics

Recognising special ethical issues can be difficult in practice. However, in food quality management some topics are rather evident, for instance working with hormones, food addition with unknown side effects and animal welfare. Gatewood (1995) described that ethical decision-making is dependent on opportunity, work relationships, and individual moral philosophy.

Opportunity refers to conditions that limit unfavourable behaviour or reward favourable behaviour. The greater the reward and the smaller the punishment, the greater the likelihood that unethical behaviour will occur.

Work relationships indicate that individual ethical behaviour is shaped by the moral climate of the organisation, or standards set by supervisors and/or the behaviour of co-workers.

A *moral philosophy* is a set of principles that describe what a person believes is the right way to behave.

b. Interest originating from the firm environment

Interests may also originate from the environment such as the stakeholders. Stakeholders who are affected by the firm's decision-making may include primary and secondary ones. Primary stakeholders, such as employees, suppliers and buyers, have formal and/or contractual relationships with the firm, whereas secondary stakeholders have less formal connection. Secondary stakeholders are for instance animal rights groups, consumer groups and environmentalists.

Relationships with stakeholders are rather complex because they are interrelated with each other. The firm is at the centre of a network of actors, which the organisation should identify and manage as far as possible (Gatewood, 1995).

c. Interest originating from the organisational culture

The third resource of interests is the organisational culture, *i.e.* the system of shared beliefs and values that are developed within an organisation and guides the behaviour of its members (Schein, 1990). Strong cultures are clear, well-defined and widely shared among members. In strong cultures there is commitment to do things within the interest of the organisation.

Figure 3.9 shows two levels of culture in organisations: the observable culture and the core culture. The observable culture is what one sees and hears when walking around an organisation as a visitor, customer or employee. A second and deeper level of culture is the core culture. It consists of core values, or underlying beliefs that influence behaviour and actually give rise to the aspects of observable culture, described in the figure. Values are often published in formal statements or corporate mission (Schermerhorn, 1999).

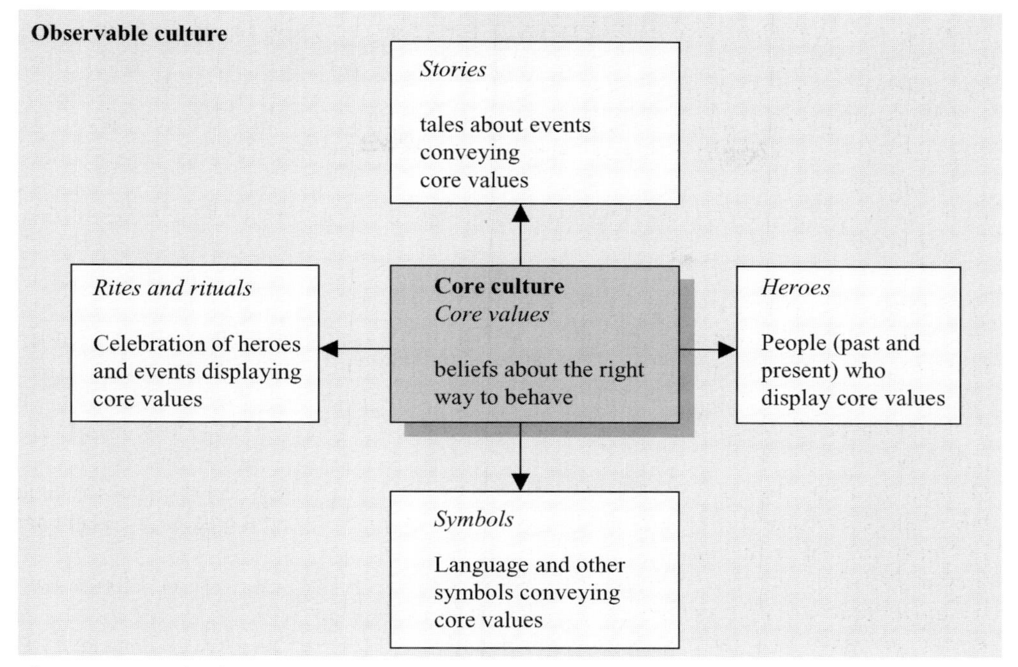

Figure 3.9. Levels of organisational culture, observable culture and core culture.

3.1.5 Group decision-making

In a group, decision-making is also the most important process. Major advantages of group decision-making are the use of more information and knowledge, and a greater acceptance and legitimacy of the decision. Major disadvantages are longer time periods for decision-making and the risk of group thinking. This is a tendency of highly cohesive groups to minimise evaluation and criticism.

Decision-making in groups can be based on the

- Authority rule, where the leader decides most times with previous discussion in the group
- Minority rule, where two or three people are able to dominate the group into making a mutually agreeable decision
- Majority rule, where formal voting may take place or members may be polled to find the majority viewpoint
- Consensus or unanimity, where discussion leads to one alternative being favoured by all members (unanimity) or most members, while the others agree to support it (consensus) (Schermerhorn, 1999).

The main difference between individual and group decision-making is the immediate interaction between participants in a group decision-making. This interaction is usually very complex because participants have different motivations, perceptions and experiences. To analyse these interactions Figure 3.7 is extended with those elements that are important in-group processes, as shown in Figure 3.10

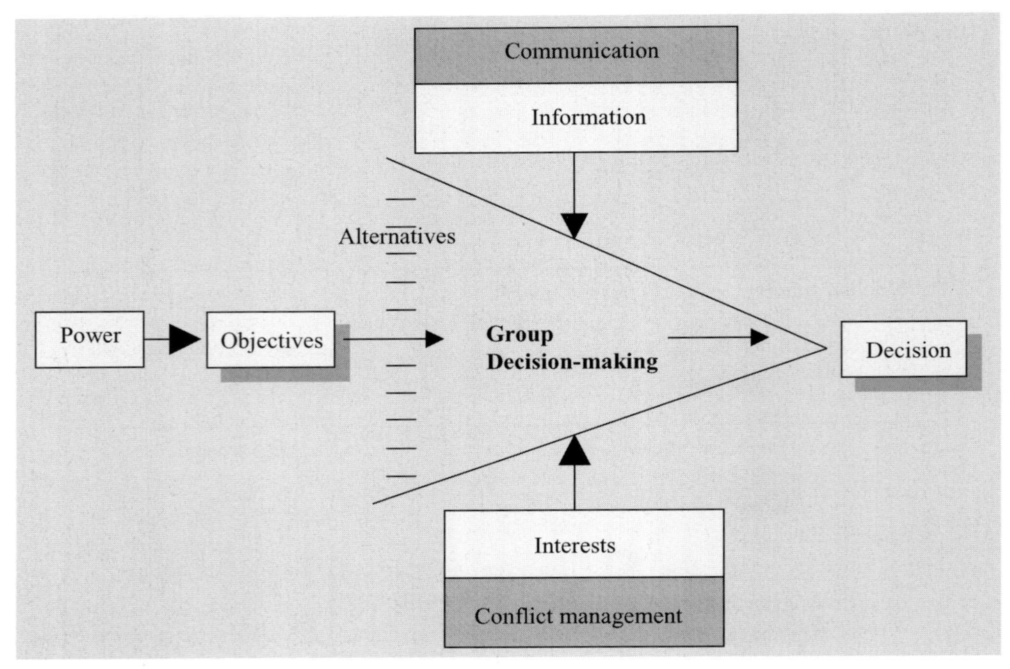

Figure 3.10. Group decision-making process.

In a group decision-making process, information and interest are also major influencing factors, but there are some differences as compared to individual processes. For decision-making in a group, additional information is used which is exchanged between group members in *communication* processes, besides the regular information input.
With respect to interests, in a group decision-making process, people often have contradictory interests and have the possibility to influence each other, sometimes causing conflicts. How groups or managers deal with conflicts is called *conflict management*.
In addition, in a group situation there are differences in power. *Power* is the ability to drive others in a direction you prefer and the way you want (Schermerhorn, 1999). Following that definition power is perhaps the most important factor in decision-making. It will direct choice among alternatives.
In short, the aspects communication, conflict management and power will be discussed below.

Communication is the process through which information and meanings are transferred from one person to another (Gatewood, 1995). The process starts with a sender, which encodes a message into symbols. Then the message arrives at the receiver, who then decodes and/or interprets its meaning (Figure 3.11). Feedback is the reverse process, whereby the receiver responds to the sender.

The receiver's interpretation almost never matches the sender's original intention. The reason is noise in the communication process. Noise is anything that interferes with the message being communicated effectively. For instance the process of encoding and decoding or inadequate channels. Robbins (1992) mentioned disturbances in communication through filtering (only

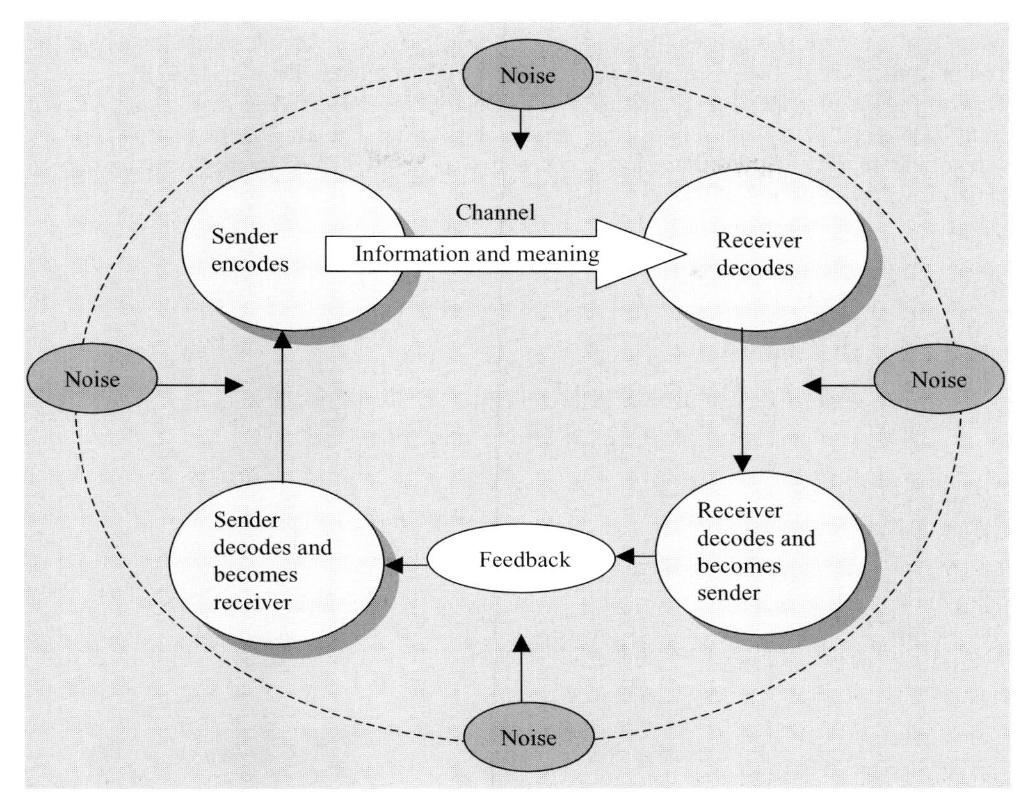

Figure 3.11. Communication process.

information is sent that is important for the sender) and selective perception (only information is used that is meaningful for the receiver).

Formal communication represents the flow of communication within the formal organisational structure and the groups and teams within it. However, also very important are the informal communication channels (the grapevine). Individuals and groups spread information to other people when they feel the need for it. Informal communication depends strongly on personal relations and the existence of informal groups.
In the past informal communication was not of great concern to managers. Today managers view the quality of communication through formal channels as being poor. Quality of informal communication also is not perfect, but managers agree that these channels are often a much better source of good and accurate information than formal channels.

Conflict management involves the way managers handle conflict situations. Conflicts can arise when different conflicting interests are at stake. There is a real conflict when someone blocks the goals, intentions or behaviours of someone else.
Conflict has traditionally been viewed as negative, however, the modern view is that conflict is both inevitable and in many cases desirable, if properly managed. Conflicts can stimulate people toward greater efforts, co-operation and creativity. Conflicts not always have the form of an

open fight between two parties (hot conflict). In fact, there is a conflict process in which the conflict, in course of time, is growing and building up to a climax (Robbins, 1992).
Kenneth W. Thomas (1976) distinguished five conflict management styles. These styles evolved from different levels of co-operation and assertiveness (Figure 3.12). Co-operation is attempting to satisfy the other party's concerns and goals. Assertiveness is attempting to satisfy own concerns and goals.

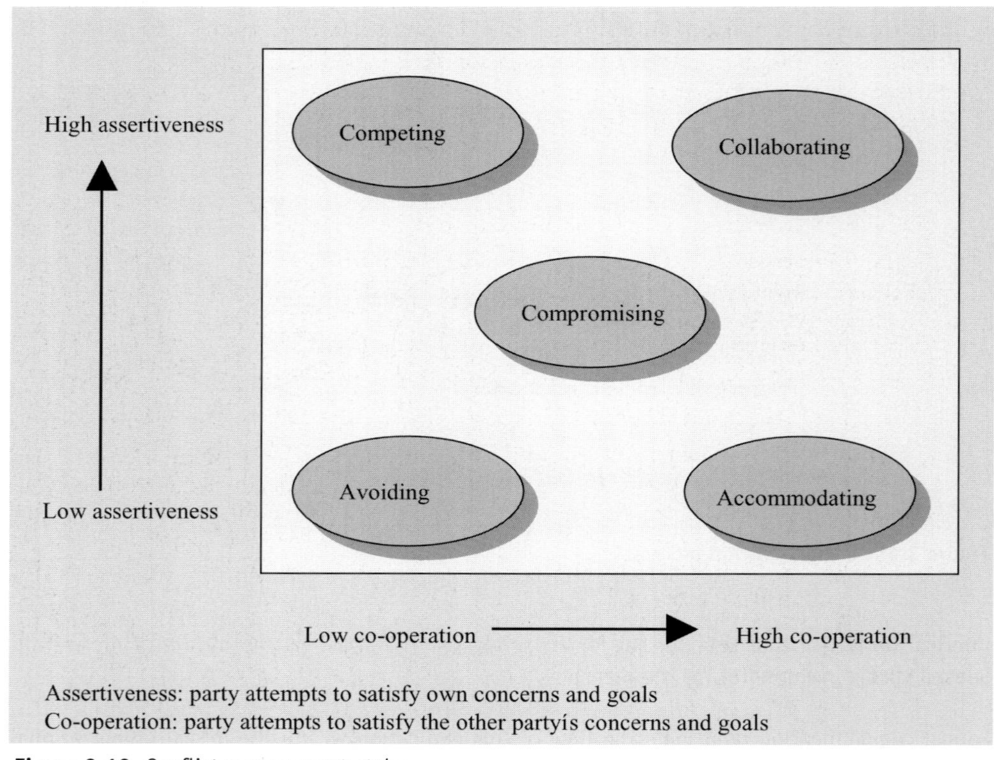

Figure 3.12. Conflict management styles.

Briefly stated, these conflict management styles involve the following behaviours (Schermerhorn, 1999):

- Avoidance, *i.e.* denying the existence of the conflict and hiding one's true feelings
- Accommodation, *i.e.* playing down the conflict and seeking harmony among parties
- Competition, *i.e.* forcing a solution to impose one's will on the other party
- Compromise, *i.e.* bargaining for gains and losses to each party
- Collaboration, *i.e.* searching for solutions that meet each other's need

Of course various conflict management styles have quite different outcomes. Collaboration or problem solving tries to reconcile underlying differences and is often the most effective style. It is a form of win-win situation whereby issues are solved to the mutual benefit of all conflicting parties.

Power is the ability to make things happen the way one wants them to happen. Power is more than just influence; it gives the possibility to get someone to do something even when he does not agree with it.
One of the most used frameworks for understanding power is developed by French and Rowen (1960). They identified five types of power:

- Legitimate power is based on the formal position in the organisational hierarchy
- Reward power stems from the ability to reward followers
- Coercive power results from the ability to obtain compliance through fear or punishment
- Referent power is based on followers' personal identification with the leader
- Expert power is based on having specialised knowledge and information.

The first three bases of power are related with the position in the organisation, whereas the last two are linked with personal characteristics (Figure 3.13).

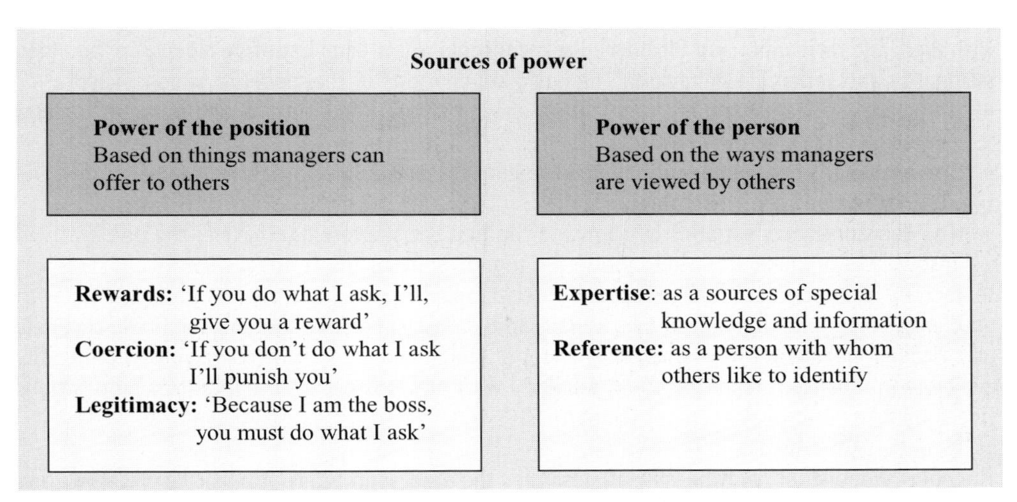

Figure 3.13. Sources of position power and personal power used by managers.

When power is actually turned into action, one can speak of political behaviour (Robbins, 1992). Political behaviour involves activities like holding back information, forming coalitions, threatening with punishments. Often political behaviour is well accepted, but there is also political behaviour that is perceived as not legitimate.

3.1.6 Decision-making models

Management theory offers a range of models of decision-making. From this broad range, three major models are described here, *i.e.* the rational model, the administrative model and the political model. These are combined in the Mintzberg process model.

The *rational* model is an approach that outlines how managers should make decisions, also called the *classical model*. The rational model is based on economic assumptions and asserts that managers are logical, rational individuals, who make decisions that are in the best interests of the organisation. The model is characterised by the following assumptions.

- The manager has complete information about the decision situation and operates under a condition of certainty.
- The problem is clearly defined and the decision maker has knowledge of all possible alternatives and their outcomes.
- Through rationality and logic, the decision-maker evaluates the alternatives and selects the optimum alternative. That is the one that will maximise the decision situation by offering the best solution (Gatewood, 1995).

In fact, the description of the decision-making process in Figure 3.6 is derived from the rational model. The main hypothesis is that analysis leads to solution of the problem.

The *administrative* model of decision-making is an approach that outlines how managers actually make decisions, also called the *behavioural model.* The administrative model is based on the work of Simon (1960), psychologist and economist, whose research and findings resulted in his winning of the Nobel Prize. Simon recognised that people do not always make decisions with logic and rationality and introduced two concepts in his administrative model. *i.e.* bounded rationality and satisfying behaviour. The basic assumptions of the model of Simon include:

- The manager has incomplete information about the decision situation and operates under a condition of risk and uncertainty.
- The problem is not clearly defined and the decision-maker has limited knowledge of possible alternatives and their outcomes.
- The decision-maker satisfies by choosing the first satisfactory alternative; the one that will resolve the problem situation by offering a good solution (Gatewood, 1995).

In fact, in this approach one is accepting that people have their own way of dealing with problems, being judging on their own behalf more or less subjective therein. Implicitly they rely on so called judgmental heuristics. They simplify search and choice procedures based on readily available information, representation (similarly with previous occurrences) and/or propositions (ideas about causal relationships) (Hogarth, 1994).

The *political* model of decision-making describes the process in terms of particular interests and objectives of powerful stakeholders (Hellriegel, 1992). The model considers uncertainty (information is limited, managers have their limitations) and potential conflict (decision-makers are different in goals and objectives, values and experiences).
The organisation consists of different coalitions striving for the accomplishment of goals that are based on their particular interests. The goals are consistent within a coalition, but are inconsistent within the organisation as a whole.
The composition of the coalitions depends on the content and circumstances of the problem or conflict. The same groups can be enemies in one conflict, whereas they can be allies in another one. The decision-making process according to the political approach consists of three steps.

- forming coalitions
- gathering information and making rules and procedures
- choice as a result of bargaining and negotiation

Cyert and March (1963) concluded that in the case of more parties in decision-making, decisions are made:

- according to decision rules giving all parties a satisfying solution instead of an optimal solution (quasi-resolution of conflict)
- using criteria derived from own points of view and objectives (local rationality)

- not taking into account all goals at the same time but one by one (sequential attention to goals)

Simon (1960) introduced bounded rationality and satisfying behaviour as important factors in individual decision-making. Considering the political model, it should be recognised that these aspects are also applicable to situations of groups or coalitions in decision-making.

Mintzberg (1976) developed a decision-making model that offers the possibility of integrating the previously described models (Figure 3.14). Mintzberg divided the decision-making process in three phases, *i.e.* identification, development and selection.

- In the *identification* phase something is recognised as a stimulus asking for a decision. This is followed by a diagnosis, *i.e.* an attempt to understand the situation in terms of cause-effect relationships
- The *development* phase starts with searching, which is trying to find existing solutions adequate for the actual problem. Searching is followed by screening, which is reducing the number of potential solutions by neglecting not "fitting" solutions. Finally, in the design step, a new solution for the particular problem is developed
- In the *selection* phase the actual decision is made, which contains three parallel decision activities, *i.e.* analysis, judgement and bargaining. By analysis, choice is a logical outcome of rational activities. By judgement, choice is made on the basis of vague criteria. By bargaining, choice is a result of decision-making activities of several decision-makers having different interest.

These activities analysis, judgement and bargaining, largely correspond with the basic ideas of respectively the rational *(i.e.* all information available, selection of optimal alternative), the administrative (*i.e.* restricted information, choosing a satisfactory alternative) and the political

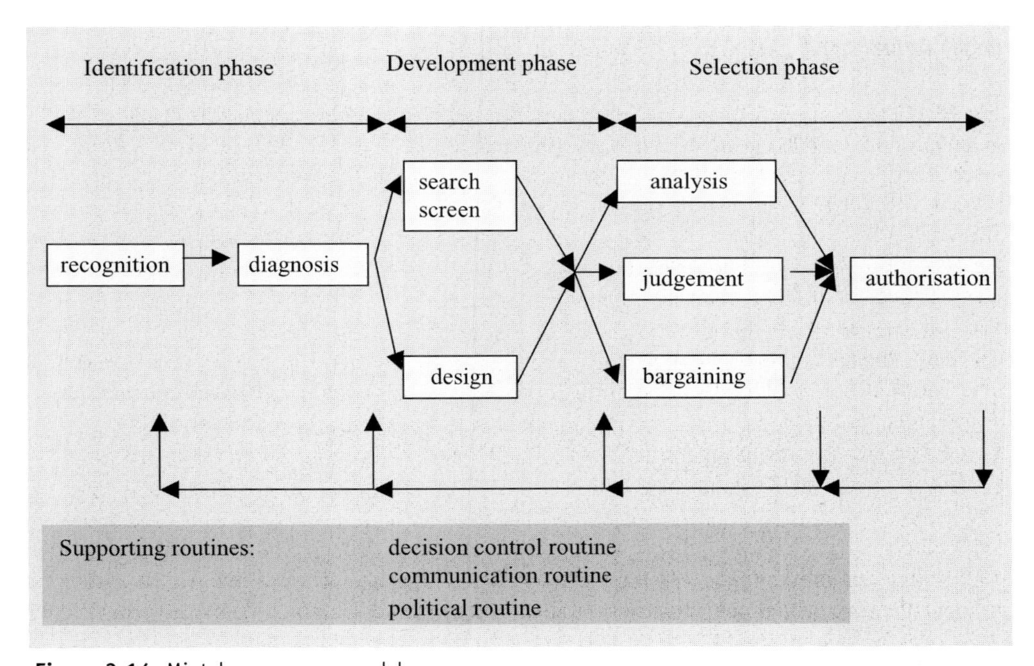

Figure 3.14. Mintzberg process model.

model (*i.e.* using criteria and decision rules giving all parties satisfying solutions). Mintzberg paid much attention to the dynamics of decision-making. Several disturbing factors can influence decision-making in such a way that phases must be repeated, phases are delayed, conflicts have to be resolved and so on.
Mintzberg (1976) defined three supporting routines, *i.e.* communication, political and decision control routines. These are rules or procedures that structure the decision-making processes. Communication routines ensure information availability. Political routines prescribe how to deal with (conflicting) interests. Decision control routines bring structure in making the actual decision.

3.1.7 Administrative concepts

On the one hand, managers make decisions, whereas on the other hand they create organisational conditions in order to influence decision-making of others. In smaller organisations, managers influence people more directly. In larger organisations, rules, procedures and structures are frequently needed in order to direct people's behaviour. In Figure 3.15 is illustrated how a manager can choose for these possibilities resulting in four administrative options, *i.e.* authority, bureaucratic, business and flexible concept.

The authority concept is the basic concept in small firms. Formalisation is not needed because a strong leader is making decisions, granting subordinates to make decisions "like he should do it". The bureaucratic concept relies on structure, rules and procedures, giving top management the guarantee that people in the organisation behave according to central goals and objectives. The top management makes all the important decisions.
The business concept allows decision-making to be dependent on market and product circumstances. Top management has influence by means of rules and procedures and by making policy statements.

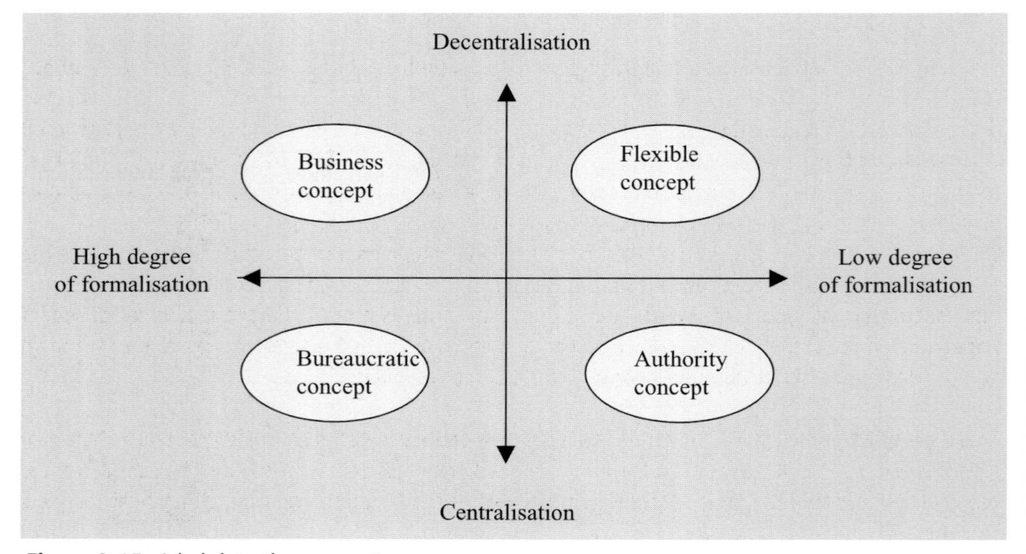

Figure 3.15. Administrative concepts.

The flexible concept accepts less influence from the top management. Top management in this case clearly defines policy and communicates it in a broad sense to all organisation members. The members, however, evaluate and weigh policy statements against other factors important at that moment and decide themselves.

3.2 Historical foundations of quality management

In this chapter a short overview of management history is given, with special attention to the development of quality management within the general evolution of management.

3.2.1 The evolution of management thought

In literature, the evolution of management thought is generally described in four approaches or viewpoints following each other in course of time, but each of them is still used in modern management (Figure 3.16). A concise description of the four approaches is given, but it is restricted to their major contributors (Gatewood, 1995).

1. The classical approach to management stresses management in a formal hierarchy of authority and focuses on task, machines and systems needed to accomplish the task efficiently. This theory emerged in the late 1800's and 1900's. This approach to management has two components, *i.e.* scientific management and administrative management.

Scientific management is a theory within the classical approach that focuses on improvement of operational efficiencies through the systematic and scientific study of work methods, tools and performance standards. Taylor (1856-1915) is considered as the founder of this theory.

Another component of the classical approach to management theory is *administrative management*, which emphasises the need to organise and co-ordinate the workings of the entire organisation.
Two important contributors to administrative management were Fayol (1841-1925) and Weber (1864-1920).
One of Fayol's main contributions to management thought was that he formulated the *functional definition of management,* which states that management is "to forecast and plan, to organise, to command, to co-ordinate and to control". Moreover, a comprehensive list of general principles of management, like division of work, authority and responsibility, was added.
Weber introduced *bureaucracy*, this theory considers management by office or position rather than by person, it is based on rational authority. In bureaucracy, it matters not who you are, but matters what you are. Weber envisioned an organisation developed around a set of impersonal and logical rules, routine, clear divisions of labour, selection based on technical qualifications and strict adherence to a clear chain of command.

These theories, indicated as classical, have formed the main roots of the study and practice of management as known today. The main criticism on the classical theories was that they focused too much on work, machines, authority structures and efficiency, ignoring the human aspects of work.

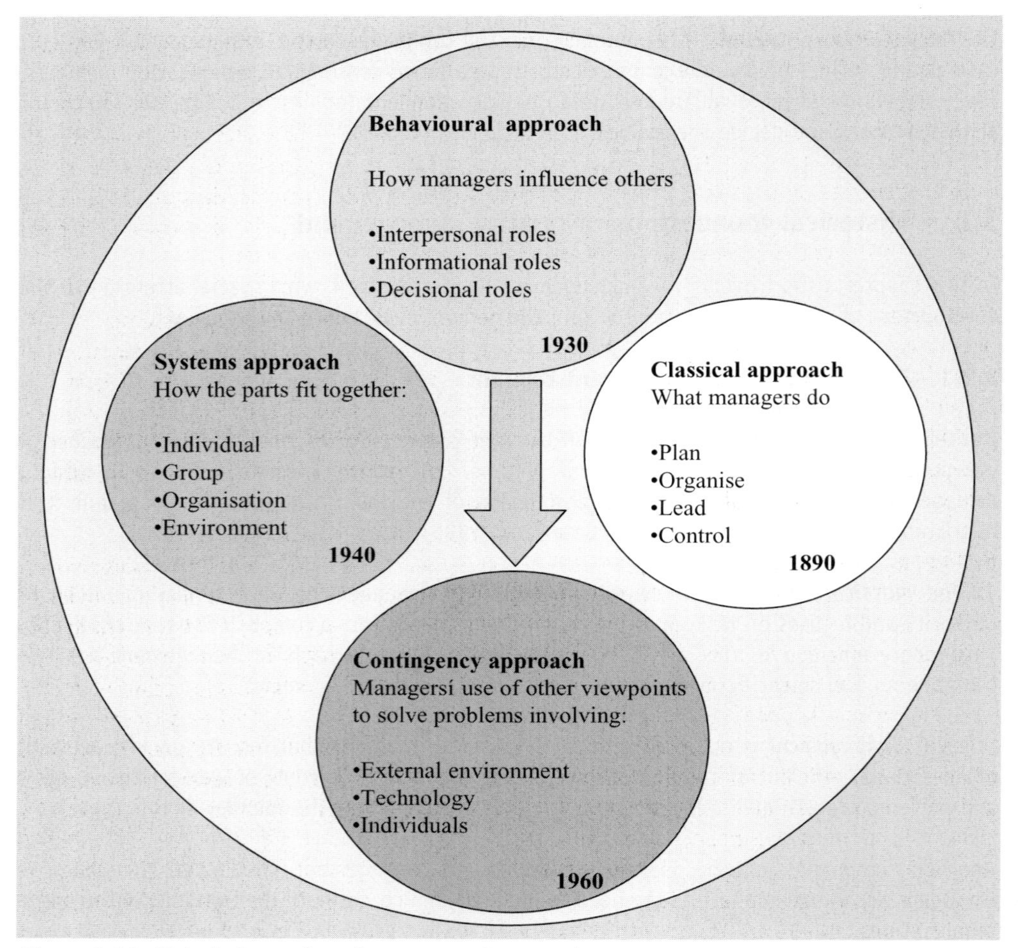

Figure 3.16. Historical overview of management approaches.

2. The behavioural approach has been initiated in the early 1900's. It was a view of management that emphasised on understanding the *importance of human behaviour, needs and attitudes* within formal organisations. Main contributors were Follet (1868-1933), the Hawthorne studies, Maslow (1908-1970) and McGregor (1906-1964).

Follet was one of the earliest management-thinkers who advocated a break from the classical management school and viewed effective management as based on the self-control and co-operation of groups of workers. She argued that work groups were one of the primary sources of influence on behaviour, even more than management's control and reward systems.

In the twenties several studies were conducted at Western Electric Company's Hawthorne plant. The general conclusions drawn from the Hawthorne studies were that human relationships and the social needs of workers were crucial aspects of business management. These studies provided the stimulus for the human relation movement within management theory and practice. They emphasised on the nature and significance of human behaviour, feelings and attitudes at work, and the role of the informal group in formal organisations.

Maslow is the developer of the *need theories of human motivation,* and advocated a humanistic approach to management. He proposed that people's behaviour at work could be explained by a need of something beyond the money essential for their basic existence. He stated that when basic survival and security needs were fulfilled by money, other needs would become important as a source of continued motivation to work.

McGregor was best known for the *theory X and theory Y.* Theory X included the assumption that people are naturally lazy, must be threatened and forced to work, have little ambition or initiative and do not try to fulfil any need higher than security needs at work. This theory presented traditional management views of directing and control. Theory Y involved the assumption that people naturally want to work, are capable of self control, seek responsibility, are creative and try to fulfil higher needs at work. These theoretic assumptions represented a new view of integration of human and organisational needs and goals.

3. The systems approach to management theory viewed organisations and the environments, within which they operate, as sets of interrelated parts to be managed as a whole in order to achieve a common goal. A system can be considered as an arrangement of related or connected parts that form a whole unit. As systems, organisations consist of inputs, transformation processes, outputs and feedback. The effectiveness of systems can be analysed according to the degree to which they are open or closed. An organisation can be considered as a closed-system if it slightly interacts with its external environment and therefore receives little feedback from, or information about its surroundings. It can be considered as open, if it interacts continually with its environment and therefore is well informed about changes in its surroundings and its position relative to these changes.

Two systems theorists will be briefly considered: Barnard (1886-1961) and Deming (1900-1993). Barnard's basic premise was that formal organisations are necessary, so that individuals can accomplish tasks they could not accomplish working alone. Therefore, organisations have to be considered as *co-operative systems*. The major mean of ensuring co-operation of individuals is to settle their believe in the organisation's purpose or goal through communication. Barnard also developed the *acceptance theory of authority*. This theory describes that in a formal organisation, authority lies in the person who receives the order. This will depend on understanding the communicated order, the believe that it is consistent with the organisations' purpose or goal, the compatibility with the personal interests, and the ability of the person to mentally and physically comply with the order. According to Barnard, these were two aspects of how organisations must be managed so that they can adjust effectively to a constantly changing external environment.

Deming's major contribution to the system approach was the integration of emphasis on quantitative methods and efficiency (typical for the classical approach), with emphasis on the *psychological and social dimensions* of the work environment (typical for the behavioural approach) into one approach. In this combined approach, all dimensions of the formal organisation and its environment are considered as part of one system.

4. The contingency approach to management theory emphasised on identifying the key variables in each management situation, and on designing an organisation that best fits in its particular situation. For example, an organisation with a highly unstable operating environment should be structured and organised quite differently from one operating in a very stable environment. This is in contrast to Weber's idea of a pure bureaucracy as the one best way of organising. Generally, the contingency approach can be considered as an evolution of the systems approach. While criticised as being inadequate as a true theoretical basis, it has inspired research and understanding on how the gap between theory and practice could be bridged.

3.2.2 Quality management history

In the first decades of the last century, statistical models of explanation had begun to be used in science. In 1931 Shewart (Shewart, 1980) published his book "Economic control of quality of manufactured products". The central topic in his publication is how to take care of data and draw conclusions from them in order to supervise and reduce the variation in the production process. In those years he already focused on customer needs. He stated that "the first step of the engineer in trying to satisfy these wants is to translate them into the physical characteristics of the thing manufactured to satisfy these wants". Shewart has had a profound influence on the modern view of quality control. If one single individual could be called the father of modern quality philosophy, that person would probably be Walter A. Shewart (Bergman, 1994).

Quality management developments can be summarised along two lines: developments in assurance and reliability and developments in attitudes to quality, as shown in Figure 3.17. The two lines of development will be illustrated now by means of a short description of the work of the four best known quality guru's Deming, Juran, Feigenbaum and Crosby. Feigenbaum is a representative of the quality assurance line at the left side of Figure 3.17, whereas Deming, Juran and Crosby should be placed in the total quality line at the right side.

W. Edward Deming and Joseph M. Juran share Shewart's statistical point of view regarding the production process. Deming in particular heavily stresses the statistical point of view.
In addition both Deming and Juran emphasise the top management role very strongly. Only if top management commit themselves fully to the quality issues is it possible to achieve continuous quality improvement. Both Deming and Juran have been ascribed great importance for Japan's successes within the field of quality. In the fifties they advised Japanese managers and had lots of success in implementing their ideas. They succeeded in turning manager's attention from costs to quality and organised a system of continuous improvement. The lessons they learned have been translated in the concept of prevention: doing things right the first time, planning and design instead of paying for failures. Thus it is not surprising that both are seen as the fathers of total quality management. In the seventies two other "guru's" in quality have also stressed these ideas: Feigenbaum and Crosby.
In the next section the core elements of the four guru's philosophies will be described in more detail.

Figure 3.18 shows the major core elements of the *Deming's philosophy* (1986). The philosophy is based on improving products and services by reducing uncertainty and variability in the design and manufacturing processes. In Deming's view, variation is the chief culprit of poor

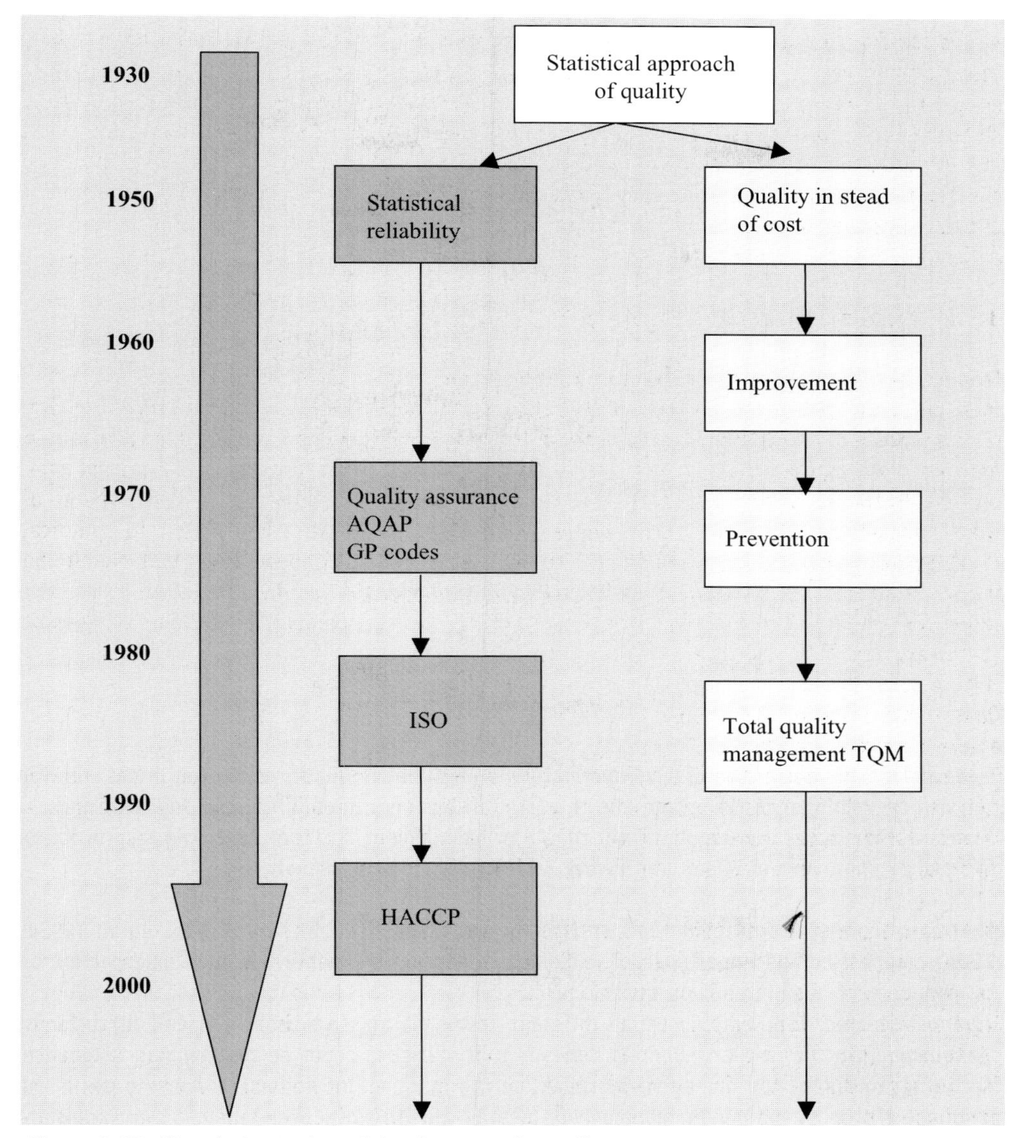

Figure 3.17. Historical overview of developments in quality management.

quality. In mechanical assemblies, for example, variation from specifications for part dimensions will lead to inconsistent performance and premature wear and failure. Likewise, inconsistencies in service will frustrate customers and so damage a firm's image.

To achieve reduced variation he advocated a never-ending cycle of product design, manufacture, test and sales, followed by market surveys, then redesign. Deming stressed that higher quality will lead to higher productivity, which in turn will lead to long-term competitive strength. In his view, improvements in quality lead to lower costs because of less re-work, fewer mistakes,

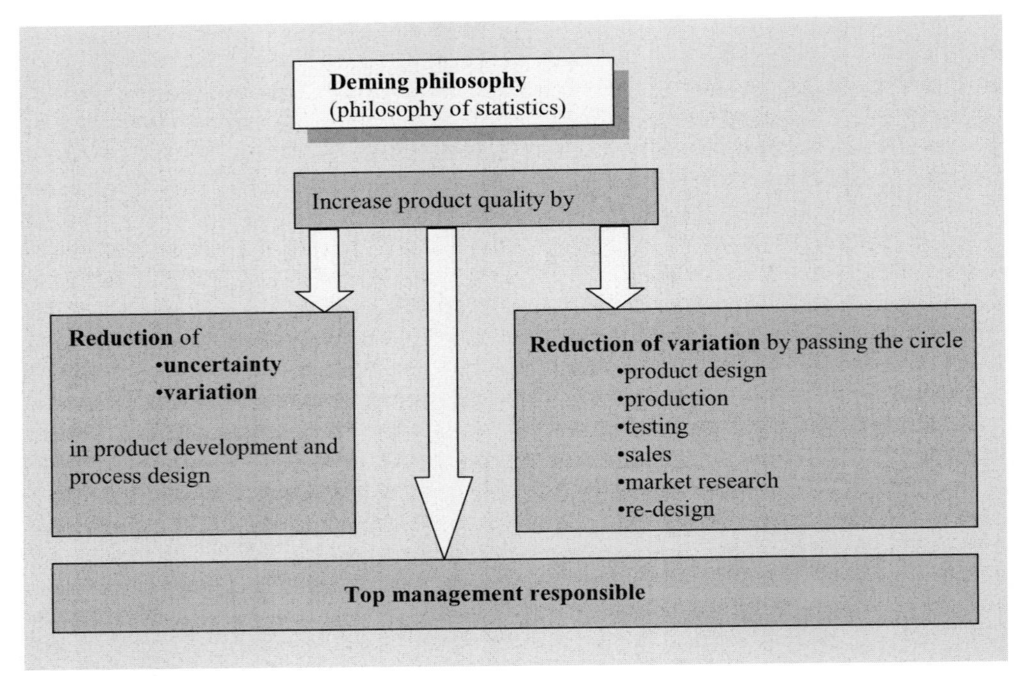

Figure 3.18. Deming's philosophy.

fewer delays and better use of time and materials. With better quality and lower prices the firm can achieve a higher market share and thus stay in business, providing more and more jobs. Moreover, Deming stated emphatically that top management has the overriding responsibility for quality improvement (Dean and Evans, 1994).

The major core elements of *Juran's philosophy* are summarised in Figure 3.19. Juran (1951, 1964) emphasised the importance of customer perception of quality and paid much attention to management commitment on quality.
Juran suggested that employees at different levels of an organisation speak in different "languages". On the contrary, Deming believed that statistics could be the common language. According to Juran, top management speaks in the language of dollars, workers speak in the language of things, and middle management must be able to speak both languages and translate between dollars and things. Therefore, to obtain the attention of top management, quality issues must be expressed in money. Juran advocated also accounting and analysis of quality costs in order to focus attention on quality problems. At the operational level, Juran proposed to increase conformance to specifications through elimination of defects, supported extensively by statistical tools for analysis. His philosophy fitted well within existing management systems. Juran defined quality as "fitness for use". This was divided into four categories: quality of design, quality of conformance, availability and field service. Quality of design related to market research, the product concept, and design specifications. Quality of conformance included technology, manpower, and management. Availability focused on reliability, maintainability and logistical support. Field service quality comprised promptness, competence and integrity.

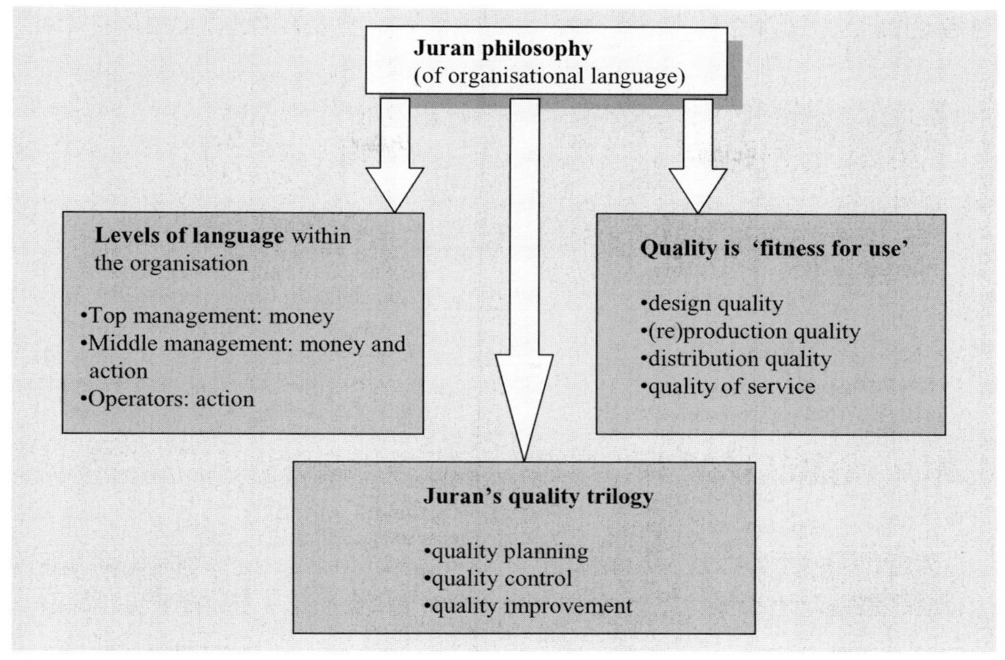

Figure 3.19. Juran's philosophy.

Like Deming, Juran advocated a never-ending spiral of activities that included market research, product development, design, planning for manufacture, purchasing, production process control, inspection and testing, followed by customer feedback. Because of the interdependence of these functions, the need for competent company wide quality management is great. Senior management must play an active and enthusiastic leadership role in the quality management process.

Juran viewed the pursuit of quality on two levels: (1) the mission of the firm as a whole is to achieve high product quality and (2) the mission of each individual department in the firm is to achieve high production quality.

Juran's prescriptions were focused on three major aspects of quality called the *Quality Trilogy* (a registered trademark of the Juran Institute). The triangle include quality planning - the process for preparing to meet quality goals, quality control - the process for meeting quality goals during operations, and quality improvement - the process for breaking through to unprecedented levels of performance (Dean and Evans, 1994).

The major principles of *Feigenbaum's philosophy* are shown in Figure 3.20. Feigenbaum (1961) suggested in the fifties that a high-quality product is more likely to be produced through Total Quality Control (also the title of his book) then when manufacturing works in isolation. Feigenbaum stated that bad quality is made in the "hidden factory" every day by producing *e.g.* scrap and rework, leading to high costs and no profit.

Feigenbaum stated also that quality should be everybody's job and everyone should feel responsible. Thereby, he emphasised Juran's message on managerial responsibility (Bounds,

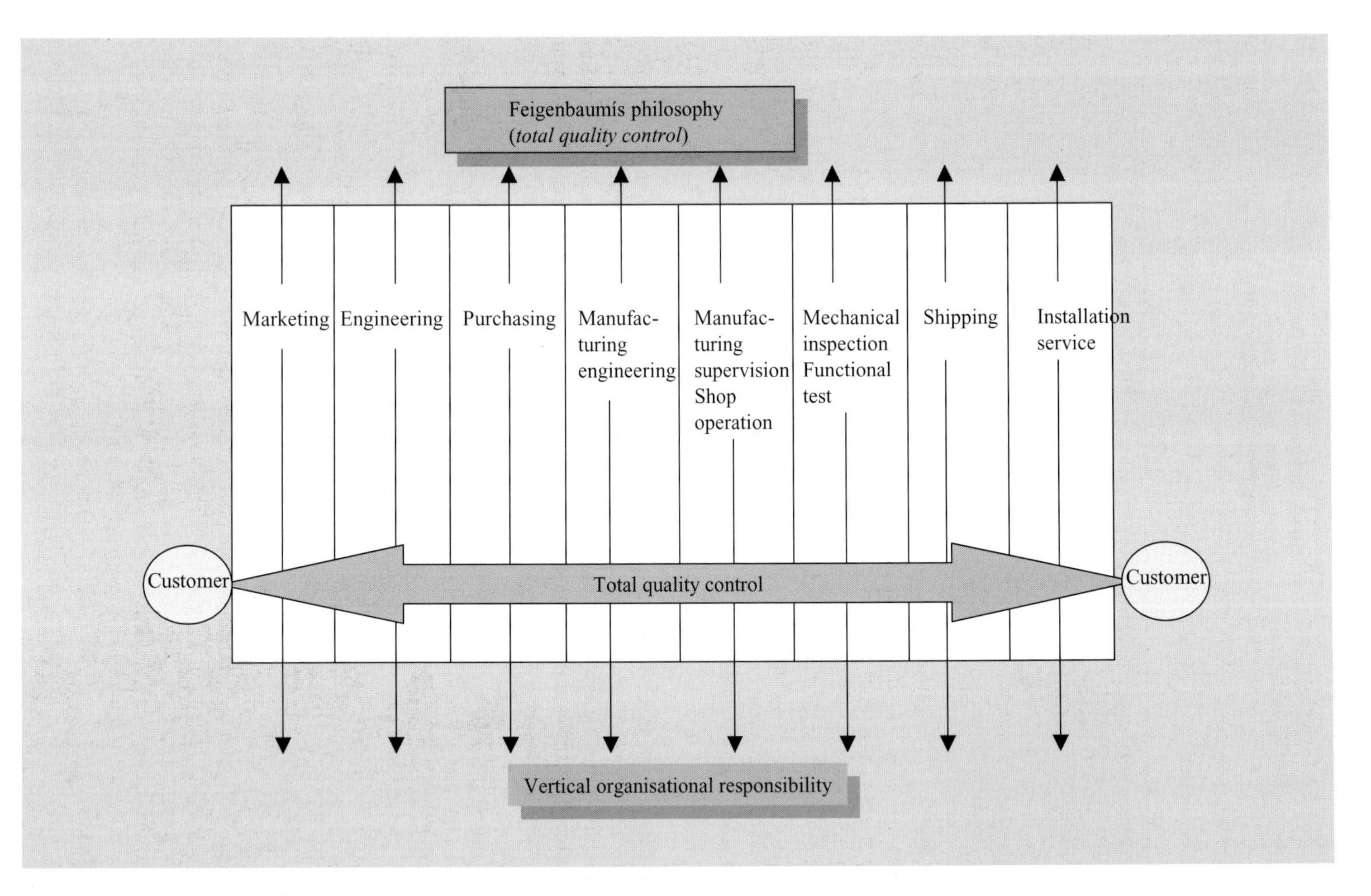

Figure 3.20. Feigenbaum's philosophy.

1994). With his company-wide system of total quality control, Feigenbaum can be considered as the father of *quality assurance*.
Total quality control is an effective system for integrating the quality-development, quality-maintenance and quality improvement efforts of various groups in an organisation. The effective system will enable marketing, engineering, production and service to operate at the most economical levels, which will result in full customers satisfaction (Feigenbaum, 1961).
Feigenbaum proposed a systems approach to quality through the definition of a quality system. It was defined as "a company and plant-wide operating work structure, documented in effective integrated technical and managerial procedures, for guiding the co-ordinated actions of work force, machines, information of the company and plant in the best and most practical ways to assure customer quality satisfaction and economical costs of quality".
Feigenbaum expressed total quality control as a horizontal concept, stretching across the functional divisions of an organisation (Kolarik, 1995).

Crosby's ideas on quality management are represented in Figure 3.21. Crosby's (1979) quality philosophy is embodied in his "Absolutes of Quality Management". The mean issues are:
Quality means conformance to requirements. Crosby dispelled the myth that quality is simply a feeling of excellence. Requirements must be clearly stated so that they cannot be misunderstood. Once a task is carried out, one can execute measurements to determine conformance to requirements. The non-conformance detected is the absence of quality. Quality problems become non-conformance problems, because it is variation in output. Setting requirements is the responsibility of management.
Crosby suggested that there is no such thing as a quality problem. The individuals or departments that cause them must identify problems. There are accounting problems, manufacturing problems, design problems, front-desk problems, but no quality problems. Quality is produced in functional departments, not in the quality department. Therefore, the burden of responsibility for such problems lies within the functional departments. The quality department should measure conformance, report results, and lead the drive to develop a positive attitude toward quality improvement.

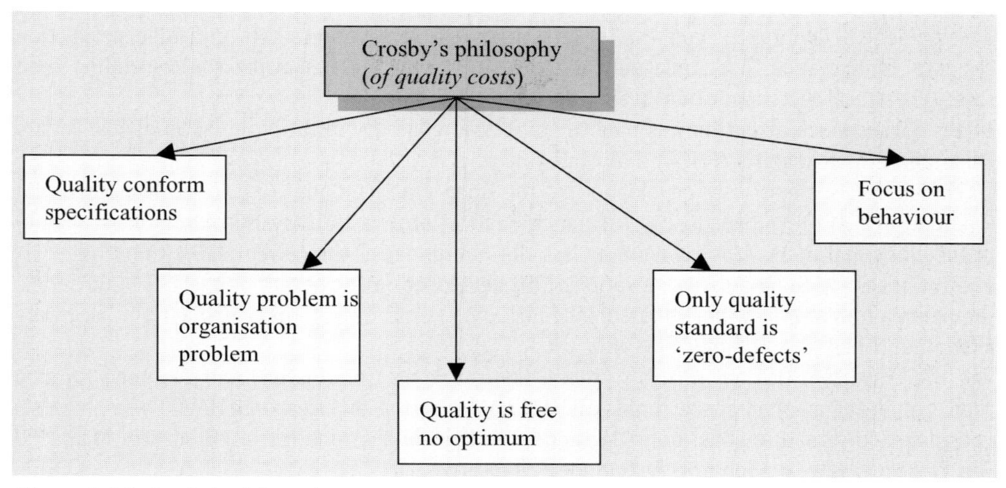

Figure 3.21. Crosby's philosophy.

Crosby also advocated that there is no such thing as the economics of quality: it is always cheaper to do the job right the first time. Crosby supported the premise that "economics of quality" has no meaning. Quality is free. What costs money are all the actions that are involved in not doing jobs right the first time. The cost of quality is the expense of non-conformance. Crosby noted that most companies spend 15 to 20 percent of their sales money on quality costs. A company with a well-run quality management program can achieve a cost of quality that is less than 2.5 percent of sales, primarily in the prevention and appraisal categories. Crosby's program stimulated to measure and publicise the cost of poor quality. Quality cost data are useful in drawing management's attention, selecting opportunities for corrective action, and tracking quality improvement over time. Such data will provide visible proof of improvement and recognition of achievement.

Crosby mentioned that the only performance standard is Zero Defects. The basic principle of zero defects is "do it right the first time", which means concentrating on preventing defects rather than just finding and fixing them. People are conditioned to believe that error is inevitable, it is human to make errors. However, the Zero Defects principle does not accept this, like we often do not accept in our personal life others to make errors.

Crosby's program is primarily behavioural. More emphasis was placed on management and organisational processes for changing corporate culture and attitudes than on the use of statistical techniques. Like Juran but unlike Deming, his approach fitted well within existing organisational structures. Crosby's approach, however, provided relatively few details about how firms should address the detailed points of quality management. The focus is on managerial thinking rather than on organisational systems.

In addition to the four described philosophers (or guru) other important names in quality management are Ishikawa, Taguchi and Imai. They contributed to specific aspects of quality management. Therefore, they will not be described here but dealt with in other specific topics (like improvement, policy) of this book.

Returning to the historical overview (Figure 3.17) it will be clear that between 1950 and 1980 the fundamentals of total quality management have been provided by the gurus mentioned above. Combined with developments in statistics they (mainly Feigenbaum) contributed to the development of assurance systems. These systems are very important in agri-food production (as it is for instance in pharmaceutical industry), being the framework for controlling food safety, health aspects and other quality aspects.

In the fifties, much attention was paid to statistical reliability resulting in methods like Failure Mode and Effect Analysis (FMEA) and Fault Analysis.

In the sixties, systems were developed like the Good Practice Codes (*e.g.* Good Manufacturing Practice) and AQAP (Allied Quality Assurance Publications), a NATO system for assuring quality of incoming materials. AQAP is more or less the forerunner of the current ISO-system.

The International Standards Organisation (ISO) developed in the seventies the ISO-9000 series, providing a framework for quality assurance including norms for external certification. They were firstly published as a standard in 1987.

The HACCP system (Hazard Analysis Critical Control Points) originated from the need for safe food supply for manned space flights. It was not until the eighties that HACCP was seriously considered for broad application in the food industry; in the nineties food industry in the EU was forced by law to apply the system (see chapter 7).

Probably in this century quality assurance systems and total quality management will be combined in flexible, though robust, innovative quality management approaches.

3.3 Quality management: planning and control

As previously mentioned, planning and control are two major functions of general management. However, for quality management, additional functions are important, like improvement and assurance. In this section, the functions planning and control are described from the quality management viewpoint. In our vision, product and resource should be considered separately. Therefore, the effect of product and resource decisions on planning and control are explained. Furthermore, the additional functions, improvement and assurance, are described in detail. Finally, our vision on the quality management functions are shortly described, *i.e.* quality strategy and policy, quality design, quality control, quality improvement, and quality assurance. These topics are the basis of this book.

3.3.1 Planning and control

There are different definitions of quality management. There is a broad acceptance of the following definition that will be used in this book. In this broad description quality management is defined as the total of activities and decisions performed in an organisation to produce and maintain a product with desired quality level against minimal costs. Quality management is that part of overall management activities that focuses on quality. When quality is defined as business performance, then quality management is a major part. When it is defined as technical product quality, quality management is a rather small part of total management activities. Nevertheless, quality management always consists of the four management functions: planning, control, leading and organising. They will be discussed in sections 3.3, 3.4 and 3.5

Planning is the process of determining exactly what the organisation will do to accomplish its objects. In more formal terms, planning is "the systematic development of action programs aimed at reaching agreed business objectives by the process of analysing, evaluating, and selecting among the opportunities which are foreseen".

The planning process consists of the following six steps, as illustrated in Figure 3.22.
1. *State organisational objectives.* Since planning is focused on how the management system will reach organisational objectives, a clear statement of those objectives is necessary before planning can begin. In essence, objectives stipulate those areas in which organisational planning must occur.
2. *List alternative ways of reaching objectives.* Once organisational objectives have been clearly stated, a manager should list as many available alternatives as possible for reaching those objectives.
3. *Develop premises on which to base each alternative.* To a large extent, the feasibility of using any alternative to reach organisational objectives is determined by the assumptions on which the alternative is based. For example, a manager could generate two alternatives to reach the organisational objective of increasing profit. Alternative (a) is to increase the scale of products presently being produced; (b) is to produce and sell a completely new product. Alternative (a) is based on the principle that the organisation can gain a larger

share of the existing market. Alternative (b) is based on the premise that a new product would capture a significant portion of a new market. A manager should list all of the premises for each alternative.
4. *Choose the best alternative for reaching objectives.* An evaluation of alternatives must include an evaluation of the premises on which the alternatives are based. A manager usually finds that the premises on which some of the alternatives are based are unreasonable and can therefore be excluded from further consideration. This elimination process helps the manager to determine which alternative would best accomplish organisational objectives.
5. *Develop plans to pursue the chosen alternative.* After an alternative has been chosen, a manager begins to develop strategic (long-range), tactical (mid-range) and operational (short-range) plans.
6. *Put the plans into action.* Once plans that furnish the organisation with both long-range and short-range direction have been developed, they must be implemented. Obviously, the organisation cannot directly benefit from the planning process until this step is performed (Certo, 1997).

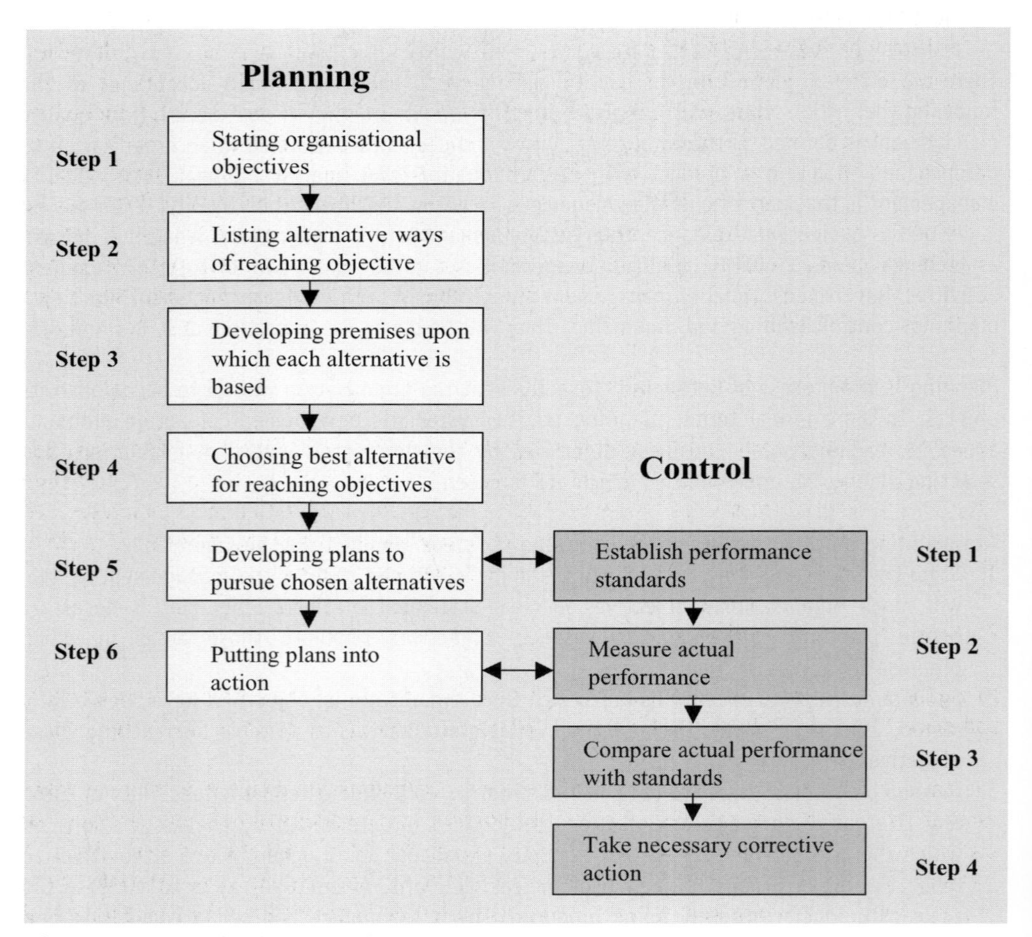

Figure 3.22. Steps in planning and control.

Controlling can be described as: "the systematic effort by managers to compare performance to predetermined standards (plans or objectives), and to determine if it is in line with these standards." Actions are taken when required, to assure that human or other corporate resources are being used in the most effective and efficient way in order to achieve corporate objectives (Certo, 1997).

The control process consists of the following steps (Figure 3.24):

1. *Establishing performance standards*. Targets set by management against which actual performance will be compared. This control is related to planning. Formal targets are formulated in plans. In the control activity these established targets are reformulated in a way that measurement is possible. They also should be clear and realistic within the framework of both the internal and the competitive environment in which the organisation operates.
2. *Measuring performance*. In many processes performance measurement is continuous and in that case often automated. But often measurement is only economic when doing it by statistical sampling. Statistics are used for determining reliable sampling procedures.
3. *Comparing performance against standards*. Once measurement results are available, they must be related to standards. Actual performance may match performance standard exactly, or it may be higher or lower than the target. Management must decide how much deviation from standards will be tolerated before considering corrective action. In fact, standards rise in most cases useless when they are not accompanied with tolerances. After comparing against standards it should be decided to take corrective action or not.
4. *Taking necessary corrective action*. Before it is assessed how one has to respond to a deviation, the reason for the deviation has to be determined. Specifically it has to be determined whether the plan was properly implemented but failed to work, or if it was not properly implemented. Then can be decided which of the three basic options is the appropriate response, *i.e.* correct deviation, change standards or maintain the status quo (Gatewood, 1995).

3.3.2 Product and resource decisions

According to Kampfraath and Marcelis (1981) a sharp distinction should be made between product and resource decisions. They stressed that decisions on resources are made long before actual manufacturing or service processes take place. In the short term, these processes and their outcome are dependent on which resources are available at that moment. Therefore, planning and control activities are quite different depending on being:

1. long-term or short-term oriented
2. resource or product oriented

In Figure 3.23, three main resources are distinguished, *i.e.* suppliers, technical infrastructure and markets, hereby following the simple input-output model of organisations (Figure 3.1).

The basic point about resources is that often large investments are required, whereas it takes time before they can be used. With respect to buildings and machinery this will be clear, but also investments are required in supplier-relations. In this case, marketing is fully aimed at creating markets by investments in market channels, promotion and customer-relations. Note that in Figure 3.23 information resources and human resources are missing. In the approach of Kampfraath and Marcelis these resources were considered as part of the administrative

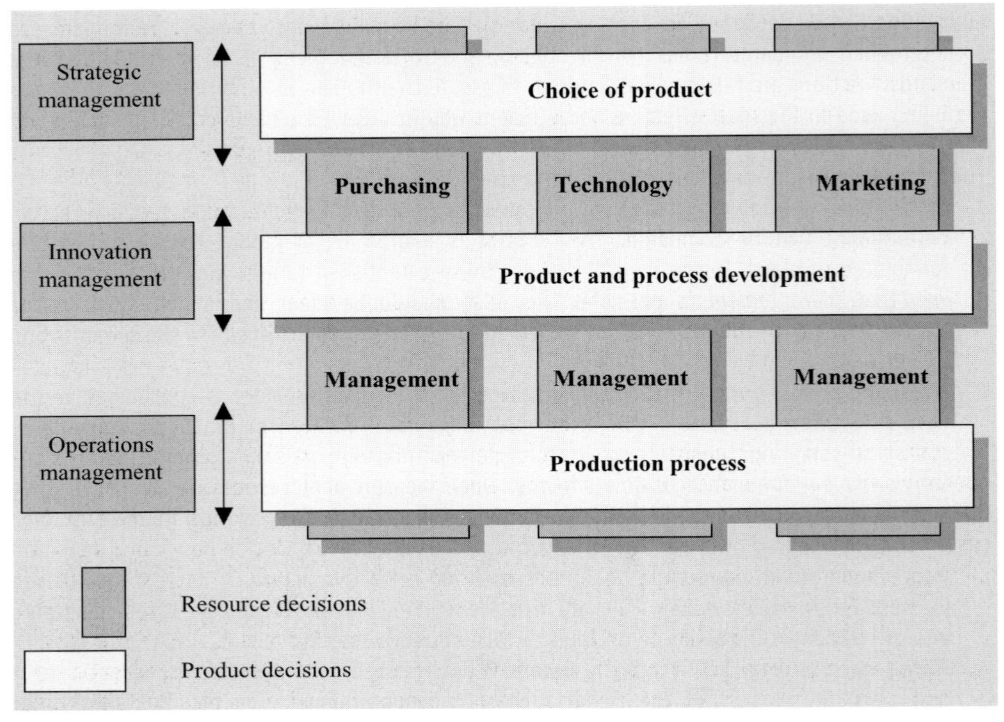

Figure 3.23. Management concerns.

infrastructure. Within this administrative structure (*i.e.* the organisation, perceived as a collective of decision-making people with their administrative support systems), the management processes have to be executed. These processes are strategic, innovation and operations management.

- *Strategic management* is the process of deciding on the organisation's mission, overall objectives, product/market combinations and major resource allocations.

Decisions on *products* deal with *e.g.* products and product-groups, segmentation of the market, and the quality level of products.

Decisions on *resources* are of course related to the product decisions. It includes decisions like supplier selection, degree of co-operation with suppliers, investments in machinery and buildings, degree of automation in manufacturing, and investments in market development and customer relations.

In strategic decision-making it is very important to look ahead and to imagine future market and production conditions. However, the actual decision will be mainly focused on resources, because these decisions are more or less irreversible and have long-term consequences.

- *Innovation management* is the process of deciding on product and/or process innovations, which are necessary to reach strategic goals and objectives, and which provide conditions and knowledge for future manufacturing.

Product decisions concern, for example, new product development, defining (new) process requirements and conditions, product testing, and developing product and material specifications.
Resource decisions concern, for example, technological process development, machinery modification, maintenance programs, detailing supplier relations, assuring purchasing channels, detailing customer relations and market development.
Innovation management is both product and resource oriented and functions as the natural link between strategic and operations management activities.

- *Operations management* is the process of deciding on customer orders and processes of supply, production, distribution of products or services.

Product decisions involve, for example, quality, quantity, delivery time, production schedules and storage and distribution schedules.
Resource decisions include, for example, the way resources are used in production processes, including necessary maintenance, hygiene in the manufacturing part of the process, and the co-operation with suppliers and customers in the remaining part of the process.
Operations management is more product-oriented than resource-oriented due to the fact that customer satisfaction is mainly dependent on product or service. Additionally, resources are hardly to change in the short term.

In most companies in the sixties and seventies, management paid much attention to operations management. In the eighties, strategic management increased strongly due to globalisation and concerns of stakeholder. Nowadays, innovation management is getting its place between them due to increasing customer-orientation and growing intensity of competition.
Based on the ideas reflected in Figure 3.23, planning and control activities appeared in different areas of management concern. While planning and control differ depending on the area of concern, they are strongly interrelated. They are most interrelated at those points where product and resource decisions cross each other.

Quality management is managing a part of the total business performance, but it addresses all areas of management concern and their interrelations. With respect to the strategic management area, quality strategy and policy are a logical part of the strategy and policy of the company. In the innovation management area quality design is perhaps the biggest part of total management's attention.
In operations management, quality control is an important part of management activity in ensuring customer satisfaction.
The broader the definition of quality (for instance including quality, delivery time, price, service, flexibility and dependability), the greater the impact of quality management in total business management.

3.3.3 Quality improvement

The four basic functions of management, planning and control, leading and organising cannot be simply translated to quality aspects. All quality philosophers pinpointed the importance of quality improvement as a cornerstone of quality management. Juran (1990) in particular focused on three major aspects of quality: the so-called Quality Trilogy (Figure 3.24).

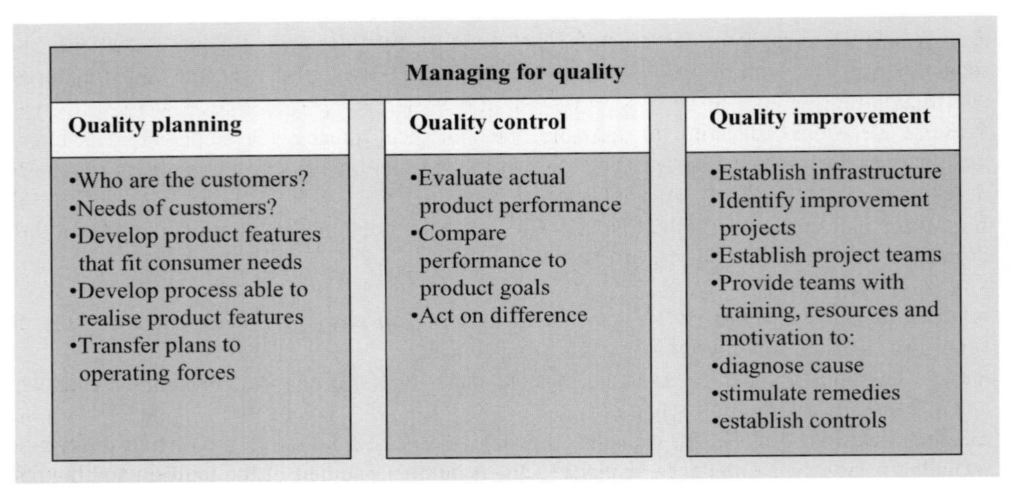

Figure 3.24. Quality Trilogy of Juran.

Juran added quality improvement to quality planning and quality control. Quality planning begins with identifying both external and internal customers, by determining their needs and developing product features that respond to these needs. Like Deming, Juran stressed employees to know who use their products, whether it is in the next department or in another organisation. Then quality goals are established, which meet the needs of customers and suppliers against minimum costs. Subsequently, a process under operating conditions must be designedwhich can deliver products that meet customers. Strategic planning for quality should be similar to the firm's financial planning process. The strategic planning should determine short-term and long-term goals, set priorities, compare results with previous plans and match the plans with other corporate strategic objectives.

Quality control involves determining what to control, establishing units of measurements so that data may be objectively evaluated, establishing standards of performance, measuring actual performance, interpreting the difference between actual performance and the standard, and taking action on the difference. Likewise, Deming emphasised on identifying sources of variation and improving the work system.

Unlike Deming, Juran specified a detailed program for quality improvement. The quality improvement process involves the following steps:

- proving the need for improvement,
- identifying specific projects for improvement and organising guidance of the projects,
- diagnosing the causes and providing remedies for these causes,
- proving that the remedies are effective under operating conditions,
- providing control to maintain improvements.

Juran's assessment of most companies is that quality control has priority among the trilogy categories. Most companies feel that they are strong in this category. Quality planning and improvement often have not high priorities and are significantly weaker in most organisations. Juran proposed that more efforts should be put in quality planning and even more in quality improvement.

Juran supported these conclusions with several case studies, in which Japanese firms used the same technology, materials and processes than American firms, but had much higher levels of

quality and productivity. He explained these findings by the fact, that since the1950s the Japanese have implemented quality improvement projects to a much larger extent than their Western counterparts. The result was that in the 1970s, Japanese product quality exceeded Western quality and continued to improve more quickly.
Massive training programs and top management leadership supported Japanese efforts on quality improvement. It encompassed training in managerial quality-oriented concepts, as well as training in the tools for quality improvement, cost reduction, data collection and analysis. The latter is one of the most important components of Juran's philosophy. Juran showed that the Japanese experience left little doubt to the significance of quality training in competitive advantage, reduced failure costs, higher productivity, smaller inventories and better delivery performance (Dean and Evans, 1994).
In fact, improvement can be perceived as an extension of planning and control. It is a kind of control in the control process, where much more attention is paid to structural causes and solutions. Structural solutions can only be found and implemented when there is more planning activity. So in this perspective, improvement can been considered as an addition to planning and control.

3.3.4 Quality assurance

Another addition to planning and control is quality assurance. Throughout the world, food manufacturing, distribution and retailing is becoming a highly complex business. Raw materials are obtained from sources world-wide an ever-increasing number of processing technologies are used, and a broad range of products is available to the consumer. Such complexity necessitates the development of comprehensive control procedures to ensure the production of safe and high quality food. In addition, consumer expectations are changing with a desire for convenience, less processed and fresher foods, with more natural characteristics. Against this background, the total food chain has to ensure that the highest standards of quality and safety are maintained. All stages of the food chain, from acquisition of raw materials through manufacture, distribution and sales, whether it would be through retail or catering outlets, must consider quality issues associated with specific products, processes and methods of handling. There are a number of reasons why, especially in the agri-food business, implementation of quality assurance systems is an issue of greatest importance.

- Agricultural products are often perishable and subject to rapid decay due to physiological processes and/or microbiological contamination.
- Most agricultural products are harvested seasonally.
- Products are often heterogeneous with respect to desired quality parameters such as content of important components (*e.g.* sugars), size and colour. This kind of variation is dependent on cultivar differences and seasonal variables, which can hardly be controlled.
- Primary production of agricultural products is performed by a large number of farms operating on a small scale.

Irrespective of size and complexity, all agri-food businesses must have an appropriate food quality assurance programme. Currently, quality assurance programmes focus primarily on food safety issues (*e.g.* Hazard Analysis Critical Control Points, HACCP), but the general principles are also applicable to the management of product quality in a wider sense.

Modern trading conditions and legislation require agri-food businesses to demonstrate their commitment to food quality and to establish an appropriate quality programme (Hoogland, Jellema and Jongen, 1998).
A quality assurance system provides extra guarantees with respect to food safety and meeting customer agreements. In terms of planning and controlling, a quality assurance system is an extra control system, initiating strategic quality plans in order to change the quality system when necessary.

3.3.5 Quality management functions

The basic functions in (food) quality management can be derived from the previous analysis. Planning and controlling on the strategic level includes decisions on quality, which can be labelled as Quality Strategy and Policy. At the innovation level decisions on quality can be summarised as Quality Design. Planning and controlling at the operations level includes decisions on quality, which can be labelled as Quality Control. These quality management functions have to be extended to Quality Improvement on behalf of the quality gurus. An extension to Quality Assurance is needed on behalf of the customer and legislative requirements concerning food safety and quality reliability. Figure 3.25 shows the different functions and their relationships.

It should be noted that each quality management function in our view has both a product-oriented and a resource-oriented part. In the next chapters, for each quality management function this distinction will be discussed. The quality management functions can be summarised as:

1. Quality strategy and policy: this is the process of deciding on
- quality goals and objectives
- quality level of products
- quality level of resources
- the quality system

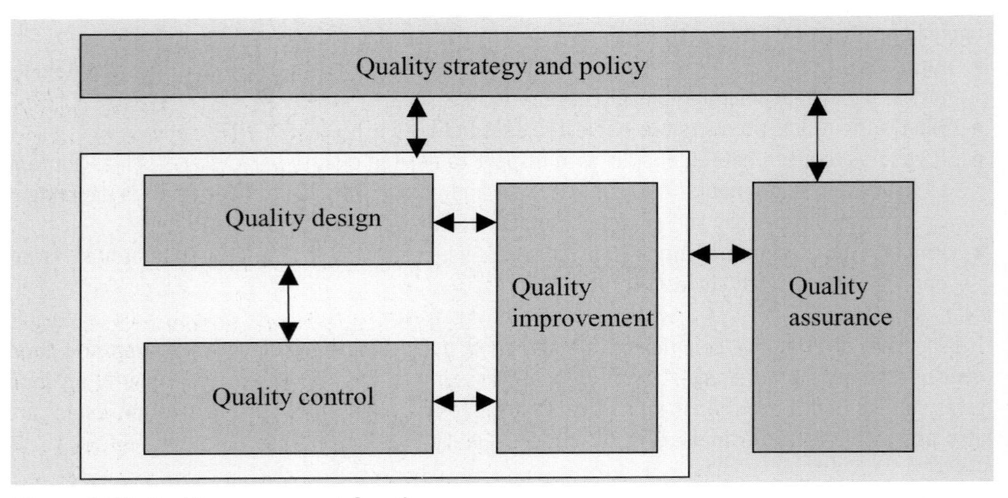

Figure 3.25. Quality management functions.

The planning activity can be characterised as long-term strategic planning, whilst the control activity can be seen as policy verification. Planning is the most important out of both.

2. Quality design is the process of deciding on
- product and material specifications
- production process requirements
- quality of resources in terms of specifications

The planning activity is mid-term oriented and can be typified as project planning, whereas the control activity is related to project control. Both are equally important.

3. Quality control is the process of deciding on
- actual product and material quality
- actual production process quality
- actual quality of process environment (production rooms, storage facilities)

The planning activity is short-term oriented and restricted to introducing quality interests in production planning. Control is the most important activity, however it is closely linked to production planning. That's why we use the term "quality control" here for what in fact is the quality planning and control of executing production. With "quality control" we use a term that is common in daily practice.

4. Quality improvement is the process of deciding on
- changes in product and material specification
- changes in production processes
- changes in quality of resources

As in quality design, the planning and control activities can be typified as project planning and project control. They are mid-term oriented.

5. Quality assurance is the process of deciding on
- requirements to the quality system
- providing the quality system related to organisation and technology
- actual performance of the quality system
- necessary changes in the quality system

Planning activity includes future developments of the quality system, as related to strategic planning. Control-activities concentrate on auditing (mostly on a yearly base) the quality system and on taking or initiating corrective action when necessary. Third parties could be involved in auditing, particularly in the case of system certification.

3.4 Quality management: leading

Management is a broad concept that encompasses activities such as planning and control, organising as well as leading. Leadership, on the other hand, is focused almost exclusively on the "people" aspects of getting a job done. That means inspiring, motivating, directing, and gaining commitment to organisational activities and goals. Leadership accompanies and complements the other management functions, but it is more concerned with coping with the dynamic, ever-changing market place, rapid technological innovation, increased foreign competition and other fluctuating market forces. In short, management influences the brains, while leadership influences the heart and the spirit (Gatewood, 1995). In this section the concept of leadership will be elaborated including the main approaches that can be found in literature. Dealing with people aspects requires an analysis of people behaviour. The most important factors influencing motivation and quality behaviour will be discussed.

3.4.1 Leadership

A treatment of leadership by Kotter (1992) compared the concept of leadership to the concept of management. According to this view, management is needed to create order in the complexity, whereas leadership is needed to stimulate the organisational change necessary to keep up with a changing environment. Kotter (1992) differentiated leadership from management by contrasting their central activities. While management begins with planning and budgeting, leadership begins with setting a direction. Setting of direction involves creating a vision of the future and a set of approaches for achieving the vision. To achieve goals, management practices organising and staffing, while leadership works on aligning people, *i.e.* communicating the vision and developing commitment to do it. Management achieves plans through controlling and problem solving, whereas leadership achieves its vision through motivating and inspiring (Dean and Evans, 1994).

Referring to the model of group decision-making in this book (Figure 3.10) it will be clear that effectiveness of leadership depends besides personal characteristics of the leader, also on the use of power, the communication process and the way of dealing with interests and conflicts. Effective leadership creates clear values that respect the capabilities and requirements of employees and other company stakeholders. It sets high expectations for performance and performance improvement. It builds loyalties and teamwork based upon the values and the pursuit of shared purposes. It encourages and supports initiative and risk taking of subordinates in the organisation. It avoids chains of command that require long decision paths.
Although quality management principles and practices may differ among firms and industries, there is unanimous agreement on the importance of leadership in implementing quality management. According to Juran, it cannot be delegated. Strong visionary leaders are an important element of a quality management approach (Ross, 1999).

3.4.2 Leadership approaches

Three different approaches to leadership, *i.e.* the trait, the behavioural and the contingency approaches, have tried to answer the question why some persons perform very well as a leader, whereas others do not (Schermerhorn, 1999; Robbins, 2000).

The *trait approach* (Figure 3.26) claims that effective leadership depends on personal traits of the leader. However, researchers were unable to establish a defined profile of traits that consistently account for leadership success.
The *behavioural approaches* tried to determine which leadership style, *i.e.* the recurring pattern of behaviours exhibited by a leader, worked best. Blake and Mouton (1964) proposed a managerial grid based on the styles of "concern for people" and "concern for production" (Figure 3.27). They concluded that managers perform best under the team management style compared with the other styles. However, there is little substantive evidence to support the conclusion that the team management style is most effective in all situations (Robbins, 2000).

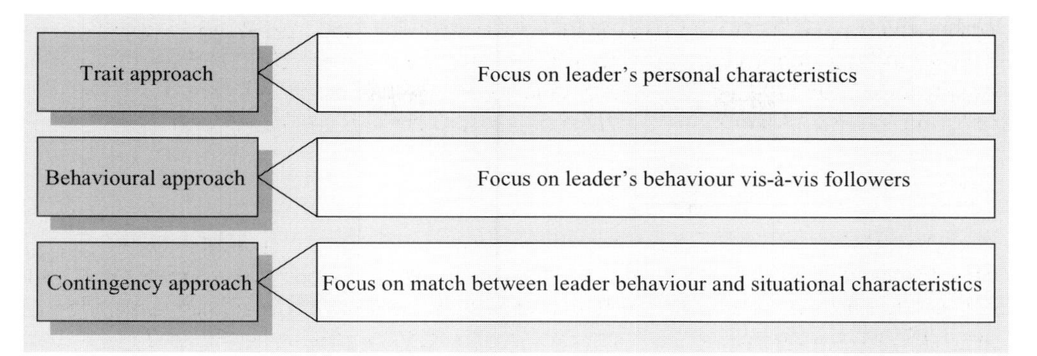

Figure 3.26. Overview of leadership approaches.

The *contingency approaches* attempted to understand the conditions for leadership success in widely varying situations. A good example is the situational leadership theory of Hersey and Blanchard (1985), whose premise was that a leader's style should be contingent on subordinates' competence and commitment (Figure 3.28).
According to this theory managers assess an employee's development level on a continuum as shown in Figure 3.28 (D1 to D4). They can find the corresponding leadership style out of the four styles (S1 to S4) differing in degree of supportive and directive behaviour.

Another leadership model is the Transformational Leadership Theory. This is a model that fits in modern quality management. According to this theory, leaders who wish to have a major impact on their organisations must take a long-term perspective. They have to stimulate their organisation intellectually, invest in training to develop individuals and groups, take some risks, promote a shared vision and values, and they should focus on customers and employees as individuals (Dean and Evans, 1994).

3.4.3 Motivation

Motivation is the inner state that causes an individual to behave in a way that ensures the accomplishment of some goals. In other words, motivation explains why people act as they do. The better a manager understands the behaviour of organisation members, the more able the manager will be to influence their behaviour to make it more consistent with the accomplishment of organisational objectives. Since performance is a result of the behaviour of organisation members, motivating them is the key to reaching organisational goals. Several theories about motivation have been proposed over the years. Most of these theories can be categorised in two basic types, *i.e.* process and content theories.

Process theories include explanations of motivation that emphasise on *how* individuals are motivated. They focus on the steps that occur when an individual becomes motivated.
Content theories include explanations of motivation that emphasise on people's *internal* characteristics. One of the best-known content theories is the Maslow's Hierarchy of Needs (Certo, 1997). Siegall (1987) described an integrated model of motivation. Gatewood (1995) adapted this model, in which the complex relations in motivation are shown (Figure 3.29).

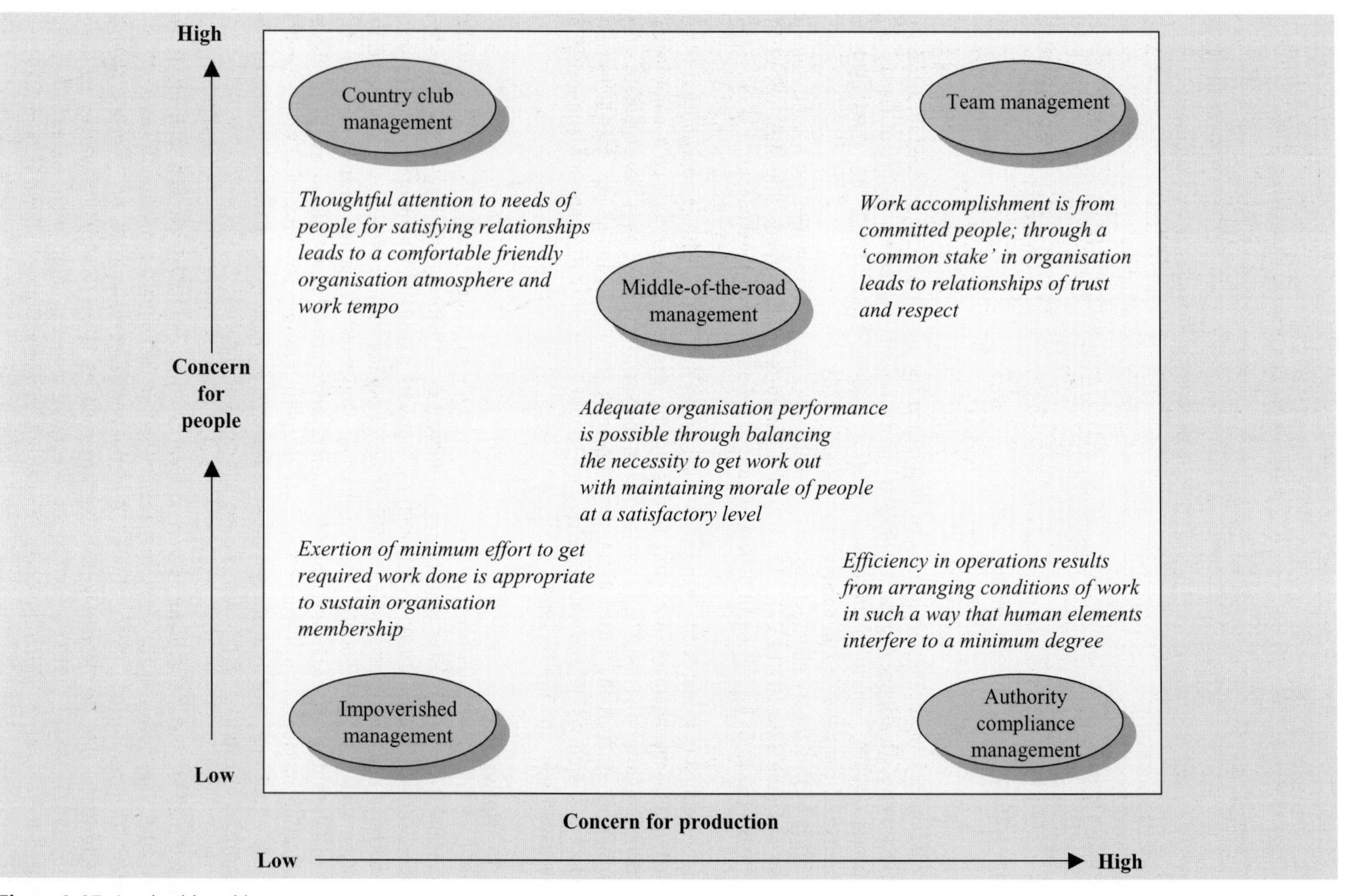

Figure 3.27. Leadership grid.

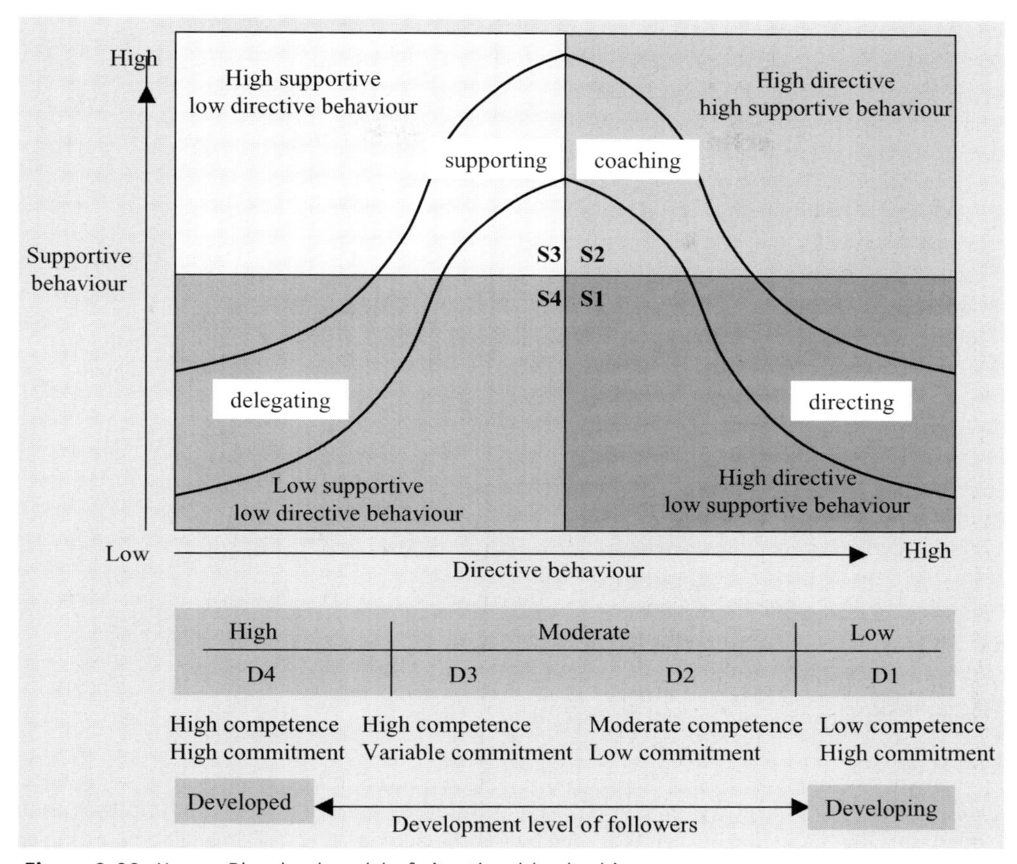

Figure 3.28. Hersey-Blanchard model of situational leadership.

Figure 3.29 shows that effort directly influences performance, whereas effort itself is affected by goals. Performance will be high when employees experience a relationship between their performance level and rewards. Although effort directly affects performance, ability has a moderating effect. It means that the more training, experience or basic talent, the greater the performance level. Performance level is directly related to outcome rewards. Performance evaluation must be fair and equitable for employees to view the outcome rewards favourably. The rewards serve to stimulate the behaviour or performance level. Inequitable or low-level rewards may inhibit maximum performance of employees. Growth-need strength moderates the relationship between outcome rewards and individual goal. It is the extent to which an employee desires activities that provide personal challenges, a sense of accomplishment and personal growth. Individual goals and individual need for achievement determine how goal-directed or motivated the individual is. The higher the employee's need for achievement, the greater the goals and the more defined the goal-directed behaviour. The need for achievement is an internal drive and individual goals may also take into account external needs.

Motivation is a process that affects all the relationships within an organisation and influences many areas such as pay, promotion, job design, training opportunities and reporting

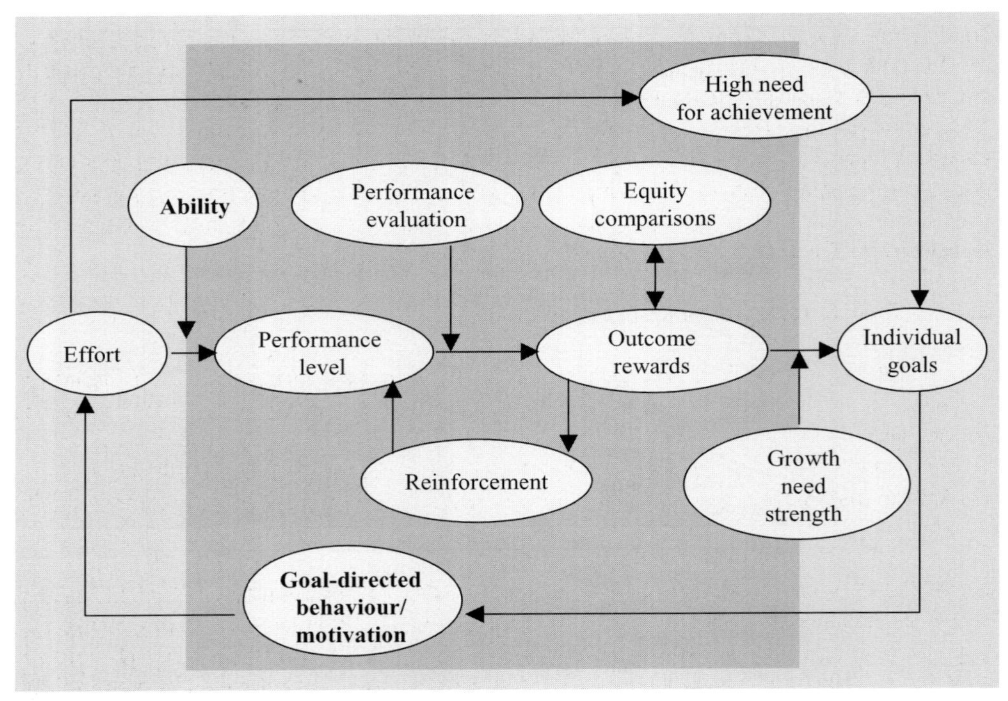

Figure 3.29. Integration of motivation theories.

relationships. Fundamentally, employees are motivated by a satisfying relationship with their supervisors, by the nature of their jobs, and by the characteristics of the organisation, as shown in Figure 3.30 (Gatewood, 1995).

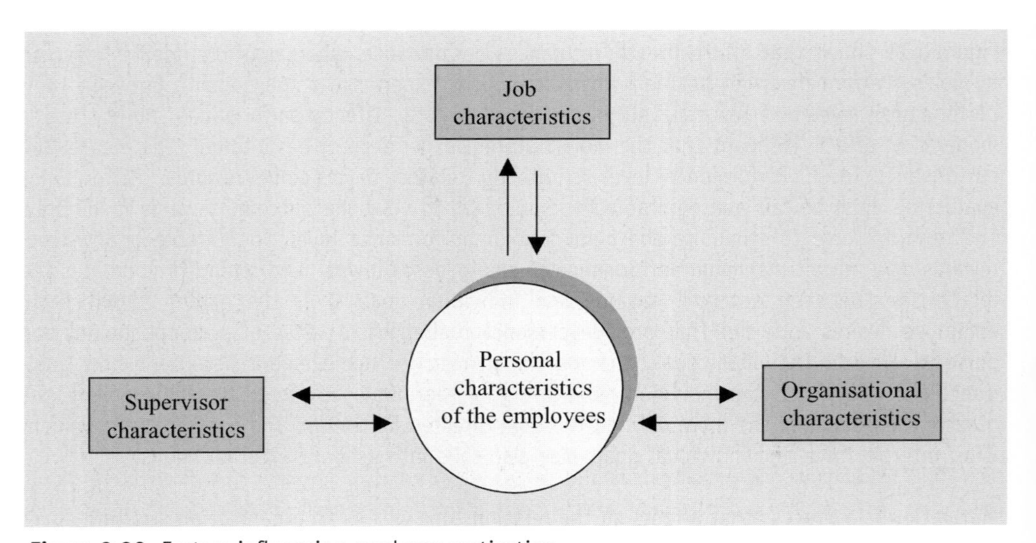

Figure 3.30. Factors influencing employee motivation.

The importance of motivation is certainly recognised in relation to quality. A firm committed to the principles of quality management frames its core question about how to motivate employees as follows: "How do we enable workers to feel a natural sense of pride in their work and to be self-motivated". This approach to motivation is based on the assumption that employees inherently want to do a good job. In the quality based management approach, the primary responsibility of management, in terms of employee's performance, is to create and maintain a motivation system that support each employee's natural ambition and pride of workmanship (Ivancevich, 1994; Deming, 1986; Crosby, 1979).

3.4.4 Quality behaviour and empowerment

Gerats (1990) made in his dissertation on quality behaviour an analysis based on the mental incongruent theory. This theory argues that in analysing behaviour, two conditions should be considered:
1. Disposition, that is the employee's own disposition to behave in a certain direction
2. Ability, that is the objective opportunity to behave in a certain direction, *i.e.* the activity area

Gerats (1990) used this theory in his research and made a translation to quality behaviour as shown in Figure 3.31. The research was carried out in slaughterhouses and was focused on hygienic working behaviour. It was concluded that 60% of the workers did not comply with the conditions concerning disposition and ability. It was concluded that the activity area (ability) for hygienic working behaviour was mainly limited by shortcomings in management with respect to hygiene, by low hygiene standards amongst workers, by low hygiene standards of first line supervisors, and by shortcomings in the hygiene facilities at the workplace. Disposition to hygienic working was mainly limited by the low knowledge level of bacteriological contamination mechanisms, by the restricted social support from colleagues, by the low interest towards hygienic working of supervisors, and by the limited opportunities for hygienic working.

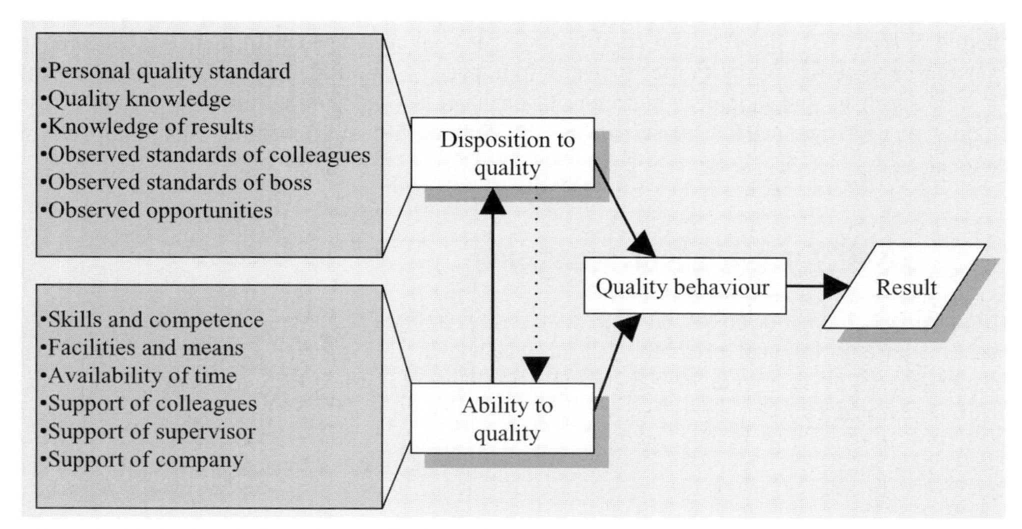

Figure 3.31. Factors influencing quality behaviour.

The research results indicated that for improving hygienic working behaviour, firstly the activity arena should be enlarged.

Ivancevich (1994) added a third factor, namely quality focus. In his view, commitment to quality includes three ingredients.

- Quality intelligence: employees must be aware of the acceptable quality standards and how these standards can be met,
- Quality skills: employees must have the skills and abilities to achieve the quality standards set by management,
- Quality focus: from top management to operating employees, everyone must sincerely believe that quality of all outputs is the accepted practice.

Commitment to quality can be obtained with motivational programs. Numerous approaches are available. They include job enrichment, goal setting, positive reinforcement and team development. Participating management involves employees in important management decisions.

A more recent development, in the context of quality management, is empowerment. Empowerment is the process through which managers enable and help others to gain power and achieve influence. Effective leaders not only accept participation, but also empower others. They know that when people feel powerful, they are more willing to make decisions and take actions needed to perform their jobs. They also realise that for gaining power it is not necessary for others to give it up. Success of an organisation may depend on how much power can be mobilised throughout all ranks of employees (Schermerhorn, 1999). Empowerment fits with the administrative concept of decentralisation (Figure 3.15). It has drastic consequences for the organisational structure. Employees will get responsibilities and authority for making decisions, and therefore will act according to those decisions, thus getting a sense of commitment and self-control. Empowerment satisfies basic human needs for achievement, a sense of belonging, and enables team members to increase their potential (Hellriegel, 1992).

Figure 3.32 shows a hypothetical continuum of team empowerment with global corresponding amount of responsibility/authority shown in four levels (Gatewood, 1995). When empowerment is well organised and supported by top management it can provide the flexibility needed in the highly competitive market place through facilitation of communication and reduction of bureaucracy.

3.5 Quality management: organising

Organising is the management function that creates an administrative infrastructure providing best conditions for goal oriented decision-making. In this section different organisational structures will be described. Special attention will be paid to the quality assurance department. Furthermore the factors that influence organisational design will be discussed.

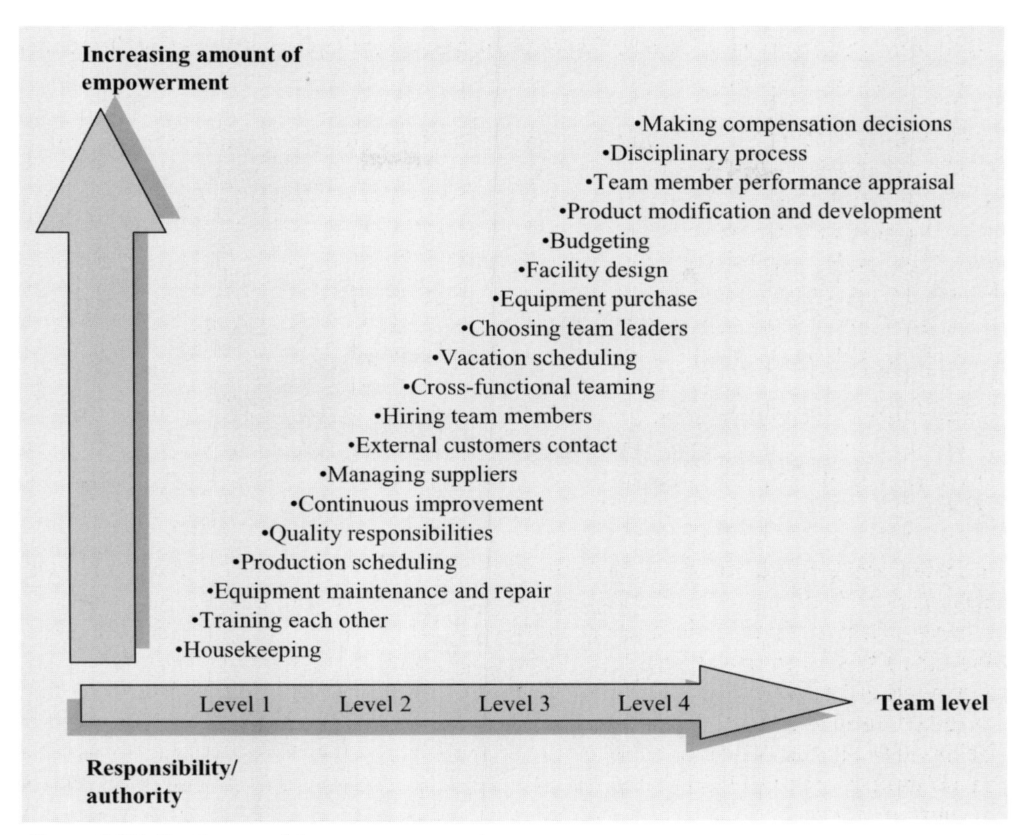

Figure 3.32. Continuum of team empowerment.

3.5.1 Organisation

Organising is the process of arranging people and other resources to work together in order to accomplish a goal. Organising involves creating a division of labour tasks to be performed and then co-ordinating results to achieve the common purpose (Schermerhorn, 1999).
As stated earlier, an organisation is a collection of people working together to achieve a common purpose (Figure 3.33). People are coming into action when they have decided about objectives and ways to accomplish them. Organising creates conditions for these decision-making processes in terms of:

- people: it includes attracting, developing and maintaining a quality workforce
- information: it involves ensuring information at the right moment and place
- system: it comprises providing resources to collect, organise and distribute data for decision support
- organisation structure: it includes defining tasks, responsibilities and authorities, rules and procedures

Today it is generally recognised that there are two prerequisites for a 'Quality organisation'. The first is a quality attitude that is spread through the entire organisation. Quality is not just a

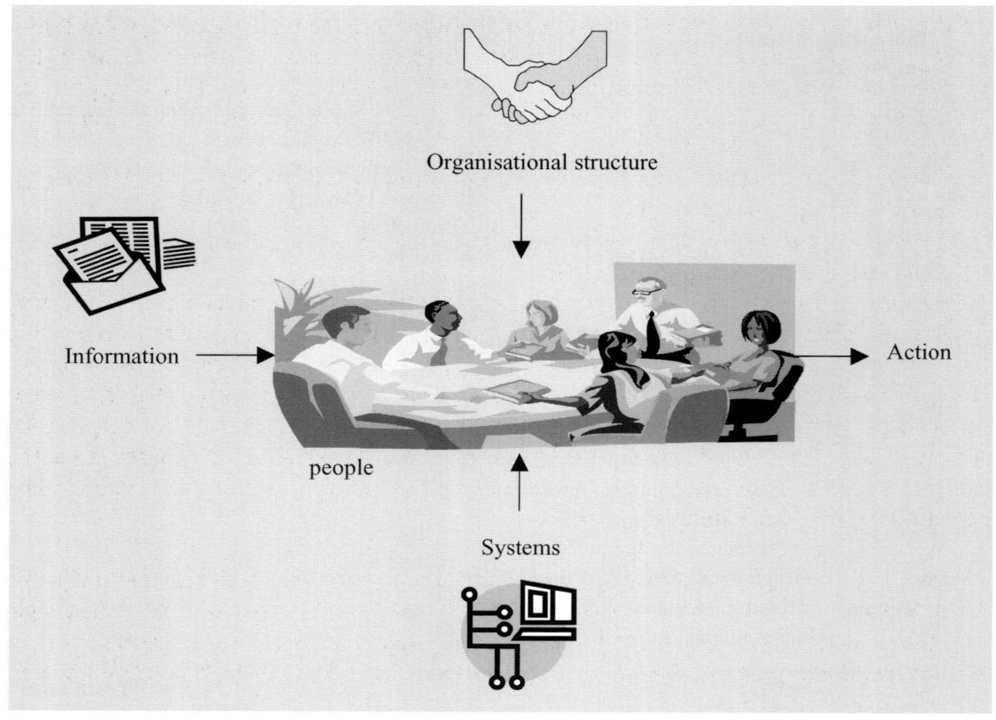

Figure 3.33. The organisation.

special activity supervised by a high-ranking quality director. The second prerequisite is an organisational infrastructure to support this attitude. People must be aware of the importance of quality and they must be trained to accomplish the necessary tasks (Ross, 1999). In one sentence: the organisation is a condition for quality behaviour.

In quality management the *human resource focus* is very important, *i.e.* how the company enables employees to develop and utilise their full potential, aligned with the company's objectives. In Human Resource Management much attention is paid to:

- *Work systems*
  How all employees contribute to achieving the company's performance and learning objectives through the company's work design, compensation and recognition approaches.
- *Education, training and development*
  How the company's education and training support the accomplishment of key company action plans and company needs. It includes building knowledge, skills and capabilities, and contributes to improved employee performance and development.
- *Employee well-being and satisfaction*
  How the company maintains a work environment and work climate that supports the well-being, satisfaction and motivation of employees.

The concept of Total Quality Management (TQM, see chapter 8) has drastic implications for the management of human resources. It emphasises self-control, autonomy and creativity among employees and requires more active co-operation rather than just compliance (Ross, 1999).

Information is a critical enabler for quality management. More and more successful companies agree that information technology and information systems serve as keys to their quality success. Three categories of information are mentioned to be critical for quality management.

- *Operational information* with emphasis on process management, action plans and performance improvement.
- *Comparative information* related to competitive position and best practices, both having an operational and strategic value.
- *Information that relates process management to business performance* providing insight in cause/effect relationships (Ross, 1999).

Information technology has many effects, including improving performance and affecting organisational structure. It helps firms to retrieve information faster and more conveniently, which serves decision-making. Moreover, it may help to improve organisational co-ordination. However, the fast developments in information and communication technology (ICT) resulted in an enormous amount of information sources and a tremendous volume of information, which are accessible through public (Internet) and private (Intranet) networks. As a consequence, the problem of technical access to information has been replaced by a problem of routing and selecting the right accurate information.

Organisational structure is a formal system of relationships that both separates and integrates tasks. Separation of duties makes it clear who should do what. Integration of duties tells people how they should work together. Organisational structure includes four basic elements:

- *Specialisation*: identifying particular tasks and assigning them to individuals or work groups who have been trained to do them.
- *Standardisation*: developing procedures in such a way that employees perform their jobs in a uniform and consistent manner.
- *Co-ordination*: developing procedures that integrate the activities performed by separate groups in an organisation.
- *Authority*: assignment of the right to decide and act to employees or work groups.

Specialisation and *standardisation* together are the basis for departmentalisation *i.e.* subdividing work and assigning it to specialised groups within the organisation.
However, in order to achieve organisational objectives managers also need to *co-ordinate people*, projects and tasks. Besides using procedures, managers also should have a limited number of employees reporting directly to them (span of control) to ensure a clear and unbroken chain of command.
Authority is the right to act or make a decision and implies responsibility and accountability. Responsibility is an employee's obligation to perform assigned tasks; accountability is the expectation that each employee will accept credit or blame for results achieved in performing assigned tasks. Delegation of authority is the process by which managers assign the right to act and make decisions in certain areas to subordinates. They assign a task to a subordinate along with adequate authority to carry it out effectively (Hellriegel, 1992).

It is worth to note in this context that behind every formal structure lies an *informal structure*. This is a "shadow" organisation, which is the unofficial, but often critical, working relationship between organisational members. If the informal structure could be drawn, it would show who talks to and interacts regularly with whom, regardless of their formal titles and relationships. The lines of the informal structure would cut across levels and move from side to side. The informal organisation would show people meeting for coffee, exercise in groups or friendly cliques, among other possibilities. It is important to realise that no organisation can be fully understood without gaining insight into the informal structure as well as the formal one. One should recognise that informal structures can be very helpful in accomplishing needed work. This is especially true during times of change when out-of-date formal structures may simply not provide the support people need to deal with new or unusual situations. Because it takes time to change or modify formal structures, this is a common situation. Through the emergent and spontaneous relationships of informal structures, people benefit by gaining access to interpersonal networks of emotional support and friendship that satisfy important social needs. They also benefit in task performance by being in personal contact with persons, who can help them get things done. In fact, what is known as informal learning is increasingly recognised as an important resource for organisational development. As a matter of fact, informal structures also have potential disadvantages. They exist outside the formal authority system, therefore the activities of the informal structures can sometimes work against the interests of the organisation as a whole. They can also be susceptible to rumour, carry inaccurate information, breed resistance to change and even divert work efforts from important objectives (Schermerhorn, 1999).

3.5.2 Organisation concepts and organisational structures

Administrative concepts (see section 3.1.7) describe alternative approaches to decision-making and administrative processes. Organisation concepts can be seen as the organisational answers to the requirements resulting from the administrative concepts (Figure 3.34).

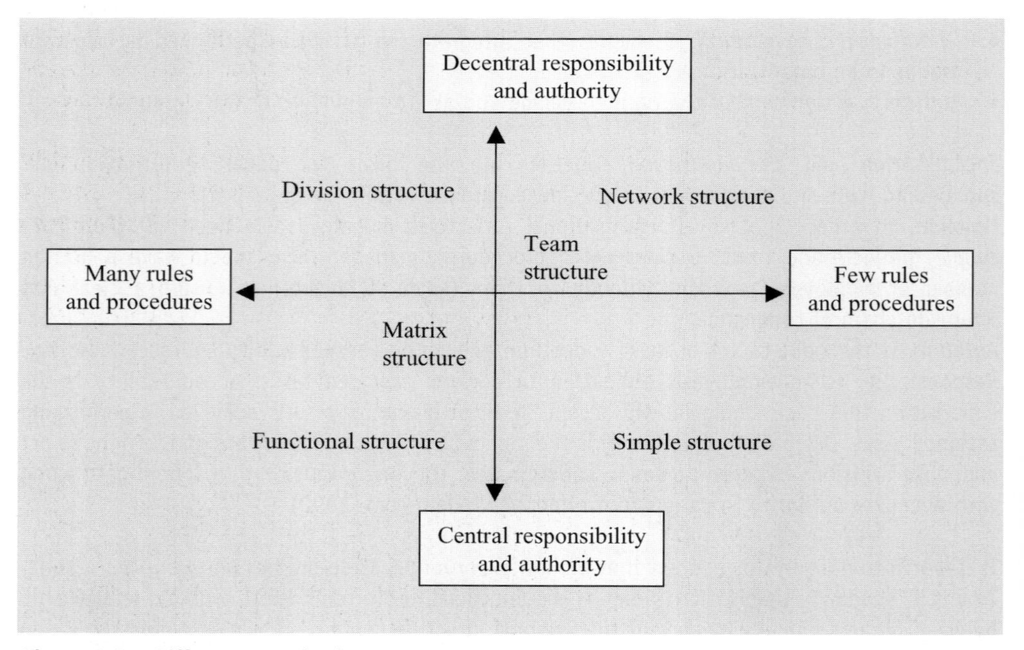

Figure 3.34. Different organisation concepts.

Decentralised responsibility and authority is the organisational answer to decentralisation. Rules and procedures often accompanied with formalised information systems are the organisational answer to formalisation as administrative principle.

The more authority is decentralised, the more delegation is needed. Delegation is the process of distributing and entrusting work to other persons (Schermerhorn, 1999). There are three steps in delegation.

In step 1, the managers assign *responsibility* by carefully explaining the work or duties to someone else who is expected to do the work. This responsibility is an expectation for the other person to perform assigned task.

In step 2 the manager grants *authority* to act. Along with the assigned task, the right to take necessary actions (for example to spend money or direct the work of others) is given to the other person. Authority is a right to act in ways needed to carry out the assigned tasks.

In step 3 the manager creates *accountability*. By accepting an assignment the person takes on a direct obligation to the manager to complete the job as agreed upon.

When formalisation is low (Robbins, 2000), job behaviours are relatively non-programmed and employees have a great deal of freedom to exercise discretion in their work. An individual's discretion on the job is inversely related to the amount of behaviour in that job that is pre-programmed by the organisation. Therefore the greater the standardisation, the less influence the employee has into how his or her work has to be done. Standardisation not only eliminates the possibility of employees' engaging in alternative behaviours, but it removes the need for employees even to consider alternatives.

According to Figure 3.34, basic alternatives for the organisation structure can be distinguished. The *simple structure* is best characterised by what it is not, rather than what it is. It has a low degree of departmentalisation, wide spans of control and little formalisation. The simple structure is a flat organisation; it usually has only two or three vertical levels in hierarchy, a loose body of employees and one individual where decision-making authority is centralised. It is most widely practised in small businesses in which the manager and the owner are one and the same. However, it is also the preferred structure in time of temporary crisis because it centralises control (Robbins, 2000).

In *functional structures* people with similar skills and performing similar tasks are formally grouped together. Members of functional departments share technical expertise, interests and responsibilities (Figure 3.35). The key point is that members of each function work within their areas of expertise. If each function performs its job properly, the expectation is that the business will operate successfully (Schermerhorn, 1999). Major advantages of the functional structure include efficient use of functional resources and high-quality technical problem solving capacity. The major disadvantages are lack of communication and co-ordination between different functions and loss of the total system perspective.

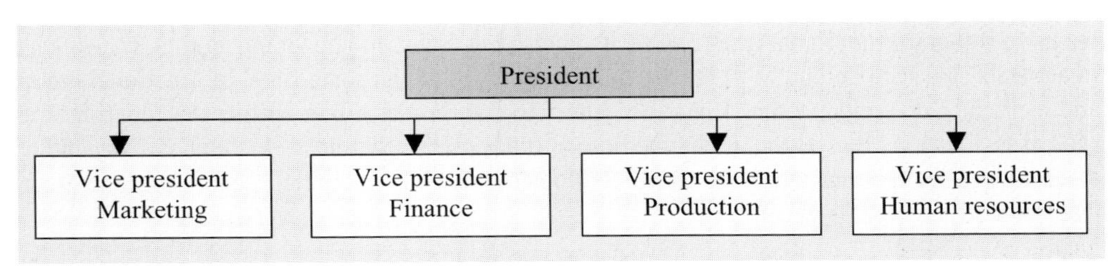

Figure 3.35. Functional structure in a business firm.

The *divisional structure* groups people who work on the same product or process, serve similar customers, and/or are located in the same area or geographical region. As illustrated in Figure 3.36 divisional structures are common in complex organisations that have multiple and differentiated products and services and/or operate in different competitive environments (Schermerhorn, 1999). The major advantages are product and customer focus and co-ordination, whilst the major disadvantage lies in increasing costs through the duplication of resources.

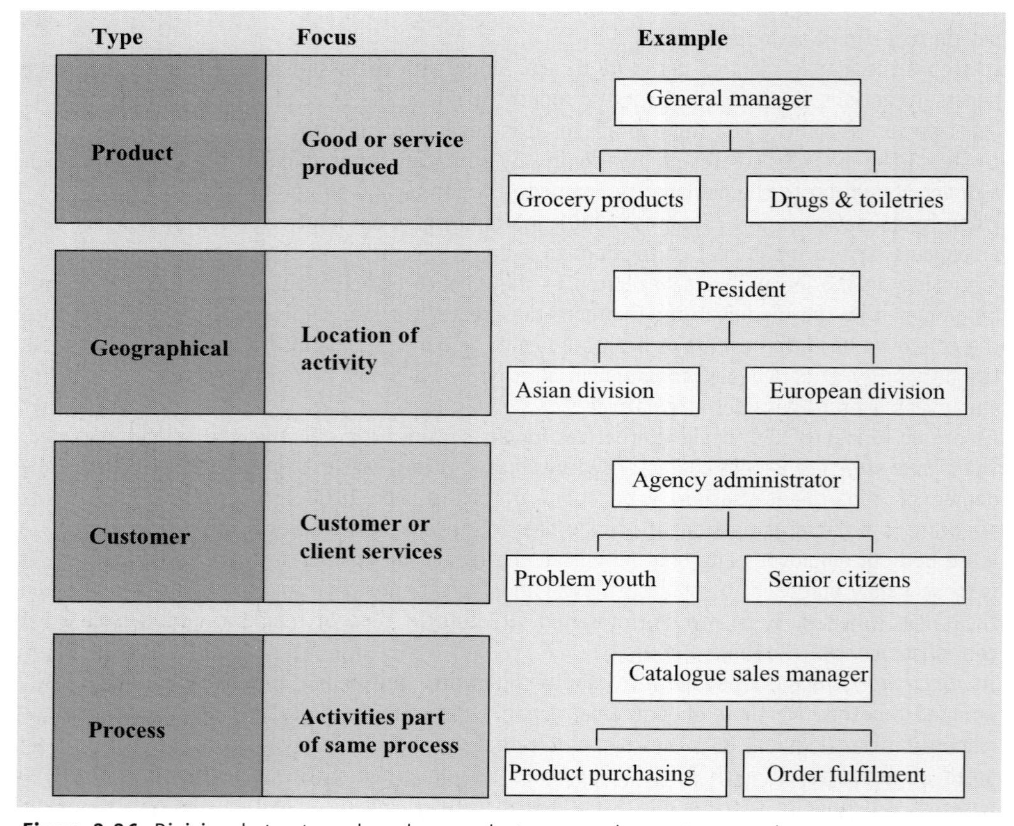

Figure 3.36. Divisional structures based on product, geography, customer and process.

The *matrix structure* is a combination of the functional and the divisional structure. In fact, it is an attempt to gain the advantages of the two structures by using permanent cross-functional teams to integrate functional expertise with a divisional focus. As shown in Figure 3.37, workers in a matrix structure belong to at least two formal groups at the same time - a functional group and a product or project team. They also report to two bosses - one within the function and the other within the team (Schermerhorn, 1999). The two-boss system is the major disadvantage being susceptible to power struggles between functional supervisors and team leaders, both trying to get authority.

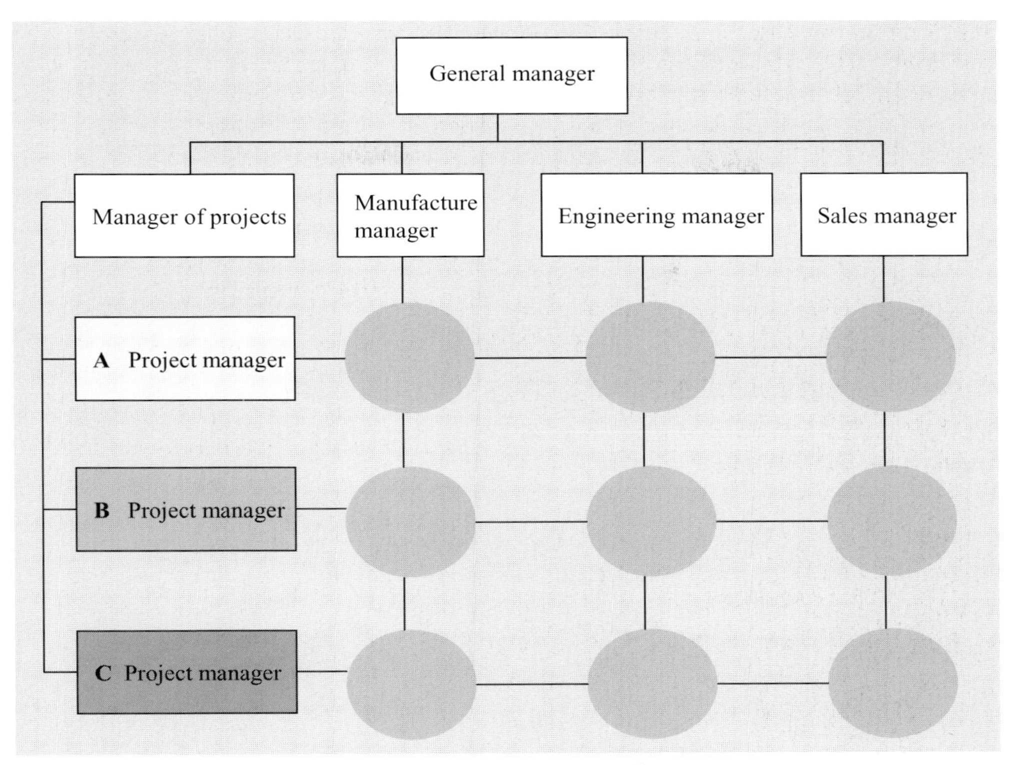

Figure 3.37. Matrix structure in a small multi-project business firm.

In organisations that operate with *team structures* both permanent and temporary teams are used extensively to accomplish tasks. Importantly, these are often cross-functional teams composed of members from different areas of work responsibility. The intention is to break down functional barriers and create more effective lateral relation for ongoing problem solving and work performance. As illustrated in Figure 3.38 a team structure involves teams of various types working together as needed to solve problems and explore opportunities, either on a full-time or part-time basis (Schermerhorn, 1999). Major advantages include the speed and quality of decision-making and breaking down functional barriers. Disadvantages are potential conflicts in loyalty to both team and functional assignments and time-consuming meetings.

A *network structure* operates with a central core that is linked through "networks" of relationships with outside contractors and suppliers of essential services (Figure 3.39). They are engaged in a shifting variety of strategic alliances and business contracts that sustain operations without the costs of "owning" all supporting functions (Schermerhorn, 1999). Network structure can be used for the firm as a whole but also for parts of it, for instance a front office in a new market having great decision authority not hampered by rules and procedures. New information and communication technology, together with network structures, are offering advantages in flexibility and operating efficiency. Disadvantages lie in difficulties in control and co-ordination.

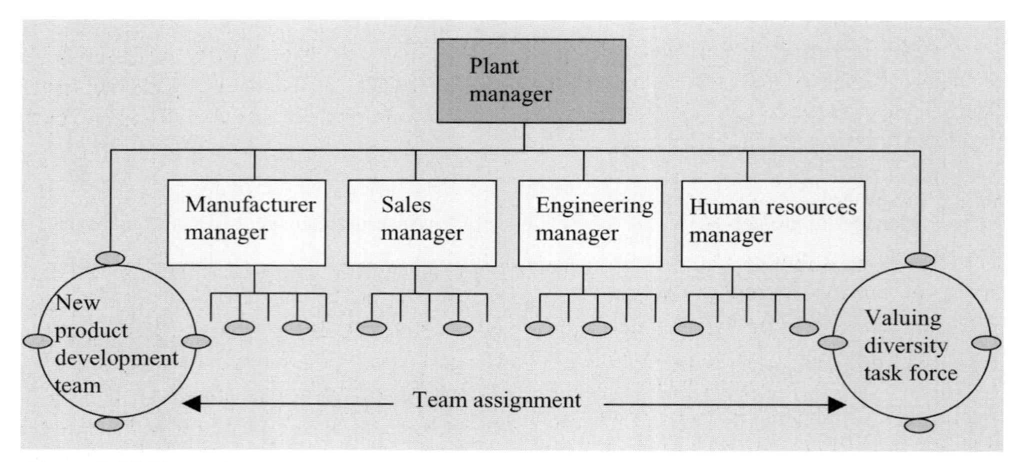

Figure 3.38. Team structure.

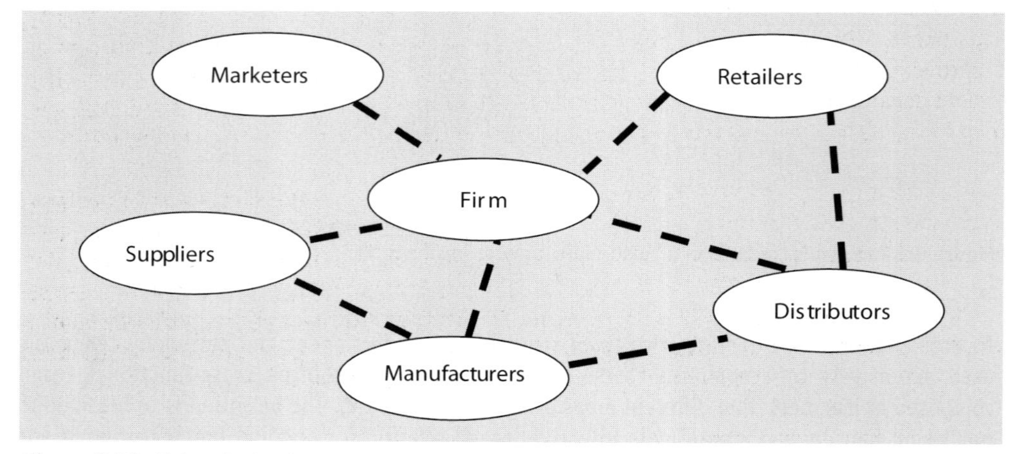

Figure 3.39. Network structure.

Considering the organisation structures, Dean and Evans (1994) mentioned several inadequacies of the functional structure:

- it separates employees from customers because it tends to insulate employees from learning about customer expectations and their degree of satisfaction with the service or product the firm is providing.
- it inhibits process improvement because no organisational unit has control over a whole process, although most processes involve a large number of functions.
- it often has a separate function for quality; this may send a message to the rest of the organisation that there is a group dedicated to quality, so it is not their responsibility.

They suggest internal customer-supplier relations, team-based organisation, reduction of hierarchy and a high-level steering committee as good conditions for quality management. From that viewpoint, they consider decentralisation and low formalisation as favourable conditions for quality.

In chapter 6 on quality improvement, the team concept will be discussed, including self-managing teams as the most far-going concept. In Chapter 4 (quality design), the 'team concept' will be illustrated by discussion of the role of project teams in new product development.

3.5.3 Quality assurance department

In a typical organisation, *line authority* belongs to managers who have formal authority to direct and control immediate subordinates who perform activities essential to achieving organisational objectives. Line authority thus flows through the primary chain of command. On the other hand, those with *staff authority* direct and control subordinates who support line activities through advice, recommendations, research and technical expertise. Staff managers and employees support line functions with specialised services and information. Line and staff conflicts can arise due to differences in roles and personal characteristics and to differences in formal and informal authority.

With respect to quality control and assurance, most companies have their own quality department. Typical functional organisations organise quality knowledge in a quality control department, being responsible for quality control of products and services. This type of quality control department will have all disadvantages typical for the functional organisation structure (like no connection to customers, no improvement opportunities, lack of responsibility etc).

In many companies, however, each individual within the firm is responsible for assessing and improving the quality of the process for which he or she is responsible. Because some managers may lack the technical expertise required for performing needed statistical test or data analysis, technical specialists, usually in the quality assurance department (QA-department), assist management in these tasks. Nevertheless, it must be emphasised that a firm's quality assurance department cannot assure quality in the firm. Its proper role is to provide guidance and support for the firm's total effort toward this goal (Evans and Lindsay, 1996).

Tasks which can be conducted by the QA-department include:
- quality assurance and control by design, improvement and control of the quality control processes
- collecting data and providing information (feedback) with regard to quality performance
- assessing process analysis to establish effective and efficient quality control, such as HACCP
- providing support to other departments in trouble-shooting, in setting complaints or recall procedures
- keeping up the knowledge base with regard to quality management
- providing support in improvement programs, such as process and product improvement
- providing support in defining and setting a quality policy

A QA-department can be structured in different ways as shown in Figure 3.40.
- *product structure:* a structure which is based on the expertise of certain (semi) products. The advantage is the total approach with input of a variety of disciplines. A disadvantage is splitting of expertise.

- *expertise or functional structure:* a structure which is based on the similarity of expertise. The advantage is centralising similar expertise in one section. A disadvantage is that sections may not have any knowledge of other processes.
- *system structure:* a structure which is based on a process approach. The advantage is bringing together interrelated expertise that forms a whole unit. A disadvantage is that it may become less effective in problem solving if feedback of other units is poor.

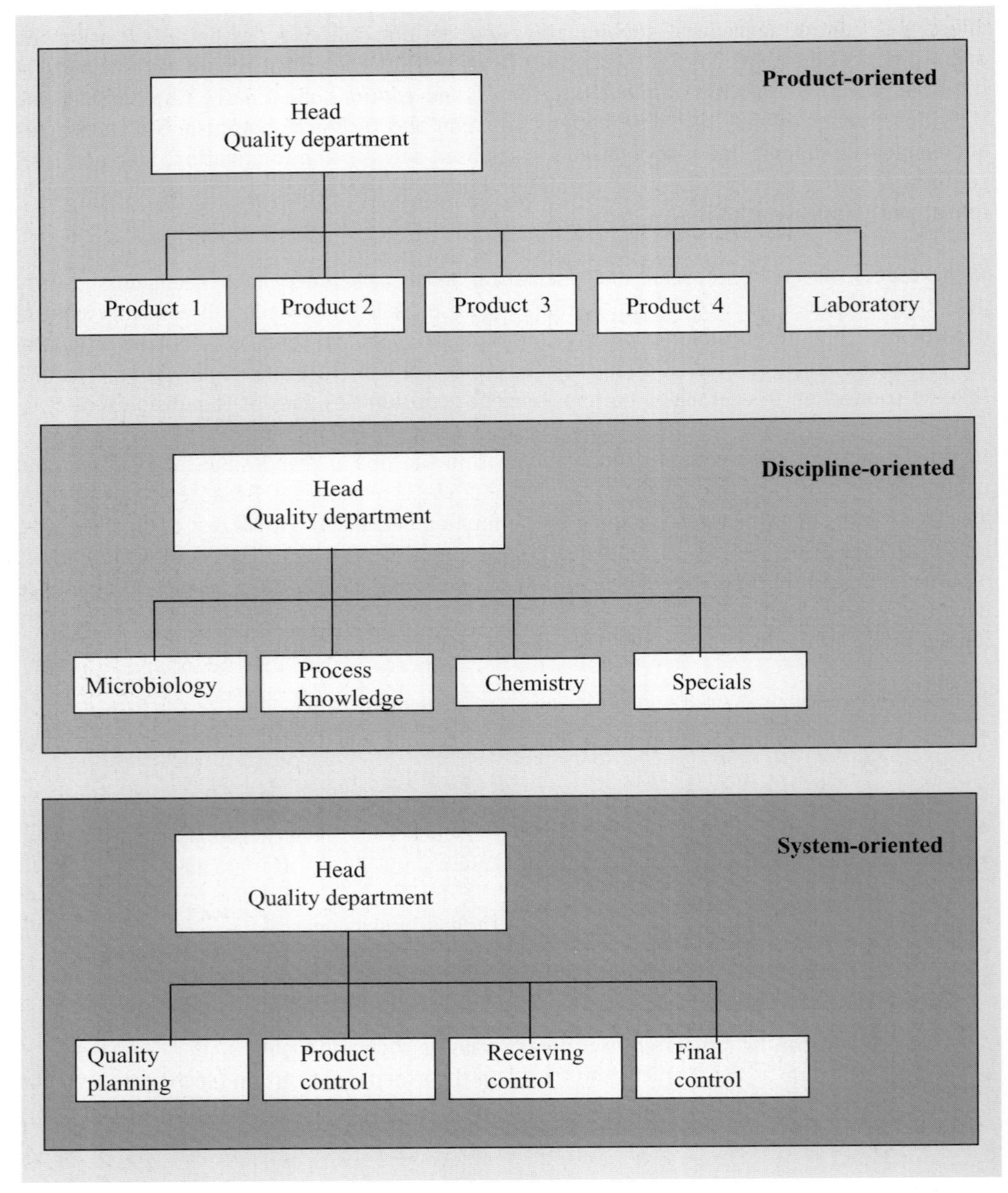

Figure 3.40. Different forms of organisation of the quality department.

As discussed, the QA-department can be structured in several ways, but can also be positioned in several ways within the firm (Figure 3.41).
It can be *staffed as an advisory department* for the management board. This position suggests that quality is of importance, that it has a major advisory function and also a great influence upon the management board. However, it is not qualified to take major decisions.
The QA-department can also be *positioned equally* as other departments, such as production, procurement and marketing. It implies that the QA-department has equal influence upon issues as other departments have.
Finally, it can be positioned, for example, as *staff under the production department*. In this case it is an advisory department for production and its influence upon issues is limited to this department. The advantages in this case lie in the acquirement of detailed knowledge of the products and processes and the commitment of staff employees to their "own" production process.

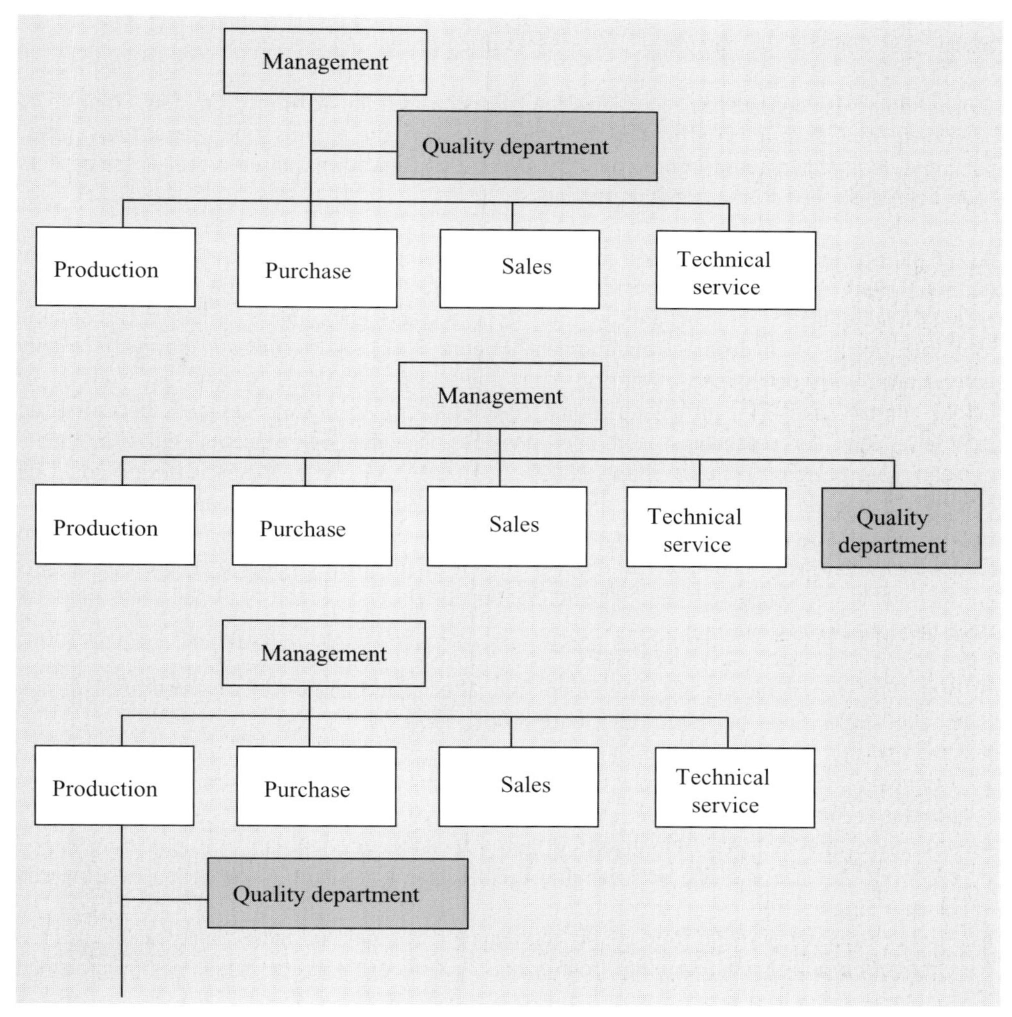

Figure 3.41. Position of the quality department within the organisation'.

3.5.4. Organisation design

Designing an organisation implies that choices on three different dimensions must be made, *i.e.* complexity, centralisation and formalisation (Ivancevich, 1994; Gatewood, 1955).
The degree of *complexity* describes how much differentiation there is between structural units. Therefore, how specialised are the organisations' jobs and how geographically dispersed and large is the organisation. If the firm has many vertical levels, has a high degree of division of labour and is a multinational co-operation, then it is a complex organisation.
Lawrence and Lorsch (1967) stressed that increased differentiation among organisational subsystems creates the need for greater integration. However, integration becomes harder to achieve as differentiation increases (Schermerhorn, 1999).
Mintzberg (1993) described different co-ordinating mechanisms than can be used to obtain horizontal or vertical integration. It includes direct supervision, standardisation of work, skills or norms and mutual adjustment. Gatewood (1995) mentioned that rules and procedures, committees and "liaison personnel" are aspects of co-ordinating mechanism.

Centralisation is defined as concentrating decision-making authority in few, high-level positions. *Formalisation* is defined as the degree to which the organisation's procedures, rules, personnel requirements and information systems are written down and enforced. These aspects have been described in the previous section.

Burns and Stalker (1961) concluded in their research among 20 manufacturing firms that two quite different organisational forms could be successful, depending on the nature of a firm's external environment. The mechanistic form fits in a stable environment, whereas the organic form fits better in a dynamic environment (Figure 3.42). A *mechanistic design* is highly bureaucratic with centralised authority, many rules and procedures, a clear-cut division of labour, narrow spans of control and formal co-ordination. An *organic design* is decentralised with fewer rules and procedures, more open divisions of labour, wide spans of control and more personal co-ordination (Schermerhorn, 1999). In Figure 3.34 the mechanistic design could be placed in the lower part on the left, and the organic design in the upper part on the right.
Organic designs work well for organisations facing dynamic environments that demand flexibility to deal with changing conditions. Therefore, for quality management these designs are favourable, although high requirements in food safety and reliability at the same time ask for a more mechanistic design.

In literature several factors are mentioned that affect the decision of organisational design (Hellriegel, 1992; Schermerhorn, 1999; Gatewood, 1995), *i.e.*

- *environment*, which can be dynamic and uncertain, or stable and predictable
- *technology*, which can vary from small-batch technology to a high-automated continuous-process technology
- *strategy*, which can vary from growth oriented to stability oriented
- *organisation's size and life cycle*, varying from a small and young organisation to a big and mature organisation
- *human resources*, varying from young and dynamic to older and more conservative
- *information processing*, varying from small flexible systems to huge, pre-programmed systems

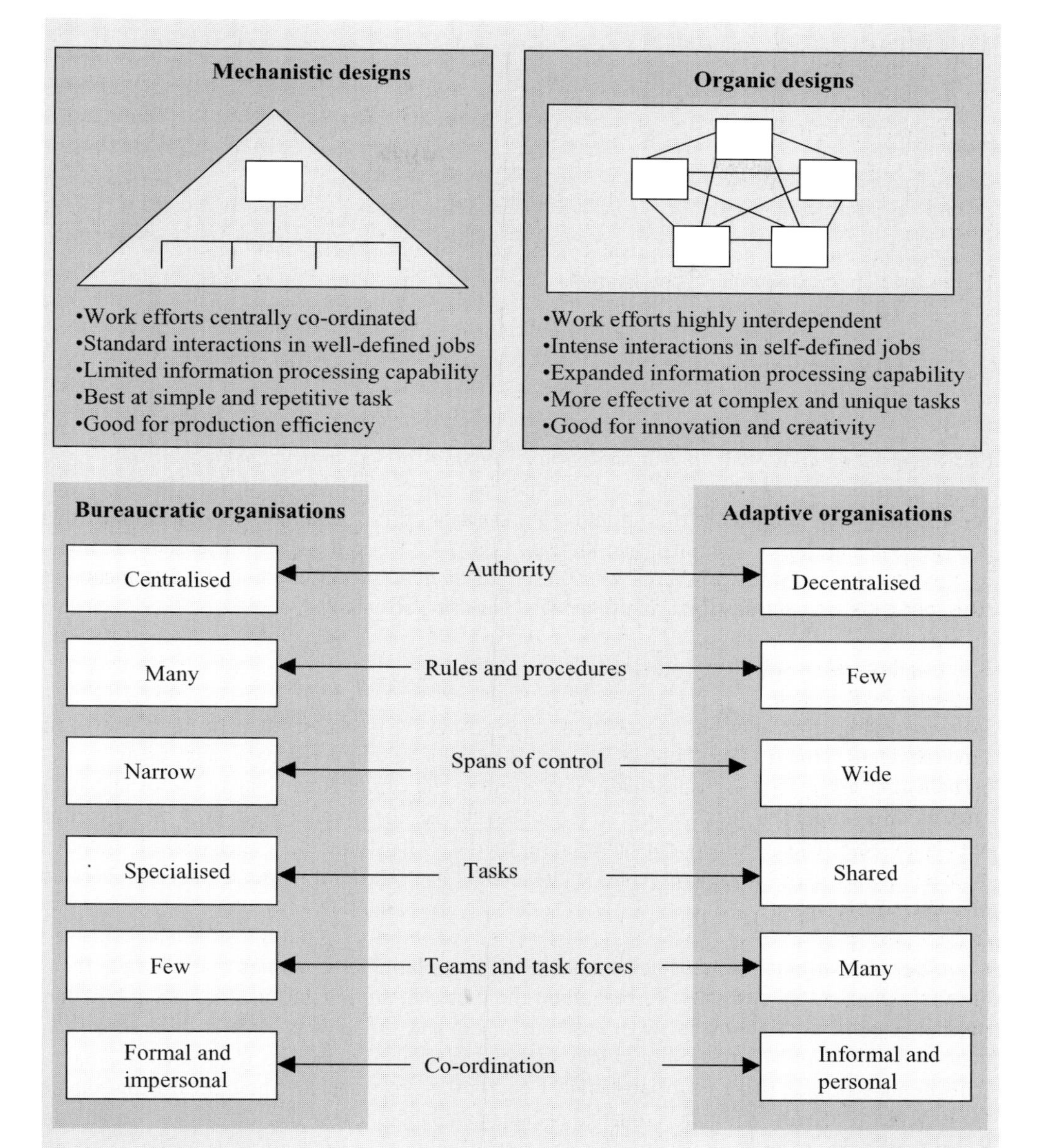

Figure 3.42. Continuum of organisation design alternatives.

Concerning the last factor, it is worth noticing that Gailbraith (1973) researched the influence of information processing strategies on organisation design (Figure 3.43). Instead of increasing the organisation's ability to process information, often it is better to reduce the need for information. In the last case organisations could be more decentralised and less formalised and therefore more flexible and adaptive.

General approach	Strategy
Increase the organisation's ability to process information	1. Create vertical information systems 2. Create lateral relations
Reduce the organisation's need to process information	1. Create slack resources 2. Create self-contained tasks or departments

Figure 3.43. Information-processing strategies.

3.6 Chain management

In the future a far-reaching integration with customers and suppliers will come up. The traditional value-chain will in many situations be transformed into value-creating networks, where the strict borders between suppliers and customers will become unclear. Business will become knowledge intense. In this situation a great deal of trust between the interested parties is of utmost importance (Bergman and Klefsjö, 1994).
In this section a restrictive overview is presented of quality related chain topics, including customer-supplier relationships, chain management and partnerships.

3.6.1 Customer-supplier relationships

From the quality management perspective, every company is a part of many long chains of suppliers and customers. One implication of this concept is that a company must focus on both their immediate and subsequent customers in the chain. Food companies, for instance, work hard to satisfy the needs of both the people who use their products, and the retailers that sell them, labelling the former 'consumers' and the latter 'customers'.
Companies should also try to establish similar productive relationships with their suppliers as with their customers.

By developing partnerships, customers and suppliers can build relationships that will help them to satisfy their shared customers further along the customer-supplier chain. The idea of creating mutually beneficial relationships with both customers and suppliers is a major difference from the traditional approach to customer and supplier relationships. In the traditional approach, self-interested competitors negotiate against each other, to maximise their slice of the pie at the expense of others. In modern quality management approaches, the focus is on expanding the pie rather than arguing over its division, thus developing a win-win relationship based on trust (Dean and Evans, 1994).

3.6.2 Supply chain management

Food and agribusiness firms, in general, are confronted with rapidly changing markets, new technologies and an almost world-wide competition. A retailer purchasing internationally perishable foodstuffs needs reliable partners. Also due to legislation on liability and food safety, only joint investments and co-operation can effectuate the development and introduction of new products. New distribution techniques enable large-scale production and frequent deliveries, although it can increase the risk of getting out of stock. In agriculture, technological developments contributed to specialisation and large-scale production, but it also increased the need for and availability of capital (Zuurbier, 1996). These developments affected thinking about quality, which exceeds the individual organisation and requires a supply chain approach. Supply chain is defined then as all stages of the entire process from raw materials to final product.

The forces, which can be considered as the drivers for the supply chain management paradigm, can be divided into four groups, *i.e.*

1. Force of *consumer demand*, which concerns demands on processing methods (*e.g.* animal welfare, additives, radiation), food safety (free from chemical, microbiological and physical hazards), food quality (*e.g.* freshness, convenience, taste and variety) and environment.
2. Force of *productivity and technology*, which concerns efficiency and effectiveness of processes (*e.g.* information technology, logistics technology, biotechnology, information and communication technology, measuring and monitoring technology).
3. Force of *government*, which concerns regulations and policies (*e.g.* shift of responsibility from government to business community, shift from domestic market protection to international market access, changes in traditional forms of subsidies, trend towards increased harmonisation).
4. Force of *resources*, which concerns demands on capital, resources and labour (*e.g.* economies of scale and size, resource and labour sustainability).

As a consequence, the world of farming, food manufacturing, distribution and retailing is becoming a dynamic and complex business, which requires management at the supply chain level. Specific market and production characteristics of food and agribusiness are additional motives for this approach. They include:

- restricted shelf-life of produce
- variability of quality and quantity of supplies of farm-based inputs
- differences in lead time between successive stages, complementary of agricultural inputs (available in joint packages only)
- stabilisation of consumption of many food products
- increased consumer awareness concerning both product and processing method
- degrading intrinsic quality (intrinsic quality of raw material is highest attainable quality for fresh products)
- availability of capital

(Zuurbier, 1996).

The restricted shelf-life of many products put great demands on duration and conditions of storage, processing and transportation at all stages of the supply chain. Therefore, the existence of assured markets is very important to suppliers of perishable products. Moreover, capital intensive processing facilities put high demands on a continuous flow of supplied inputs to buyers. Differences in lead-time between stages require efforts to match these stages. For example, pigs need time to grow, to reach optimal productivity and they cannot be stored alive. Meat production is a process that will inherently lead to a wide range of final products. In addition, some products cannot be made in isolation. For example, hams can not be produced without pork chops. As a consequence, it is not always possible to produce exactly what is demanded because of side-products inherent to the required product. This complicates the supply.

Nowadays, the mentioned developments increased the awareness of companies that effectiveness, efficiency and quality (as perceived by the customer) is served by a supply chain approach. It contributed to the developments of concepts such as Efficient Consumer Response (ECR) and Supplier-Retailer Collaboration (SRC). The idea of these concepts is to improve the performance of the entire supply chain by integrating marketing and logistic decisions and optimising activities between links.

3.6.3 Partnerships

Within the supply chain every firm conducts several specific activities in the process of transforming raw materials into final consumer products. This concept has been made more explicit by Porter's (1985) concepts of the 'value added chain' and the 'value system'. If a single firm performs activities, such as design of products, purchasing, production, packaging and shipping, it adds value. This value adding chain is part of a larger system, which embodies the value system, or supply chain.

Every firm is part of a value system and by means of co-operation the entire performance of the value system can be improved. To establish value-added partnerships (VAP) is very relevant for firms operating in agribusiness and the food industry. The basic idea is to co-operate with partners that fit best to the firm's own competencies in order to create synergetic effects. Partnerships in the perspective of a VAP is defined as co-operation among firms to achieve common objectives and to meet customer's needs at a maximum of added value and against a minimum of costs. Basically a VAP is temporary and partial. Its structure and organisation are results of joint activities, exchange of information, people and means (Zuurbier, 1996). If markets or technologies change, it might imply that partnerships are dissolved. The success of the partnership becomes apparent in aspects such as higher quality products, increased market access and more efficient processes. In other words, the supply chain's overall quality performance.
Very often the decision to co-operate is inspired by the existence of mutual interdependencies (Figure 3.44). Several interdependencies are at issue, like technical or technological knowledge and social interdependencies (Kamann, 1989; Zuurbier, 1996).

Activities of actors are linked within a partnership. They are continued in the scope of efficiency, in such a way that interdependencies will be developed and/or will be strengthened. The structure and organisation of a partnership will act as a control mechanism, which makes some changes easier and others more difficult. Besides the mutual interdependencies, the success of a partnership is also affected by:

1. *bounded rationality of actors*. Firms are restricted in obtaining and processing all available information
2. *opportunist behaviour of actors*. Firms may take advantage of their position and impose others by providing incomplete or incorrect information deliberately, which is also affected by uncertainty and complexity of the business environment and degree of market concentration
3. *rigidity of organisational structures*. Within partnerships routine structures may be developed, which impede organisational learning. Firms may withhold or manipulate information to retain gained positions

4. *balance of power between actors*. Relations can vary from symmetric to asymmetric. A relation is asymmetric if control is unbalanced. These types of relations can be found in a dual production system, where a dominant firm ties a group of dependent firms. As a consequence, new partnerships may be developed to create countervailing power
5. *appropriation of resources*. The inability to account costs and benefits to each of the actors may hamper the development or duration of partnerships.

Basically there are four key factors which determine successful partnerships (based upon Hughes, 1994):
1. clear benefits for all participants
2. a good strategic fit of partners
3. the involvement of all management levels
4. (organisational) flexibility.

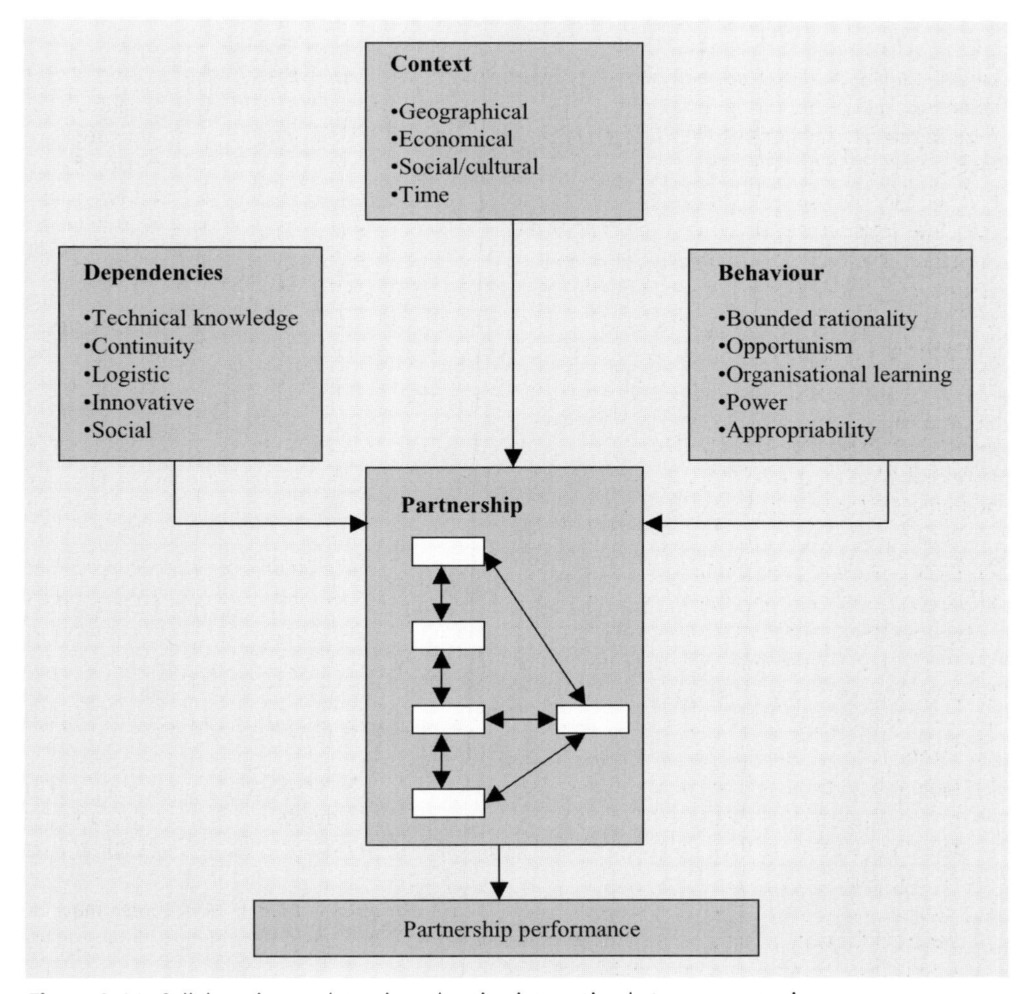

Figure 3.44. Collaboration as dynamic and active interaction between companies.

Partnerships will only last if they are able to meet the key factors of successful partnerships. Changes occur especially during economical recession in order to maintain continuity and control of gained positions. In addition, functional, strategic and political issues should also be involved. If common objectives are splitting up and co-ordinating actors are losing their influence, then the partnership might decline.

4. Quality design

One of the aspects that contribute to the realisation of quality includes quality of design. According to Juran (1992), Evans and Lindsay (1996), high quality can only be achieved in complex products by starting at the source of the production cycle. In other words, quality should be built in at the design stage of products and corresponding manufacturing processes. Quality is generally summarised as 'meeting or exceeding customers expectations'. Therefore, it is essential to incorporate consumer requirements and expectations, also called 'voice of the customer', into the design. Moreover, minimal costs, optimal productivity, minimal environmental impact of product and production, and compliance with legal requirements, should be guaranteed. According to Jongen (1995), an additional requirement is that food products must fulfil a role in the maintenance and improvement of human health.

For years and years, product development and process design were technology-driven. It means that technologists invented new products, which were subsequently launched on the market supported by advertising. After a test period, consumer perception of the new product was evaluated and, if necessary, the product concept was modified or removed. This common approach to product development appeared to be expensive and time-consuming, and it often resulted in a high rate of product failures in the last decades. Up to 75-80%, of all product-ideas never reached the market stage and 30-40% of the finally launched products did not survive the test market (Cooper, 1993; Kotler, 1994; Keuning 1994). Dekker and Linnemann (1998) have described this process as the innovation funnel from ideas to successful products (Figure 4.1).

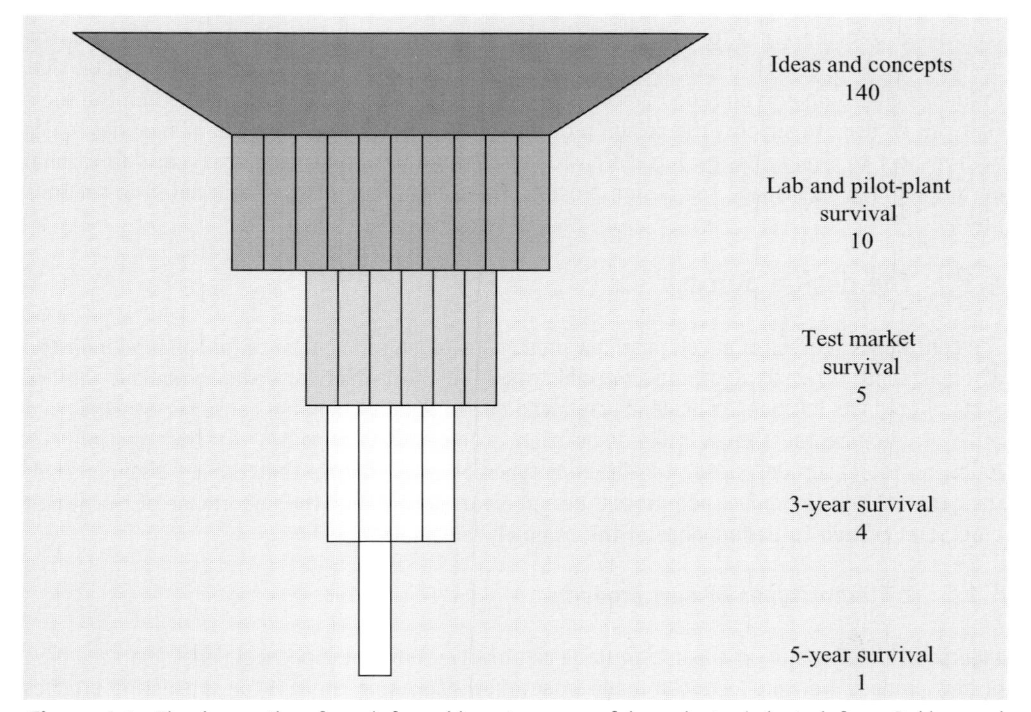

Figure 4.1. The innovation funnel from ideas to successful products (adapted from Dekker and Linnemann, 1998).

Several studies focussed on assessing causes of failure and success in product development. Van Trijp and Steenkamp (1998) summarised determinants for success and failure of new products, which were analysed in different studies (Cooper and Kleinschmidt, 1990; Urban and Hauser, 1993; Grunert *et al*, 1996). They distinguished determinants related to consumers, organisational aspects, competition and marketing. The latter three factors have been also called 'the voice of the company'.

- Key factors to success, related to consumers, are proper defining of product concepts and adding a higher value for the consumer to a product.
- Proper structuring of the design process, appropriate embedding in the organisation and commitment from top management, are typical organisational key factors for success in product development.
- Competition related factors include *e.g.* competitive activity and turbulence, size and attractiveness of the market.
- Accurate assessment of market size and desired positioning are typical marketing related determinants.

Several concepts, tools and methods have been developed, which can support consumer-driven product development and process design. Nowadays in the food industry, there is an increasing interest in these concepts and tools (Charteris, 1993; Grunert *et al.*, 1996; Bech *et al.*, 1997 Dekker and Linnemann, 1998; Van Trijp and Steenkamp, 1998; Costa *et al.*, 2001). However, to apply these concept, tools and methods, the typical characteristics of food and its production, as well as managerial aspects of the respective firm should be taken into account.

This chapter will focus on relevant technological and managerial aspects that should be considered to achieve quality in food products and process design. Firstly, general steps of the design process will be outlined. Subsequently, typical technological and technical variables that influence quality in agri and food products and process design will be explained. Then the tools that can be used to support the design process are shortly described. Typical managerial topics with respect to quality design include customer-oriented design management, cross-functional design and how to manage the design process. These topics are discussed in detail in sections 4.5-4.7.

4.1 The design process

The continuous process of developing new products is of major importance for the food industry. Whereas about 10 years ago products used to stay in the market for years, nowadays the life cycle of products has been considerably reduced due to rapidly changing consumer demands and other environmental factors (like legislation, competitive markets). In this section the following topics are described, *i.e.* different types of new products that can be distinguished, the general steps in product and process design are described, and the importance of the design process as related to performance of the company is explained.

4.1.1 Different types of new products

Incorporation of quality at the design stage can involve different aspects depending on the type of new product. New products can range from a new packaging concept for an existing product to a completely new product that has not been manufactured before; the latter is called innovation. New products can be classified into different categories; each puts special demands on the design process. Fuller (1994) distinguished seven categories of products and described their major characteristics with respect to the design process.

- *Line extensions* are new variants of an established (food) product. Typical examples are new flavours for existing products (*e.g.* flavoured potato chips) or new tastes in a family of products (*e.g.* new taste in series of dried soups). The design process of these products can be characterised by relatively little effort and development time, small changes in the manufacturing process, little change in marketing strategy and minor impact on storage and/or handling techniques.
- *Repositioned existing products* are current products that are again promoted in order to reposition the product. For example, by the increased attention for health products, a margarine brand was repositioned because of its natural high content of tocopherol (vitamin E). The development time for repositioned products can be minimal and only the marketing department should put efforts in capitalising the niche market.
- *New form of existing products* are products now in existence, which are transformed to another form (*e.g.* solved, granulated, concentrated, spreadable, dried or frozen). An illustration is the transformation of frozen ready-to-eat pizzas to pizzas that can be stored in the refrigerator; the latter have a much shorter shelf life. These types of products may require an extensive development time because physical properties of the product change drastically. In the case of pizza it also affected storage and distribution conditions. A frequent cause of failure is that the consumer not always appreciates the change. The change must have an added value.
- *Reformulation of existing products* involves current products with a new formula. There can be several reasons for reformulation, like reducing costs of ingredients, bad availability of raw materials, or new ingredients with improved characteristics. Examples are products with better colour, improved flavour, more fibres, less fat etc. The design process for reformulation is usually inexpensive and needs a relatively short development time. However, for food products minor changes in composition might have great consequences, for *e.g.* the chemical or microbial shelf life. Therefore, a profound knowledge of food technology is required to estimate and anticipate to possible changes in advance.
- *New packaging of existing products* involve current products with a new packaging concept. For example, the technique of modified atmosphere packaging created many opportunities for extension of product shelf life. With respect to the design process, new packaging concepts may require expensive packaging equipment, but also products may be reformulated for the new application (*e.g.* microwave packaging).
- *Innovative product* is defined as one resulting from making changes in an existing product otherwise than described above. The changes must have an added value. The design process is generally longer and more expensive, when more product changes are required (higher innovation degree).Marketing may also be costly because consumers may have to be educated to the novelty. However, in some cases time and costs of innovation were relatively little, *e.g.* assembling frozen vegetables and a frozen pastry on a tray resulted in a very successful innovative ready-to-cook product.
- *Creative product* is described as one newly brought into existence, *i.e.* the never-before seen product. Typical examples are novel protein foods (or meat replacers) that are based on vegetable proteins. Creative products commonly require extensive product development, tend to be costly (much marketing efforts, new equipment) and have a high failure chance. However, once a product appeared to be successful, imitators will rapidly flood the market.

Before starting the development process, it is useful to assess the character of the new product in order to consider the consequences for different development activities (*e.g.* technological, marketing efforts, packaging technology).

4.1.2 Steps in the design process

Different product development procedures are being used in practice and various ideas exist about how it should be conducted. Some consider product development as a procedure of sequential steps *e.g.* where screening is depicted as second step after idea generation (Oickle, 1990; Graf and Saguy, 1991). Others suggest a more iterative process, where screening occurs along each step of product development. The product is getting more and more refined using feedback information at each step, from technologists, marketing, finances, production, quality control etc. (Blanchfield, 1988; Fuller, 1994).
In traditional product development much attention was paid to testing of concepts and prototypes, while neglecting previous steps of product development, such as collecting information on consumer demands. According to Schmidt (1997), these approaches do not integrate marketing science, quality management and product design concurrently. Additionally, many companies make a fatal mistake by starting with preconceived ideas of what they *think* consumers will want (Fuller, 1994).
Ideally, the design process should be an interactive process of product development and process design. Products-in-development set requirements on the process design, whereas process requirements limit or facilitate product development opportunities.
In fact, the design process should include both product development and design of processes and equipment. Product development can be described as all activities carried out to translate the voice of the customer into a product that can be effectively manufactured. Process design involves not only design of process equipment, but also planning of physical facilities and development of information and control systems required for manufacturing of the product (Evans and Lindsay, 1996).

A common procedure for product development and process design is depicted in Figure 4.2. The different phases are described below:

- The design procedure should start with an *inventory* of consumer requirements and defining the focus group (*i.e.* for whom the product will be designed). Also boundary restrictions, such as the companies' strategy and objectives, legislative regulations and technological opportunities should be considered. At this stage the 'voice of the consumer' and the 'voice of the company' must be clarified.
- In the *concept phase* product ideas are generated and screened by the product development team based on the inventory of consumer requirements and other boundary restrictions. Then product ideas are embodied in a product statement or concept, which includes all attributes as desired by the defined 'group' of consumers (*i.e.* the expected end users), but also by other customers like distribution and retail.
- In the *prototype phase,* (food) technologists usually start with a basic recipe. These recipes can be obtained from, for example, cookbooks, ingredient suppliers or analytical data of competitor's product. Subsequently, basic recipes are modified with ingredients to obtain desired product attributes (Fuller, 1994). After preparation of several prototypes, the quality attributes must be evaluated and screened by objective testing. Objective testing includes quantitative measures like *how much* sugar, *how* fruity the taste etc.

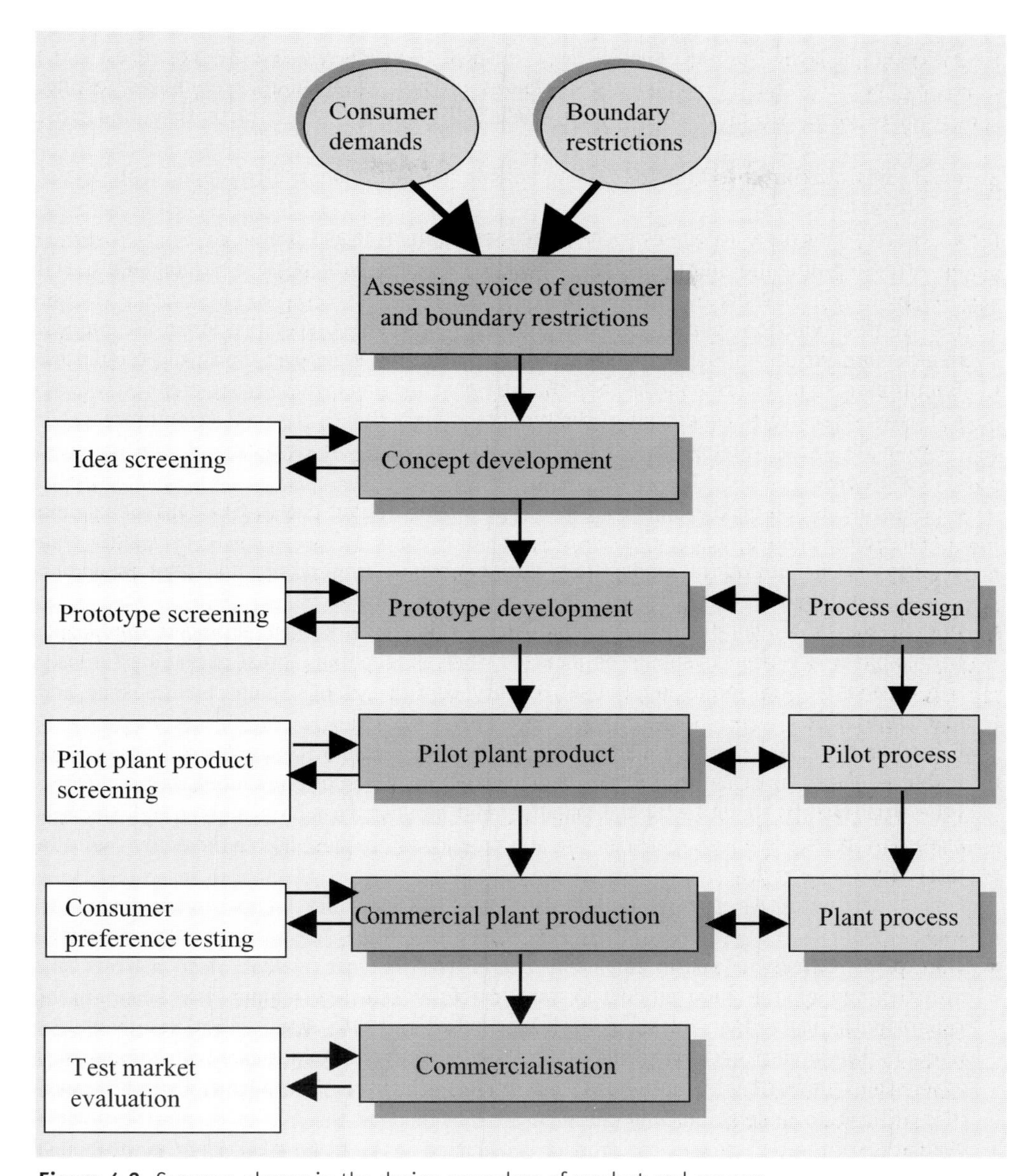

Figure 4.2. Common phases in the design procedure of product and process.

The prototype development provides input for process engineers to design the process, like which handling (*e.g.* cutting, mixing), preservation techniques (*e.g.* heating, drying, packaging) and process conditions are required. On the other hand process engineers must give feedback to the product developers on design restrictions.

- In the *pilot production phase* the product is made in a real process line. Commonly, the packing process is included at this stage. Typical aspects that should be considered in the pilot production phase include:

   - Determination of product shelf life with respect to safety and sensory properties.
   - Subjective evaluation of quality attributes of pilot products by preferences testing, where panellists must state which product they prefer, which taste they find most acceptable etc.
   - Finding of reliable sources for *e.g.* ingredients, raw materials and packaging materials at acceptable costs.
   - Establishing other resources such as required equipment and tools.
   - Assessing potential hazards in the production process, with respect to food safety, and how to control them (*i.e.* quality control).

  The pilot production phase provides input for in-process specifications. All process treatments that can influence quality attributes must have specified limits and tolerances. Typical process parameters in the food industry that require specifications are:
   - Time-temperatures conditions of the product at each step in the production process. For example, heat treatments, flow rates, and product viscosity affect the required conditions.
   - Shearing action on the product (*e.g.* during mixing action, due to flow rate, depending on pipe diameters etc.).
   - Pressure changes on the product (*e.g.* during mixing action, due to flow rate, depending on pipe diameters etc.).
- In the *commercial plant production*, the product is made under the actual process conditions. In this phase new aspects should be included, such as labelling, actual packaging (primary and secondary), transportation, labour, quality control systems, plant maintenance and plant sanitation.
  Furthermore, recipes must be modified to make them suitable for commercial production. Problems due to up-scale from pilot to commercial plant should be reduced to a minimum. The actual quality attributes of the manufactured product must be evaluated by consumer preference tests. Moreover, real shelf life and product safety must be confirmed by shelf life tests.
  Finally, the definitive product and process specifications and product price should be established.
- The *commercialisation phase* commonly starts with a test market. Test markets are introductions of the new product(s) into carefully selected regions for a variety of geographical, marketing and company reasons (Fuller, 1994). Basic questions should be *where* (is the product peculiar to the test market?), *when* (*e.g.* which season) and *how* (*e.g.* advertising, promotions) to introduce. Test market results are often misinterpreted because it is difficult to understand fully consumer's behaviour and the role of competition. After evaluation of the test market, the product is launched at the real market place and the sales are monitored.

Product development and process design are not once-in-a-life-time events but should be a major and continuous process in the organisation. As a matter of fact, in Western countries, product innovations in food production have become an important competitive tool to maintain or increase market share in a saturated market (Jongen and Meulenberg, 1998). Major motives for innovation of food products include (Fuller, 1994; Dekker and Linnemann, 1998):
- The *decrease of product life cycles*. Each product has a life cycle, which consists of the following stages: (1) product introduction, (2) increasing of sales by new consumers and repeated buyers, (3) declining growth because of market saturation, (4) stable sales period

and (5) finally decline due to competitive products. The time between introduction and product decline has been reduced in the last decades, therefore continuous innovation is required.
- Also *market conditions* force companies to change their product concepts, in order to anticipate to the constantly changing requirements. For example, currently several forms of retail outlets have been developed, such as internet shopping, shop-around-the-clock, and convenience food shops at fuel stations, to anticipate to changing market demands.
- *Company's policy and strategy* may adopt an aggressive growth program to comply with long-term set goals. The strategy may therefore stimulate product development.
- *Technological developments* may offer opportunities for new product concepts, which were not possible before. For examples, current developments on new mild preservation techniques (like High intensity light HIL, and ultra high pressure UHP) might offer new opportunities for producing healthy products with a fresh character.
- Various *external sources*, such as legislative changes, health programs, and/or agricultural policy may compel product innovation. For example, the packaging convenant of the European Union requires that the amount of packaging waste must be reduced within a set period. This convenant stimulated the food and packaging industry to redesign their packaging concepts.

Thus on the one hand, product development and process design are important for incorporating quality in an early stage; on the other hand, it is a competitive tool for food companies to survive in a dynamic market with many competitors.

4.1.3 Design and business performance

Customers are demanding continuing improvements in quality and more customised products. Furthermore, they want a quick delivery and a pleasant smile. They compare each product's performance with its costs to assess the value. Innovation and design activities are, of all activities, perhaps the most important ones influencing business performance. Until recently product design was confined largely to introduction and growth stages of the product life cycle (Noori and Radford, 1995). Once a product had matured, firms standardised product design and began to pursue economies of scale. Specialised production equipment was put in place where possible, and few if any fundamental changes in the product design were made from then on. The more costly the effects of redesign, the less likely the redesign. Nowadays, firms are focused on design and update of design of more and more products each year to stay competitive, even if they are not following a multiple-product strategy. Almost all firms increase the speed with which they design products as well as the quality of those products.
Mass customisation is a strategy that strives for satisfying a market's desire for a customised product while retaining the economies of scale of mass production (Melnyk and Denzler, 1996; Noori and Radford, 1995). It combines economies of scale with economies of scope. In food markets mass customisation will be a strategy that is getting more and more attention. Several design principles and methods will be necessary to underpin this strategy. Perhaps the most important is Modular Product Design. In a modular product-architecture each module performs an essential function, including for example existing, off-the-shelf components for some functions. This type of design may speed up product development and lower its cost. On the other hand careful attention is required to design interfaces that connect functional units.

Modular design offers a great advantage when technologies of the different modules are changing at uneven rates. Some modules can be upgraded without changing the others.
Design for Manufacture and Design for Assembly are techniques that seek to integrate the activities of product engineers with those of designers of manufacturing or assembling processes. The main objective is then to match product design and existing or new manufacturing and assembling technologies. Design for Service and Design for Recycling are more recently developed techniques.

4.2 Product development

In order to develop products that meet or exceed customers' expectations it is necessary to know which factors contribute to the realisation of quality. Furthermore, it must be recognised which factors can be affected by technological and/or by managerial measures. Different factors that contribute to realisation of product quality are (physical) product quality, quantity (or availability), and costs. Service, flexibility and reliability are additional factors that are important for the overall quality of production as realised by the firm. Technological design-variables mainly affect (physical) product quality, whereby a proper distinction between intrinsic and extrinsic attributes should be made. Typical intrinsic attributes are safety, health and sensory properties, whereas *e.g.* choice of production system and environmental impact of a product are extrinsic attributes. Product costs and availability and service, flexibility and reliability of the firm, can be greatly influenced by management activities. In this section are described the technological variables, which are relevant for product development and process design of food products.

4.2.1 Technological product design-variables

Food products are distinctively different from non-perishable products. The typical product and production characteristics put special demands on product and process development. Major product characteristics are described below.

Stabilisation of intrinsic quality attributes
A typical aspect of food is the phenomenon of decay processes, which starts immediately after harvesting (*i.e.* fresh materials) and processing (*i.e.* manufactured products). Several decay processes, like chemical, microbial, physiological, enzymatic and physical reactions, can cause a decline of intrinsic quality (*e.g.* changes in colour, odour, taste, texture and appearance, or vitamin C degradation) (see Chapter 2). In order to maintain desired intrinsic attributes during the required shelf life, it must be established which and how quality attributes must be stabilised. Stabilisation can be obtained by modifying the composition (*e.g.* regulating water activity (a_w), acidity (pH), use of additives), optimising process circumstances (*e.g.* appropriate time-temperature conditions), and/or selecting suitable packaging concepts (see Chapter 2). Examination of the stabilisation of intrinsic attributes should start as early in the development process as possible.

Possibilities and limitations for stabilising quality attributes can have a great impact on further processes in the food chain, like choice of the distribution channel. For example, fresh export beef is commonly transported by air, but development of innovative packaging concepts enabled a much cheaper alternative, *i.e.* transport by ship. Distribution conditions can also be driven by microbial stability properties, *e.g.* chilled food must be transported through a cooled chain to maintain the safe microbial state of the product. Therefore, also distribution requirements should be considered in an early stage of product development because they can be set by stability properties of the product.

Safety aspects

Care for product safety, and thus of human health, is characteristic for food (and also for pharmaceutical products) and is increasingly important for consumers. Product safety can be affected by growth of pathogen micro-organisms, by toxic compounds (*e.g.* microbial toxins, environmental contaminants or pesticide residues), by foreign objects and calamities (see Chapter 2). Microbial safety should therefore be evaluated and assessed along each step of product development from concept to final product. All factors should be considered that can influence product safety, such as initial bacterial contamination (of raw materials), composition (*e.g.* below pH 4.5 seldom growth of pathogens), appropriate process parameters (*e.g.* correct core temperature of product), hygienic operation and hygienic process design. The latter three factors need to be considered during process design as well.
Moreover, a safe product, produced by a company, can be turned into an unsafe product due to *e.g.* inaccurate distribution or storage conditions or due to misuse by consumers (*e.g.* a too high temperature for a longer period can favour the growth of pathogen bacterial spores). Therefore, during product development risks of potential hazards should be examined that might occur due to production and/or due to factors outside the company (*e.g.* distribution, consumers).

Product complexity

Food products and especially manufactured products are often very complex. They consist of a wide variety of chemical compounds, which can affect each other and thus also the final intrinsic quality. Besides, complex chemical, enzymatic, physical and physiological reactions may occur simultaneously. As a consequence, modifications applied to the composition (*e.g.* reformulation of existing products) may result in surprising changes. For example, the water activity (a_w) of dried powders was increased to prevent lipid oxidation. However, the product turned brown due to the Maillard reaction, which was favoured by the increased water activity (see Chapter 2). So profound knowledge on reactions and synergistic effects in food are required for product development.

Product-packaging interactions

Another typical aspect for food is product-packaging interaction, which may negatively influence intrinsic quality attributes. Interactions can be divided in three major mechanisms *i.e.* (1) diffusion of food compounds into packaging material, (2) diffusion of foreign compounds through the packaging materials into the product, and (3) diffusion of packaging compounds into the product. A typical example of diffusion of food compounds is the diffusion of flavours into polymeric packaging materials, which can modify product flavour and in some cases even changes packaging properties. Diffusion of foreign compounds through the material depends on the packaging characteristics. Each material has its specific diffusion properties; some are permeable for water vapour whereas others are more permeable for oxygen (Paine and Paine, 1983). Improper use of packaging materials may result in rapid loss of intrinsic attributes (*e.g.* softening of cookies due to water diffusion or desiccation of soft products). Also packaging compounds, such as colour additives, plasticisers and heavy metals, may diffuse into the food product and so affect flavour or even product safety. It should be noted that migration limits for packaging compounds are enacted in the European Directive (90/128/EEC) for food contact materials. So, the choice of the packaging concept should be tuned on product properties and legislative restrictions as early in the development procedure as possible, and not at the end of the design process.

Variability and availability of raw materials

In contrast to most other industries, the availability of raw materials in the agri-business and food industry can be seasonal. As a consequence raw materials must be obtained from other countries, or stored under proper conditions, to gap the time between harvest and use. For example, some fruits and vegetables can be stored frozen or dried until processing, whereas others can be stored under controlled conditions to extend their shelf life. However, storage conditions and different origins can affect the composition of raw materials, influencing intrinsic quality attributes of the final product. For product quality it is important to keep these attributes constant, therefore the effects of changes in raw materials should be taken into account during product development. A typical solution for this problem is the development of blends with raw materials from different origins. This principle is commonly applied for grained coffee.
Also, variability of raw materials (*e.g.* due to whether and seasonal conditions) may put special demands on the recipe. Variability of raw materials should be considered when establishing product specifications and tolerances. The contribution of variation of raw materials to variation of the end product may differ, depending on recipe and process conditions. This variation must be considered during product development as well.

Production method and environmental impact

As previously mentioned also other attributes, like choice of production method or environmental aspects, can contribute to quality perception of food products. Therefore, the possible effects of production system characteristics (*e.g.* use of genetically modified food, use of innovative packaging concepts, origin of raw materials) on consumer perception should be evaluated and/or considered during product development, *e.g.* by acceptance studies. For food, the environmental impact is commonly reflected in the selected packaging concepts (*e.g.* minimal packaging, recyclable, reusable), which must be considered besides the functional aspects of packaging (*e.g.* diffusion properties).

The relevant technological variables of food production, described above, should be considered during the product development procedure.

4.2.2 Technology tools supporting product development.

At each step in the product development procedure decisions must be made with respect to screening and selection of the product-in-development. For this purpose, the (food) technologist and quality control manager can use several techniques that can support in assessing quality attributes, shelf life characteristics and/or potential hazards during product and process development. Some common techniques and methods are described below. Another tool, which can be used in product development, is quality function deployment (QFD). It is a technique that can support in structuring the development process while assuring that consumer requirements are incorporated in the product and process design. QFD is not only a tool to support the technologists, but also other participants of the product development team are closely involved. The principles of QFD are described in the following section.

A. Technology supporting 'decision techniques'

The methods described below can help in making decisions with respect to the choice of new food concepts, pilot products and/or test market products, from a technological point of view. Other constraints like consumer demands, legislative regulations, finances, and actions of competitors can also have a considerable influence on the selection procedure during product development.

Sensory techniques

Sensory tests can be used to evaluate the sensory properties (like taste, colour, appearance, odour, texture) of the product-in-development. Sensory techniques can be roughly divided in objective and subjective testing. In objective testing, difference methods are used, whereby a well-trained analytical panel determines quantitative differences between samples. Subjective testing involves a representative group of untrained consumers, who indicate liking, preference or acceptability of test products.
The difference methods are mostly used in the prototype and pilot product screening, whereas consumer panels evaluate commercial plant products.

Shelf life tests

Shelf life tests are storage experiments to determine the actual storage life of a product, by evaluation of the quality attributes. The shelf life is often assessed by the just noticeably difference (JND), which means the earliest time when a difference of test and control samples can be detected by a trained sensory panel. This criterion can be applied to specific quality attributes or to the overall perceived product quality (Singh, 1994). Prior to applying shelf life tests, the testing criteria (*e.g.* colour changes, sensory changes, bacterial count, loss of nutrients etc.) and the acceptable loss should be defined. Two typical shelf life tests can be distinguished *i.e.* long-term and accelerated tests (Ellis, 1994).
In *long-term tests,* products are stored under the conditions that will be applied in practice. In *accelerated tests*, products are subjected to more extreme conditions (*e.g.* higher temperature, high oxygen concentration etc.) to speed up the ageing process and so reducing experimental storage time. In the latter experiments, the actual shelf life must be confirmed by long-term tests or it can be estimated by using predictive models. The accelerated test conditions must be selected very carefully because normal ageing or spoilage processes can be changed by the extreme conditions. These tests are often used to compare shelf life of prototypes with existing products, whereas long-term testing should be used to assess the actual shelf life of final products.

Expert systems

Expert systems are computer-based systems that enable technologists to run formulation simulations and rapidly calculate, for example, the theoretical shelf life. An expert system should include all relevant relationships between product composition and quality attributes like *e.g.* microbial growth, taste, colour etc. For example, in the UK a 'Cake Expert System' has been developed for the baking industry. This expert system calculates the equilibrium relative humidity of a certain formulation, and based upon this information the theoretical mould-free shelf life of the specific formulation is estimated (Walker, 1994). Such expert systems can be used very early in the product development procedure, even before actual prototypes are being developed.

Challenge testing
Challenge testing can be used to estimate risks due to potential pathogens and spoilage bacteria. Challenge testing comprises the direct inoculation of a food with specific micro-organisms. For this purpose are selected those spoilage or pathogen micro-organisms which are expected to cause problems in the new product. These tests, however, are very time-consuming and require laboratory skills, which are often not present (Shapton and Shapton, 1991).

Microbiological models
Microbiological models can be used to predict potential hazards due to growth of pathogens and spoilage bacteria. These models can be used to get an impression of the microbial stability or safety of *e.g.* prototypes with different compositions, even before development of the prototypes. Moreover, it can be used in the risk analysis of the production process, *i.e.* what are the risks with respect to survival of pathogens under given process conditions? For detailed information one is referred to literature (Walker, 1994; Zwietering, *et al.*, 1996; Wijtzes, 1996).

Hazard analysis critical control points
Another important method that should be applied during product development is the Hazard Analysis and Critical Control Points (HACCP). The objective of HACCP is to evaluate potential hazards and determine where to control them in the production process, before actual production. Preparation of a HACCP plan comprises a structured hazard analysis of the production process. It contains a detailed flow chart of the production process with an overview of potential hazards and critical points in the process, which must be monitored and controlled to assure food safety in production. The developed HACCP plan should provide input for the design process, *e.g.* at which stage must control points be incorporated and which hygienic requirements should be set for process design. Details about the HACCP procedure will be discussed in Chapter 7.
The techniques described can be used to make decisions during product development with respect to food safety and quality from a technological point of view.

B. Principles of quality function deployment (QFD)

One method that is attracting attention in the food industry as supporting methodology in product development is quality function deployment QFD (Charteris, 1993). QFD was developed in Japan in 1972 as a planning tool for the development of electronics, computer software, defence and health care. More recently it is also being applied for food product development (Kogure and Akao, 1983; Hofmeister, 1991; Dalen, 1996; Bech *et al.,* 1997; Dekker and Linnemann, 1998).

Quality function deployment is a concept that provides means to translate customer (or consumer) requirements into appropriate technical (or technological) requirements for each stage of product development, including marketing strategies, planning, product design and engineering, prototype evaluation, production and sales (Sullivan, 1986; Fortuna, 1988; Cohen, 1995). In fact, it is a planning process that ensures that quality is incorporated into a product and its process at the design stage (Charteris, 1993; Dekker and Linnemann, 1998). In the QFD process one or more matrices, also called quality tables, can be constructed as illustrated in Figure 4.3. The matrices depict interrelationships between what's (*e.g.* what do consumers want) and how's (*e.g.* how can these consumer demands be translated to physical measurable units). Moreover, it shows the priorities of the 'what's' and the 'how's'.

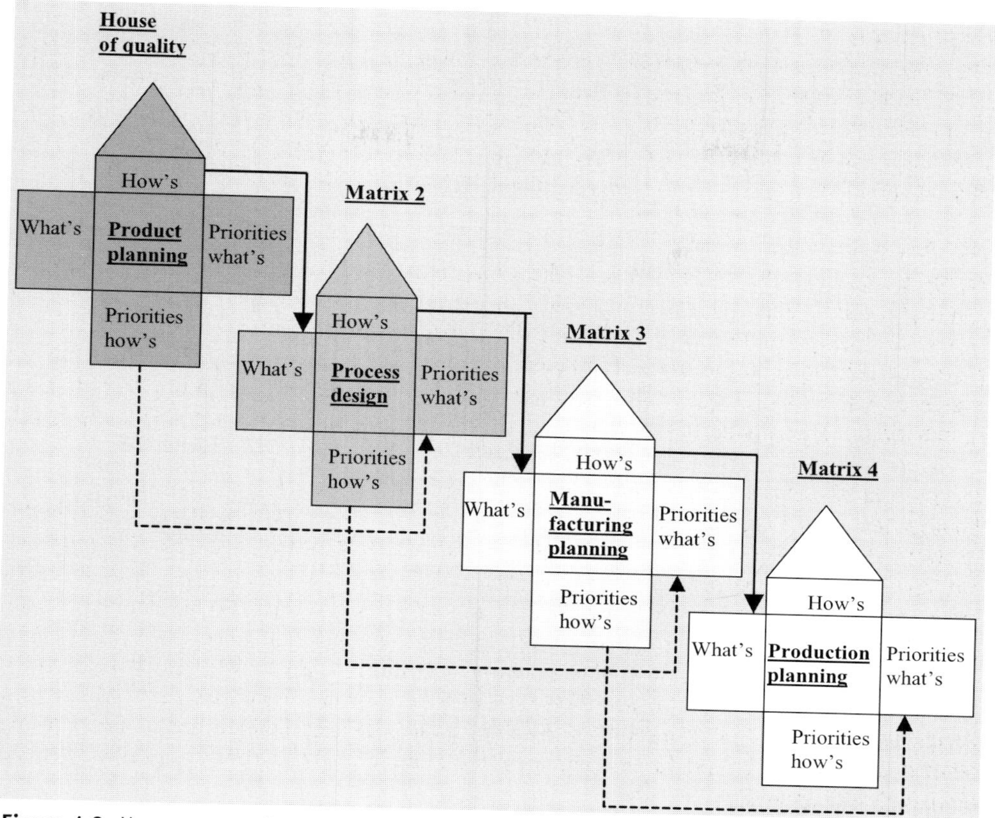

Figure 4.3. Houses or matrices of the Quality Function Deployment (QFD) process.

The first of these houses or matrices is called the 'House of Quality' as shown in Figure 4.4. The house of quality consists of six sections as described below (Cohen, 1995; Govers, 1996; Dekker and Linnemann, 1998).

- *Section 1*: The 'WHAT's' in the house of quality is a *qualitative part* and contains a structured list of consumer (or customer) wishes and needs. Consumers often provide different types of needs, ranging from sensory properties and health aspects to product convenience. These needs must be sorted, classified and structured. The needs can be sorted by asking questions like, what, when, where, why and how. For example, *what* means convenience for you: Easy-to-open, a short preparation time or others? Needs are then classified, for example, in function (*e.g.* packaging properties), benefit (*e.g.* healthy aspects) and targets (*e.g.* maximum amount of fat is 5%). The final result of this section is a detailed, clearly defined and structured list of consumer requirements.
- *Section 2*: The 'HOW's' are the measurable characteristics described in the words of the (food) technologist. So the consumer language (*i.e.* the wishes and needs) must be translated to the technological language, *i.e.* the product specification requirements. In this section, the following aspects are considered: unit of the specification (minutes, kilo's?), direction of the specification (the more the better or a specific target) and how it can be measured.

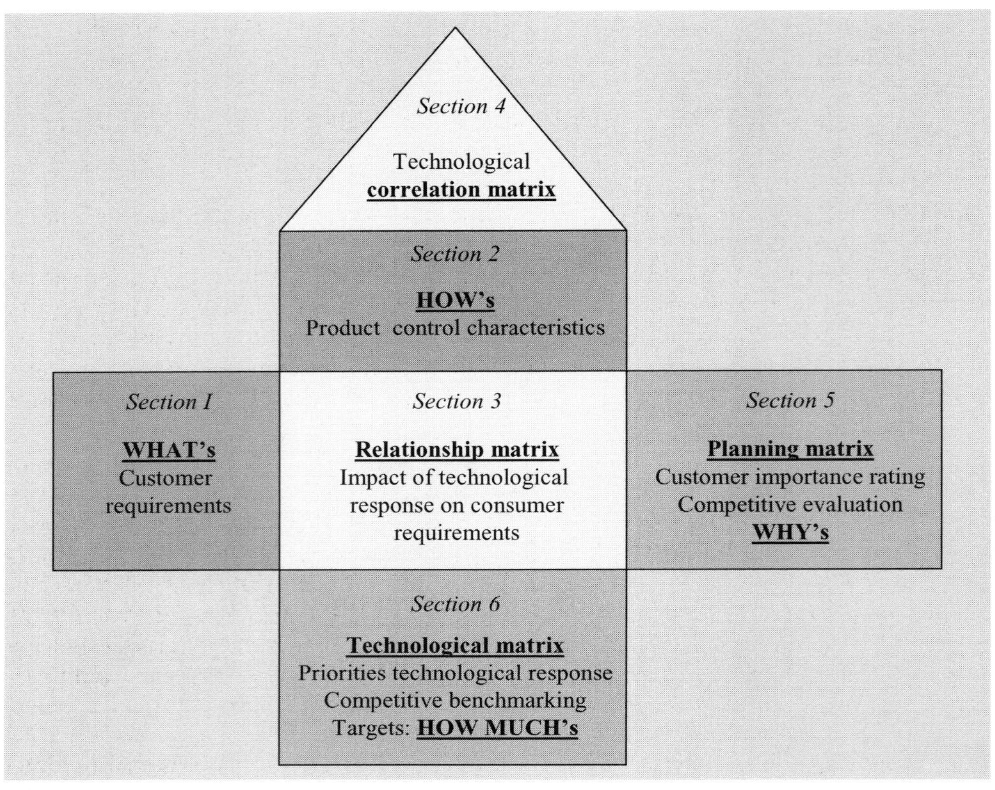

Figure 4.4. House of quality.

- *Section 3*: The relationship matrix shows the strength of the relationship between the consumer needs ('WHAT's') and technological product specifications ('HOW's'). The relationship can be strong, medium or weak. Thus, the relationship matrix indicates to what extent the product specifications fulfil specific needs or wishes of consumers. For example, the percentage of fat in the product (HOW) is strongly negatively related to the need for a healthy (low fat!) product, but is often positively related to desired sensory properties ('WHAT's').
- *Section 4*: The correlation matrix is the roof of the house and shows interrelationships between different product requirements; changing of one can affect the other requirements. For example, sugar content is strongly related to sensory perceived sweetness. Decreasing the sugar content will, therefore, weaken the perceived sweetness. If consumers prefer a strong sweetness and a low sugar content, then special technological modifications are required because the product requirements are negatively related. The correlation matrix can provide insight in areas that require more technological research to fulfil consumers' wishes and needs, which are not yet technologically feasible.
- *Section 5*: The planning matrix ('WHY's') forms the *quantitative part* of the *'WHAT's'*. It contains quantitative marketing data indicating the relative importance of the different needs to the consumer. In this part of the matrix consumers' wishes and needs are ordered from high to low priority. Furthermore, the matrix contains information about to what

extent the company's product and the competitor's product fulfils specific needs and requirements. It gives insight in which product scores best on which needs and requirements?

- *Section 6*: The technological matrix with target values ('HOW MUCH's') forms the *quantitative part* of the *'HOW's'*. It contains three types of information, *i.e.* technological priorities, comparative information and performance targets.
  The technological priorities are based on the relative importance of attributes (*i.e.* needs and requirements) as set by consumers (section 5), and the relationship between these attributes and the technological response (section 4). Thus, a technological measurement having a strong relationship with an important consumer need (high priority score) receives a high technological priority score. Comparative information includes analysis of the competitor's product and comparison with own product performance. Performance target values ('HOW MUCH's') are established based on the technological priorities and comparative analysis data.

After completing the house of quality, additional matrices can be constructed. Figure 4.3 shows the cascade of four matrices, which is widely applied in the USA (Cohen, 1995). The house of quality (*i.e.* product planning matrix) forms the input for the process design deployment. In this second matrix the process design characteristics are established and rated according to their relative importance (*i.e.* priorities). The outcome of this matrix can be used as input for the manufacturing planning-matrix, where the process parameters are set and rated. This forms in turn the input for production operations planning. In practice, only a few use the additional matrices beyond the house of quality.

QFD can thus support the product development team in incorporating consumer needs in the design process. It is a structured method, which helps in translating consumer needs to the different technological/technical languages (*e.g.* of the food technologist, process engineer, production planning manager etc.). QFD forces to prioritise, *e.g.* which attributes are rated highest by consumers, how does the company's product score as compared to the competitor's, and what might be the unique selling points for the company. Furthermore, QFD provides insight in the technological barriers (see correlation matrix), which should be overcome to fulfil consumer requirements in the future.

However, to become a widely applied tool in the food industry, more knowledge is required on relationships between needs (like sensory attributes) and technological measurements, to fulfil these consumer needs properly. More information is also required about the complex interactions between ingredients and components in the food-matrix to fill in the correlation matrix (section 4). Costa *et al.* (2001) made a comprehensive evaluation of the opportunities and bottlenecks for QFD applications in the food industry.
Another, more general, comment on QFD is that the approach is focused on an initial diagnosis of what consumers want, which is transformed to performance measures, whereas during further development process there is no feedback to the consumer. QFD designs for and not with the consumer (Kaulio, 1998).

4.2.3 Taguchi method for product design

In order to estimate the true cost of quality and to improve cost-effectively quality the Taguchi method provides a tool to designers to improve the product as well as the process by which it is made (Evans and Lindsay, 1996). Taguchi asserted that the quality of a product is a function of key product characteristics referred to as performance characteristics. The ideal value (or state) of a performance characteristic is its target value. A high quality product performs consistently near target values throughout its life span and under all operating conditions. Taguchi's quality loss function estimates the loss to society from the failure of a product to meet its target for a particular performance characteristic (Figure 4.5). This loss can be incurred by the consumer (*e.g.* short product life, increased maintenance and repair costs), by the firm (*e.g.* increased scrap, rework and warranty costs, damage to reputation, lost market share) or by society in general (*e.g.* pollution, safety). Either exceeding or not reaching the target will result in a loss even when the product falls within specifications. The closer each product is to its target value, the better it fits with adjacent products, the better it looks and the better it performs its intended function. The better fit with adjacent products is especially important for assembled goods (like cars, computers etc). The quality loss function enables a firm to quantify cost savings arising from product and process improvement. Another question, which might rise, is to reduce the sensitivity of a product to environmental fluctuations. Taguchi described a three-step approach:

1. System design. Firstly, an initial product design is developed, which includes preliminary settings for parameters that affect the value of performance characteristics.
2. Parameter design. Secondly, the factors that contribute most to variation in the end product must be identified. A set of experiments must be carried out to determine the settings that minimise the sensitivity of engineering designs to the sources of variation, thus minimising the expected loss to society. Taguchi also used these statistically designed experiments to identify design parameter settings that reduced costs without hurting quality.

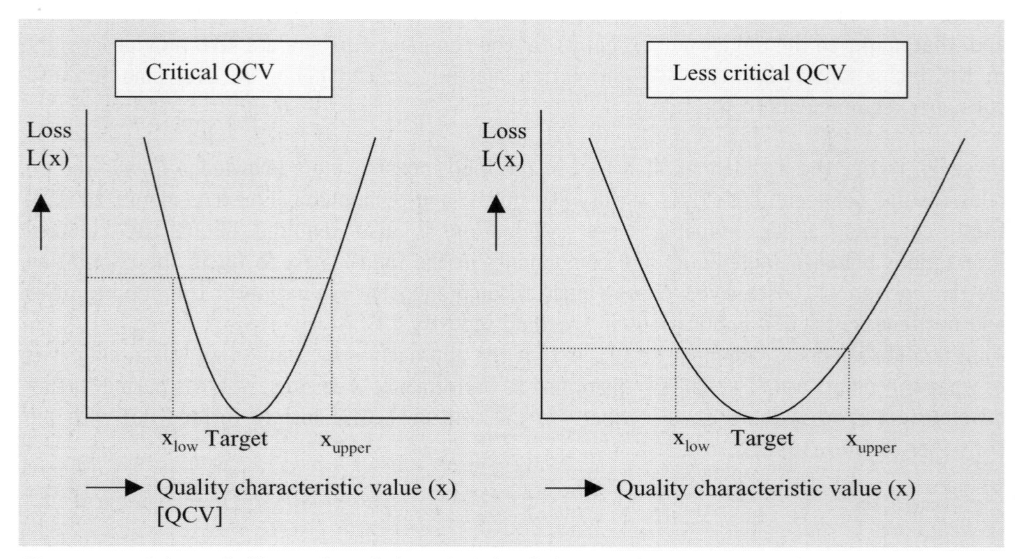

Figure 4.5. Schematic illustration of the principle of the Taguchi's quality loss function.

3. Tolerance design. Acceptable tolerances are determined around the values of the target design parameter as set in step two. Not all tolerances need to be tight, only those around design parameters whose variation will result in significant variation in the product's performance characteristics should be tight (Noori and Radford, 1995).

The resultant product will have acceptably low levels of product variation despite fluctuations in materials, manufacturing (equipment, operators) and environmental conditions. Since performance variation is minimised by reducing the influence of the sources of variation rather than by controlling them, Taguchi methods are cost effective.

4.3 Process design

Quality design of the manufacturing process not only includes process engineering, but also design of control and information systems, organisation of production, and organising material and stock control, as illustrated in Figure 4.6.
Process engineering can include several elements such as system design (*i.e.* design of equipment and tools), parameter and tolerance design, according to the Taguchi approach. *Control and information systems* must be designed as a communication tool between operators and equipment. It includes routine documents like operation sheets, route sheets and flow charts, but also the methods and procedures to control processes and flow of materials in production. These control and information systems are required for various quality issues (like process control, end inspection). *Organisation of production* involves decisions about *e.g.* use of human labour or automated equipment? How will the process be arranged, *e.g.* process-based (similar processes in the same area) or product-based (entire production lines are dedicated to specific products)? *Material and stock design* includes the organisation of the supply and internal storage of (raw) materials. Just-in-time (JIT) is a Japanese philosophy to material management, which aims at keeping the inventory as low as possible, because perishable products and internal transports cost a lot of money (Evans and Lindsay, 1996).

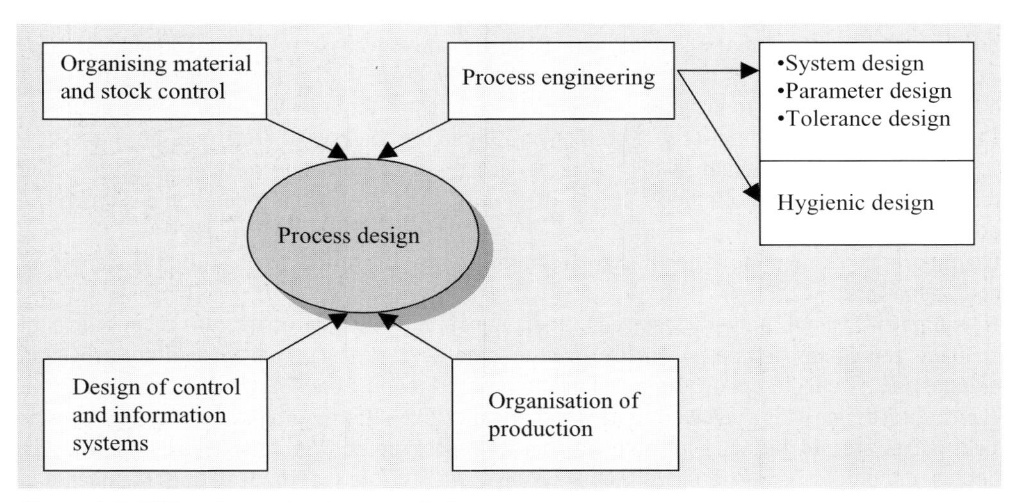

Figure 4.6. Different aspects of process design.

In the next section the technological process variables and tools will be discussed. Typical managerial aspects of the design process will be discussed in sections 4.5 and 4.6.

4.3.1 Technological process design variables

Designing of equipment for food applications require knowledge of or input from engineering on the one hand and food science on the other hand. This section is not intended to describe food-engineering details, but we have attempted to show some major points of attention for process engineers with respect to quality engineering of food processes. For food process engineering the following general aspects of food should be considered in the design:

- *Variability*. Physical properties of the agri-food materials often change during the different stages of processing, which can have major consequences for the design of each process step.
- *Complexity*. The (raw) materials and ingredients used for food production often contain many components, they have complex matrices and (many) interactions between components may exist. The materials can therefore often not be considered as a homogenous mass, which can have consequences for the equipment design.
- *Perishable character*. Agri-food materials are perishable, which requires special attention with respect to design. For example, adequate cleaning is of major importance for the safety of the products made.
- *Living materials*. Processing conditions of the raw material(s) are often determined by properties of the end product to be manufactured. For example, the heat load during pasteurisation of milk for the production of low-pasteurised milk is much more critical than for milk powder production; for low-pasteurised milk some enzyme activity must remain after the heating process. So, depending on the type of end product that will be manufactured from the same raw material, different process conditions and/or design may be required.
- *Consumer demands*. Consumers' requirements, like more fresh-like, safe food without additives, may also put specific demands on the process design.

Process engineers should thus consider the specific characteristics of raw materials and end products to be made, as well as the typical properties of raw materials and ingredients from set requirements on engineering at different steps in the process. Table 4.1 represents some major process steps in food manufacturing (column 1) and considerations that should made by the process engineer (column 3) to achieve required process characteristics (column 2). For example, proportioning is a process step commonly used in the food industry. The usual objective of this step is to dose the correct amount, in the right order with the desired accuracy. The features of the material (*e.g.* a powder or a liquid, viscosity of the material etc.) and how it changes during processing (*e.g.* change of viscosity due to pressure or temperature) set the requirements on the way of proportioning. This information must be considered in designing the equipment.

Cleaning and disinfecting of equipment requires special attention because of the danger of contamination.

If equipment is not properly designed, then remaining food materials can be a source of spoilage and toxic bacteria, which can contaminate next batches. This food safety aspect in engineering is reflected in hygienic design and is described in more detail in next section 4.3.2. Typical processing steps involved in new technologies like ultra high pressure and high electric field pulses are not included in Table 4.1, because most are not yet commercially applied. For details on process engineering of food systems one is referred to literature (Leniger and Beverloo, 1975; Zeuthen et al, 1990).

Table 4.1. Major process steps in food manufacturing and related points of attention for process engineering.

Process step	Process characteristics	Engineering aspects
Proportioning of ingredients/materials	• To dose the *correct amount* (weight, volume, density etc.) • In the *right order* • With desired *accuracy*	Product properties (*e.g.* type of product, homogeneity) and required accuracy determine the possibilities for proportioning *e.g.* • Liquids can be dosed by volume or time • Powders and granular materials can be dosed by volume, weight or time
Composing ingredient mix *e.g.* mixing, homogenisation etc.	• To obtain *equal (homogenous) distribution* of ingredients • To obtain *desired structure* and *rheological properties*	Effects of shear tension on bulk properties put demands on the design of the stirring (or homogenisation, mixing) equipment like, • Shear velocity of *e.g.* the stirrer or screw • Shape and size of stirring blades or screw
Transportation of (intermediate) product(s)	• To get the product at the *right place* • At the *right time* • And in a *proper condition*	Product characteristics determine the choice of transportation facilities • Transport via pipes *e.g.* for fluid products • Transport via conveyor belt *e.g.* for solid products And the transport conditions like • Shear tension during transport, which are affected by *e.g.* pump capacity, pipe dimensions • Temperature changes during transport • Residence time and residence time distribution (RTD)
Separation processes	• *Separation* of *desired streams* of (raw) materials	Material properties and their structure determine the level of separation and the separation mechanism, *i.e.* • Macro level, often mechanical separation by *e.g.* sieving, air circulation • Meso level, depending on particle size separation can be performed by *e.g.* filtration, hydro cyclones • Molecular level, depending on material is the separation based on *e.g.* thermodynamic properties (boiling point), molecular weight etc.

Table 4.1. Continued

Process step	Process characteristics	Engineering aspects
Chemical conversions	• To *change* the *molecular composition* *e.g.* backing and cooking, hydrogenation of lipids, or proteins hydrolyses	Depending on the product characteristics and the required conversion the following should be considered • Process conditions *i.e.* temperature, residence time and residence time distribution, which are affected by type of heat transfer, equipment design (*e.g.* length, diameter) and pump capacity • Type of (bio)catalyst required
Heating and **cooling** processes	• To obtain the product at *accurate temperature* • During the *right time* • To obtain the *suitable heat transfer process* (i.e. correct energy load per time unit)	Product characteristics (*e.g.* product tolerance for local over-heating) during manufacturing constrains process design. Important engineering aspects include: • Type of heat transfer *i.e.* direct/indirect and medium used (steam, air, cool water etc.) • Residence-time and residence time distribution, which are affected by equipment design (*e.g.* length, diameter) and pump capacity
Filling and **packaging** process	• To *dose accurately* the product • To obtain an *integer* (no leaks), *safely* (*e.g.* no glass pieces) and *correctly* packaged product	Product characteristics require certain packaging properties and both put demands on the filling/packaging equipment, such as • Velocity of filling and packing process • Accuracy and unit (weight, volume) of dose measuring device • Seal conditions
Cleaning and disinfection	• Equipment must become *microbiologically, physically* and *chemically clean*	Details in section hygienic design

4.3.2 Technological tools supporting process development

Similarly to product development, also for process design typical supporting tools are available, which enable incorporation of quality in the design phase. In this section, first are described the principles of Failure Mode and Effect Analysis (FMEA). Secondly, the principles and legislative requirements of hygienic design are outlined. Hygienic design as such is not really a tool, but it provides general criteria and detailed guidelines, which can support the design process. FMEA is a general tool applied in process engineering, whereas hygienic design is specific for the agribusiness and food industry.

A. Principles of Failure Mode and Effect Analysis (FMEA)

FMEA is a systematic analytical technique to detect potential failures at the design stage of product, service or process in order to minimise loss due to failures (Kehoe, 1995; Banens *et al.*, 1994; *Dale et al.*, 1998). There are two forms of FMEA, *i.e.*

- *design FMEA*, which involves the analysis of potential failures of a new product or service.
- *process FMEA*, which includes failure analysis of the manufacturing (or service) process.

This section focuses on the principle of process FMEA. The procedure for design FMEA is very similar and will not be explained; details have been described in literature (Kehoe, 1995; Banens *et al.*, 1994). Process FMEA must be carried out by a team. This team should include employees from relevant areas, such as process engineering, manufacturing, maintenance, sales and marketing.

Process FMEA consists of eight steps as shown in Figure 4.7. Some examples of process FMEA are given in Table 4.2. Below the different steps are shortly explained.

- At the *first step* the different components (*i.e.* steps, elements or chains) of the process are identified and established by the FMEA team.
- In the *second step*, at each process component the various potential failures are identified. For example, potential failures of pasteurisation include a too high or too low temperature combined with a too short or too long heating time (Table 4.2). Too low temperatures combined with too short heating times may result in inappropriate inactivation of pathogenic bacteria, whereas a too high temperature and too long heating time can have negative effects on sensory properties or nutritional value.
- *Thirdly*, causes of each potential failure are identified *e.g.* inappropriate heating may be due to improper functioning of the steam valves, which regulate the temperature.
- In *step four*, the impact of each failure at each process component on the internal and/or external customer is identified, *e.g.* inappropriate heating may result in reduced product safety and a short shelf life because of the potential that pathogenic and/or spoilage bacteria may survive and grow during storage and distribution.
- *Step five* includes identification of existing or proposed controls, which are applied to monitor or prevent the mode of failure. Regular inspection of the steam valves may prevent improper heating.
- At *step six,* failure modes are assessed in terms of severity [S] and occurrence [O] of the failure and the ability to detect [D] the failure. Severity, occurrence and ability to detect can be classified according to a scale, ranging from 1-10. For example, an occurrence [O] classified in scale 1 means that the chance that a defect occurs is 'very rare'. The product of severity [S], occurrence [O] and ability to detect [D] is called the Risk Priority Number (RPN) (Kehoe, 1995).

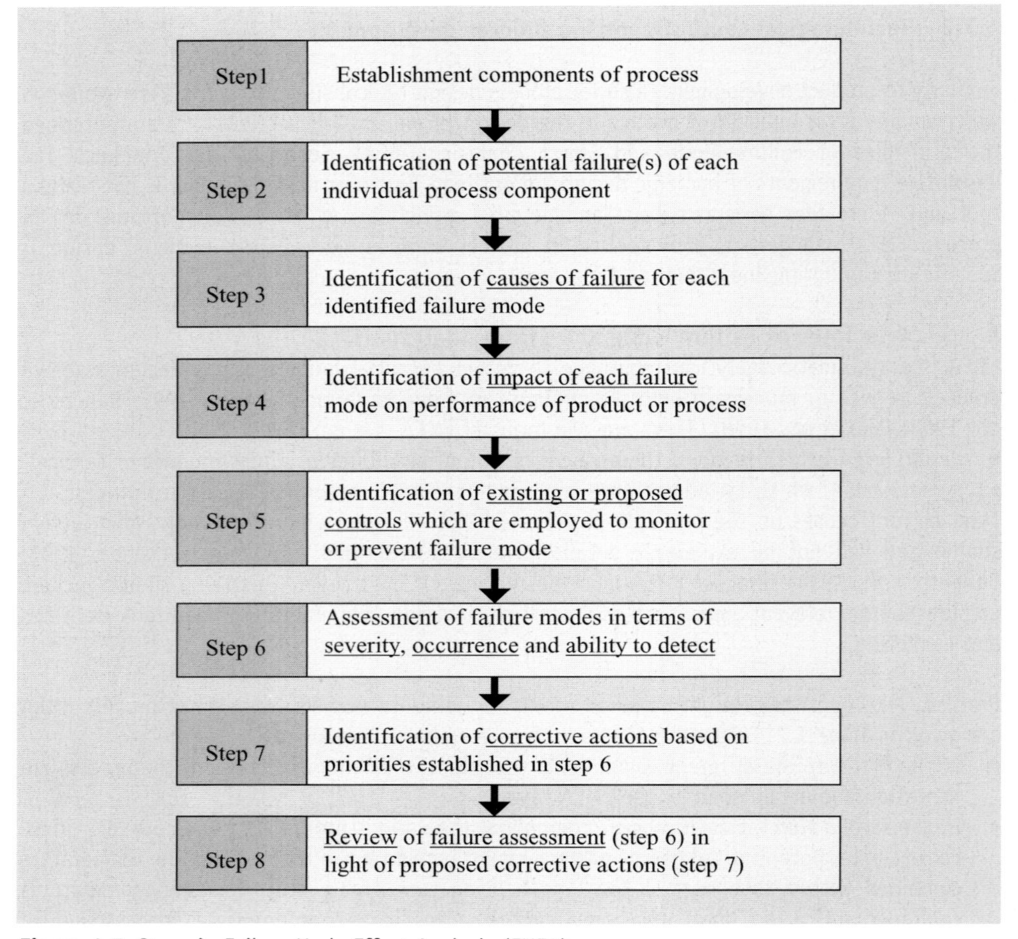

Figure 4.7. Steps in Failure Mode Effect Analysis (FMEA).

  - RPN= S x O x D
  - The ratings for severity, occurrence and ability to detect are scored on a scale from 0-10 according to descriptive and quantitative criteria
  - If RPN > 90 then a corrective action should achieve a high priority
  - The RPN gives only an indication of the importance of each process step with respect to the consequences of potential failures. FMEA is not a quantitative method and RPN just underpins estimations, knowledge and ideas of the FMEA team (Banens *et al.*, 1994)
- In *step seven* corrective actions are identified to improve significant failures based upon the relative priorities (*i.e.* failures with an RPN > 90) as set in step 6. If the RPN exceeds 90, then a corrective action requires a high priority *e.g.* by implementation of charts to control the process. By implementing a control chart the ability to detect [D] increases and the value for [D] decreases, resulting in a lower RPN as illustrated in Table 4.2.
- Finally, in *step eight*, the failure assessments of step 6 are reviewed in light of the corrective actions proposed in step 7.

Table 4.2. Example of process FMEA.

Process	Purpose	Potential failure	Causes of failure	Effects of e failur	Controls	Existing RPN=SxOxD	Recommended actions	Result on RPN=SxOxD
Pasteurisation	• Inactivation pathogens and spoilage bacteria • Shelf life extension	• Temperature too high / too low • Time too short /too long	• Steam valves do not function properly • Timing not correct	• Reduced product safety (pathogens) • Shortened shelf life (spoilage bacteria) • Sensory affects	• Inspection of valves • Calibration of time measure	5x5x*7*=175 5x3x*6*=90	Implement control charts (improve detection)	5x5x*3*=75 5x3x*2*=30
Packaging	• Prevention contamination • Shelf life extension • Quantities	• Leaks in packaging • Incorrect weight	• Sealing conditions not correct • Packaging errors (equipment or human)	• Contamination can negatively affect product safety and shelf life • Material shortages	• Specification of settings • Calibration weight scales	5x3x3=45 2x5x3=30	No actions planned	

The values of S, O and D are assessed using the 10 points classification system as comprehensively described by Kehoe (1995).

So FMEA can be a useful tool in evaluating potential failures in the process design and enables a prioritising of corrective actions and improvements that should be carried out first at specific steps of the process. For more details on FMEA principles one is referred to the literature (Kehoe, 1995; Banens *et al.*, 1994)

Process FMEA provides information about potential failure modes, which are ranked in priority, in order to identify corrective actions and improvements early in the design stage. In fact, process FMEA helps in understanding the process. A technique, which is based on the process FMEA principle, is Hazard Analysis and Critical Control points. Whereas FMEA was developed for economic reasons (*i.e.* to minimise economic loss caused by process failures), HACCP was developed from a food safety point-of-view (*i.e.* to minimise health risks). For details about the HACPP system one is referred to Chapter 7.

B. Hygienic design in the food industry

Increasing public awareness of food safety encouraged the safe and hygienic processing of food products. Hygienic design aims at designing equipment in such a way that contamination by micro-organisms or by cleaning and disinfectant chemicals can be prevented. In fact, if agri and food processing equipment is of poor hygienic design then more severe cleaning procedures, more aggressive cleaning chemicals and longer cleaning and decontamination cycles are required (Curiel *et al*, 1993).

Legislative aspects of hygienic design have been enacted at different levels *i.e.* international, European and national.

- At international level, the World Trade Organisation (WTO) is responsible for regulations with respect to food hygiene. The SPE-agreement (Sanitary and Phytosanitary measures) of WTO contains a broad range of measures focussed on protecting human and animals against food-borne diseases (see Chapter 2).
- At European level, since January 1995 all new equipment must comply with the Machinery Directive 89/392/EC. The Machinery Directive includes safety requirements for equipment with respect to the user and a hygiene section for agri and food machines. The latter includes requirements for the production of safe and hygienic products.
- In EC member states, the EC Directive for hygienic process design has to be implemented in their national regulation.

Furthermore, several international and national standard organisations produced standards for, amongst others, hygienic design of food equipment like ISO (International Organisation for Standardisation), CEN (European Committee for Standardisation), NIN (Nederlands Normalisatie Institute) and DIN (Deutsches Institut für Normierung).

In addition, Good Manufacturing Practice (GMP) codes established minimum acceptable standards and conditions for processing and storage of products. Several GMP codes contain requirements for hygienic design, construction and correct use of equipment (IFST, 1992; FDA, 1994).

The formulation of the hygiene requirements in the Machinery Directive is rather general. In practice, the requirements were often differently interpreted and implemented. In anticipation to this problem the EHEDG (European Hygienic Equipment Design Group) has been established in 1989. EHEDG is an independent consortium formed to develop *guidelines* and *test methods* for safe and hygienic processing of foods (EHEDG, 1995). The group includes representatives from research institutes, food industry, equipment manufacturers and government organisations in Europe. EHEDG reports contain detailed instructions for design engineers of agri and food machines. Hygienic design requirements as recommended by EHEDG are described in the *functional requirements* and involve that equipment:

- Must be easy to clean and disinfect, preferably cleaning in place (CIP)
- Should protect the product from microbial and chemical contamination
- Must prevent the ingress of micro-organisms in case of aseptic design
- Can be monitored and controlled at equipment's functions that are critical to microbiological safety (Curiel *et al.,* 1993).

To fulfil these functional requirements, a range of design criteria have been established for the construction of hygienic design, including criteria for construction of materials, criteria for layout and geometry and criteria for installation. Each is described in more detail below.

- *Criteria for construction of materials*. The materials that are used for the construction of agri and food machines must be non-toxic, non-absorbent and resistant to both product and cleaning and decontamination chemicals, under the conditions used. Contact surface areas must have a finish of an acceptable R_a value and should be free of pits, folds and other irregularities. Irregular surfaces are more difficult to clean and micro-organisms can remain and multiply in the imperfections and so contaminate the next batch. Several materials can be used for the construction such as stainless steel, plastics, and elastomers. Selection of stainless steel type depends on the corrosive properties of the product to be manufactured. Chloride-containing liquids may cause corrosion. When choosing plastics the effects of the product and process conditions should also be considered. For example, some plastics may swell due to absorption of lipids. Elastomers are often used for seals, gaskets and joint rings. Excessive pressure and heating conditions can damage the elastomer; therefore the compression of the elastomer should be controlled under the conditions used (Curiel *et al.*, 1993).
- *Criteria for geometry and lay-out*. All equipment and pipes must be self-draining and horizontal surfaces must be avoided (Figure 4.8). Dead ends and sharp corners should be avoided as well, because micro-organisms can remain there and multiply contaminating the next production. Joints and seals are critical parts of the design. They must be bacteria tight and free of imperfections. For example, metal-to-metal joints are critical; they must be continuously welded and free of imperfections.
- The *criteria for installation* refer to how aspects must be considered when equipment is installed, like direct seal to the floor to prevent difficult cleaning under equipment.

A summary of major hygiene design criteria as recommended by EHEDG is represented in Table 4.3. Typical examples of incorrect and correct designs are schematically shown in Figure 4.8.

However, these criteria are still rather general and, therefore, EHEDG also developed guidelines for hygienic design of specific equipment like for example:

- Microbiologically safe continuous pasteurisation of liquid foods (Lelieveld, 1992);
- Hygienic packing of food products (Mostert *et al.*, 1993);
- Hygienic design valves for food processing (Abram *et al.,* 1994).

Finally, it should be emphasised that hygienic characteristics of equipment must be proved. For this purpose EHEDG developed (and still develops) special test methods to evaluate hygienic design characteristics *e.g.*

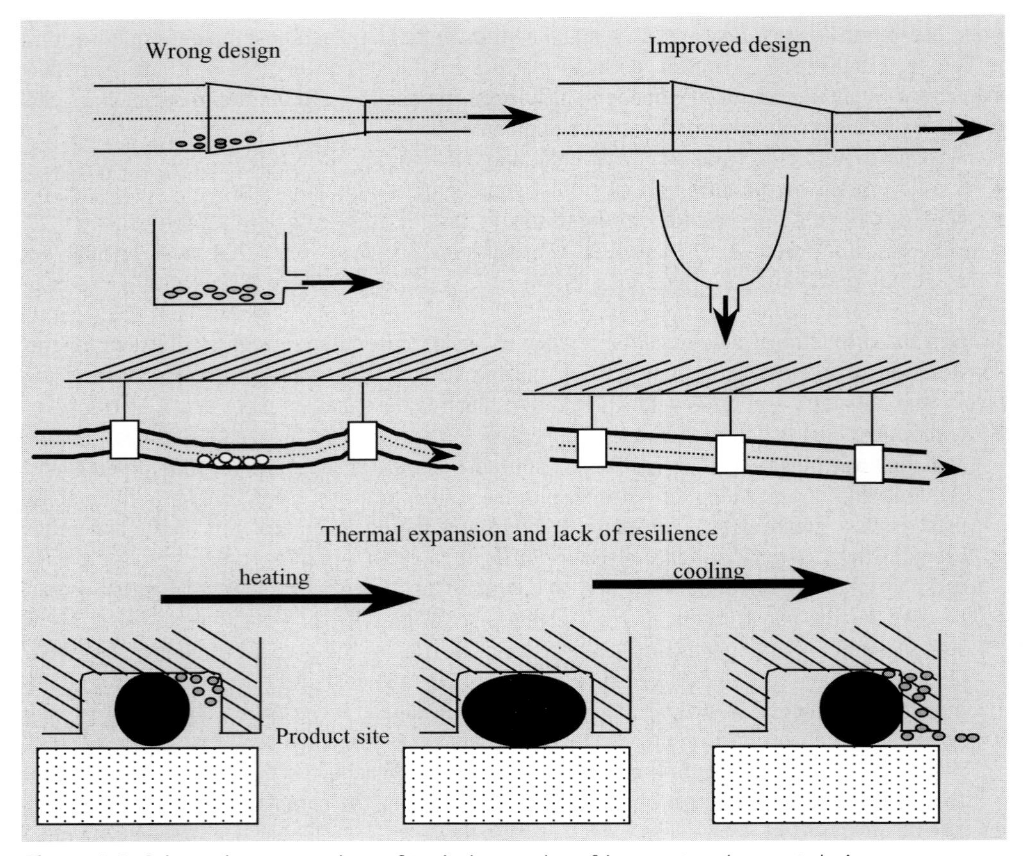

Figure 4.8. Schematic presentations of typical examples of incorrect and correct designs.

- Method for assessing the in-place cleanability of food-processing equipment (Holah *et al.*, 1992)
- Method for the assessment of bacteria tightness of food-processing equipment (Timperley *et al.*, 1993)

The manufacturer may mark equipment with the special CE (Conformité Européenne) symbol, if the equipment complies with all requirements of the EC Machinery Directive. In fact, without the CE mark and evidence that its use is justified, equipment may be neither sold nor put into use (Lelieveld, 1993).

Another important aspect is the maintenance of food-processing equipment. On regular basis, product contact surfaces must be controlled for signs of wear, like corrosion and crevices. The ageing of materials (such as compression capacity of elastomers) must also regularly be checked. Furthermore, the hygienic status of equipment can be tested by several assessment methods, as developed by EHEDG. Maintenance of equipment costs time and often the production must be put out of operation. However, the costs of equipment defects, because of poor maintenance, can be much higher then the costs of regular maintenance.

Table 4.3. Hygienic design criteria for agri and food equipment.

Criteria for construction materialat product contact side	Criteria for geometry and lay-out of equipment	Criteria for installation
• Non-toxic and food approved • Non-absorbent • Resistant to process conditions • Contact surface free from crevices and imperfections • Roughness surface $R_a \leq 0.8$ µm • Materials used: 1. Stainless steel 2. Plastics (PP, PVC, PC, PE)[1] • Elastomers, *e.g.* nitril rubber, silicon rubber • Adhesives • Lubricants shall comply with FDA regulations	• Equipment should be self-draining • No dead ends • No sharp corners • Joints and seals • Must be bacteria tight • Free of imperfections • Avoid misalignment • Metal-to-metal joints must be continuously welded • Insulation of equipment/pipes by vacuum or cladding (no ingress of water) • Construction of supports for equipment/piping such that soil and water cannot remain on surface	• Minimise risk of condensation on equipment; otherwise collect condense • Equipment directly sealed to floor/wall without gaps • Otherwise adequate space for cleaning and inspection

[1]PP= polypropylene; PVC= polyvinyl chloride; PC= polycarbonate; PE= polyethylene

Hygienic design becomes more and more important because of the increased public awareness on food safety.The implementation of EC Directive 93/43/EC on food hygiene also puts demands on hygienic design. As a matter of fact, evaluation of the hygienic design of process equipment for agri and food machines should be part of a Hazard Analysis and Critical Control Points (HACCP)-study (Burggraaf, 1998). HACCP comprises a structured analysis of potential hazards of the manufacturing process of food products. Subsequently, critical control points are established, which must be monitored and controlled to assure food safety in production (see Chapter 7). Equipment of poor hygienic design at critical process steps may be sources of pathogens and/or spoilage bacteria and therefore potential hazards.

4.3.3 Process capability and Taguchi method

As previously mentioned, designing the process not only includes system design (*i.e.* design of equipment and tools), but also parameter and tolerance design are important aspects that must be considered. *Parameter design* involves the translation from the designer's concept into concrete manufacturing specifications, which have nominal dimensions (*e.g.* gram, millilitre, °C, minutes, etc.). A typical aspect of food products and food packaging materials is that the nominal dimensions (*i.e.* parameter design) are often set by legislative regulations, whereas for mechanical products the designer usually sets the parameters.

Tolerance design includes the determination of tolerances around the nominal settings as identified by parameter design. Tolerance is the permissible variation in the dimension.

Tolerances are necessary because it is not possible to produce food products or components, which all exactly comply with specifications, due to common causes. Natural variations (*i.e.* common causes) in a production process or system can be due to different sources, including people, materials, machines and tools, methods, measurements and environment (see Chapter 5).

The assessment of the process capability is an important tool for process design. Process capability refers to the ability of the process to meet the design specifications for a product or service; parameter design. Design specifications often are expressed as a nominal value (or target) with specified tolerances (or allowance above or below the nominal value or target); tolerance design. A process is capable if it has a process distribution whose extreme values fall within the upper and lower specifications for a product or service, as shown in Figure 4.9. In addition, the capability of a process is also a boundary condition for statistical control of a process. The principles of statistical process control (SPC) and the effect of process capability are discussed in more detail in Chapter 5.

Improving process capability is possible through quality engineering, using for example the Taguchi methods, as previously mentioned for product development. It is an approach that involves the combination of engineering and statistical methods to reduce costs and improve quality (Krajewski and Ritzman, 1999). In practice, the consequences of tolerance setting on product functionality, manufacturability and economic effects are often not considered. Taguchi related tolerance design to economic implications and advocated to design narrow tolerances, preferably at critical (processing) steps. In other words, narrow tolerances must be set where exceeding the upper and/or lower tolerance limit has large economic consequences. This approach is called the Taguchi's quality loss function, as previously illustrated in Figure 4.5. The loss increases as a quadratic function when values do not meet the target value. Therefore, according to Taguchi, the only meaningful specification is being on-target.
However, for food products narrow tolerances are often not feasible, because of large natural variations in raw materials due to, amongst others, seasonal effects, weather conditions, genetic predisposition or transport conditions (see Chapter 2). Nevertheless, the principle of Taguchi's quality loss function is also applicable to food products. For example, if an expensive

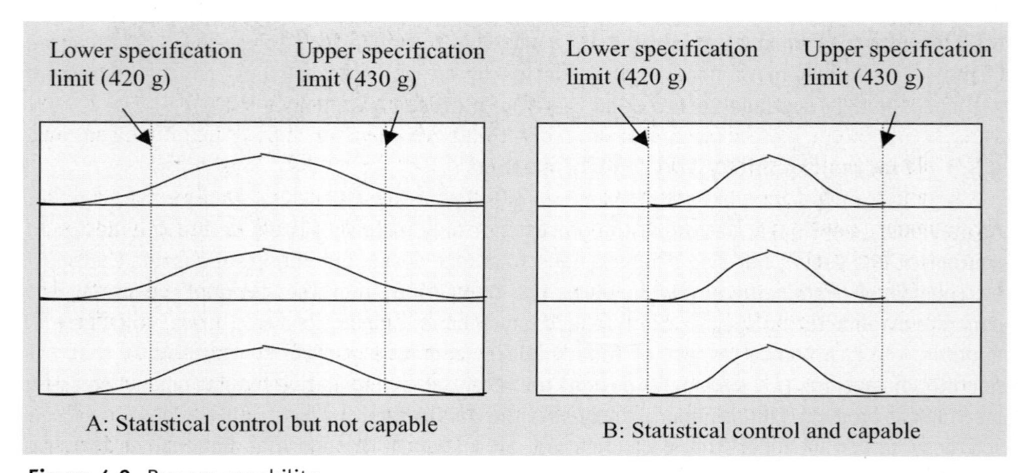

Figure 4.9. Process capability.

aroma ingredient is used, exceeding the upper limit will cost a lot of money, whereas too low levels will have less economic consequences, until it can be noticed in sensory experiments or by consumers (too low concentrations). Thus when designing the critical limits (*i.e.* tolerance design) the costs of exceeding these critical limits should be considered.

4.4 Customer-oriented design management

Although design processes in the food industry deal with complex technological problems, customer-orientation is getting more attention. In this section customer information is brought up as a vital condition for customer-orientation. Secondly the process of a customer-oriented design is described.

4.4.1 Customer information.

A major key determinant for success in product development is the degree of fit between the new product and consumer needs. As a consequence, customer-oriented design takes consumer needs as the starting point for the design process, and product and production technology as its derivative. In other words, the new product and its production technology are not a goal as such, but rather an instrumental mean to the realisation of consumer value through fulfilment of their wishes and needs (Van Trijp and Steenkamp, 1998).
To use the voice of the consumer as starting point of product development, information about their needs and wishes are required. A broad range of methods and sources, which provide input for product ideas, can be consulted as shown in Table 4.4. Some methods give *direct information* about consumer needs, like surveys, focus groups or field intelligence (Evans and Lindsay, 1996). A survey is aimed to measure consumer satisfaction but often includes questions to indicate the importance of particular quality attributes. A focus group is a panel of individuals who answer questions about a company's product as well as those of competitors. This approach aims at comparing experiences with expectations. Typical questions are: what do you like or dislike about the product? If you had the opportunity, how would you change the product? Field intelligence means that all employees, ranging from sales to receptionists, who have contact with customers or consumers, obtain useful information just by listening.
Indirect information about consumer requirements and needs can be obtained by using collective memory, analysis of complaints and evaluation of competitors' product and analysis of purchase data. Analysis of purchase data may reveal gaps in product assortment offering opportunities for product development. Collective memory includes all knowledge and experience present in the organisation. This collective memory should be organised into accessible catalogued information, which can be consulted for product development.
Furthermore, supporting information can be obtained, which can help in directing product development (Table 4.4). This information is described by Henning (1988) as research guidance; it includes topics such as

- *Societal changes*, which might have consequences for future consumer demands. For example, the reduction of family size, the increasing number of dual income families and also economic processes pushes towards the development of ready-to-eat meals.
- *Technological developments*, which may offer new opportunities. For example, new preservative techniques, like high-pressure treatment, electric field pulses, modified atmosphere and active packaging, may offer new opportunities to satisfy consumer demands for fresher, more tasty and less processed products.

- *Changes in the food production chain* may have consequences for *customer* demands in the chain and *consumer* needs at the market place. For example, a trend in retail to open shops at fuel stations might have consequences for packaging concepts, *e.g.* packaging must be easy to handle in the car, and possibly barrier properties against fuel components in the air are required!
- *Legal developments* at national and international level may have consequences for the boundary restrictions for product development. For example, the packaging convenant stimulated the packaging and food industry to develop new packaging concepts with less material in order to reduce packaging waste.

All these sources give input and boundary conditions (*e.g.* legal limits) to the product development team.

Table 4.4. Sources of information for generation of product ideas.

Direct information		**Indirect information**	
• Surveys and interviews		• Complaint analysis	
• Focus groups		• Analysis competitor's product	
• Field intelligence		• Purchase data analysis to detect gaps	
• Direct customer contact		• Collective memory of company	
Supporting information			
Societal developments	*Technological developments*	*Changes in food chain*	*Legal changes*
• Governmental publications	• Conferences, symposia	• trade literature	• governmental publications
• Research publications	• trade literature	• sales representatives	• workshops
	• scientific literature		• trade literature

4.4.2 Customer-oriented design processes

The customer-oriented design approach is not new, but is now getting more and more attention in the food industry. Juran (1988) was the first to present the design process starting with customer needs. In his 'Quality planning roadmap' the first step includes identifying customers and discovering customer needs (Figure 4.10). Juran paid much attention to translating customer needs into units of measure, thus providing the basis for the concept of Quality Function Deployment (Van den Berg and Delsing, 1999).

Urban and Hauser (1993) stated that the new product's superiority could be summarised in the "Core Benefit Proposition". This core benefit proposition can be described as the unique benefits of the product that comply with consumer needs and those benefits that are required to meet and surpass competition. Van Trijp and Steenkamp (1998) supported the new product design process model of Urban and Hauser and adapted it to the model presented in Figure 4.11.

Van Trijp and Steenkamp (1998) made a distinction between a managerial and a consumer behaviour research component, called "voice of the company" and "voice of the consumer", respectively.

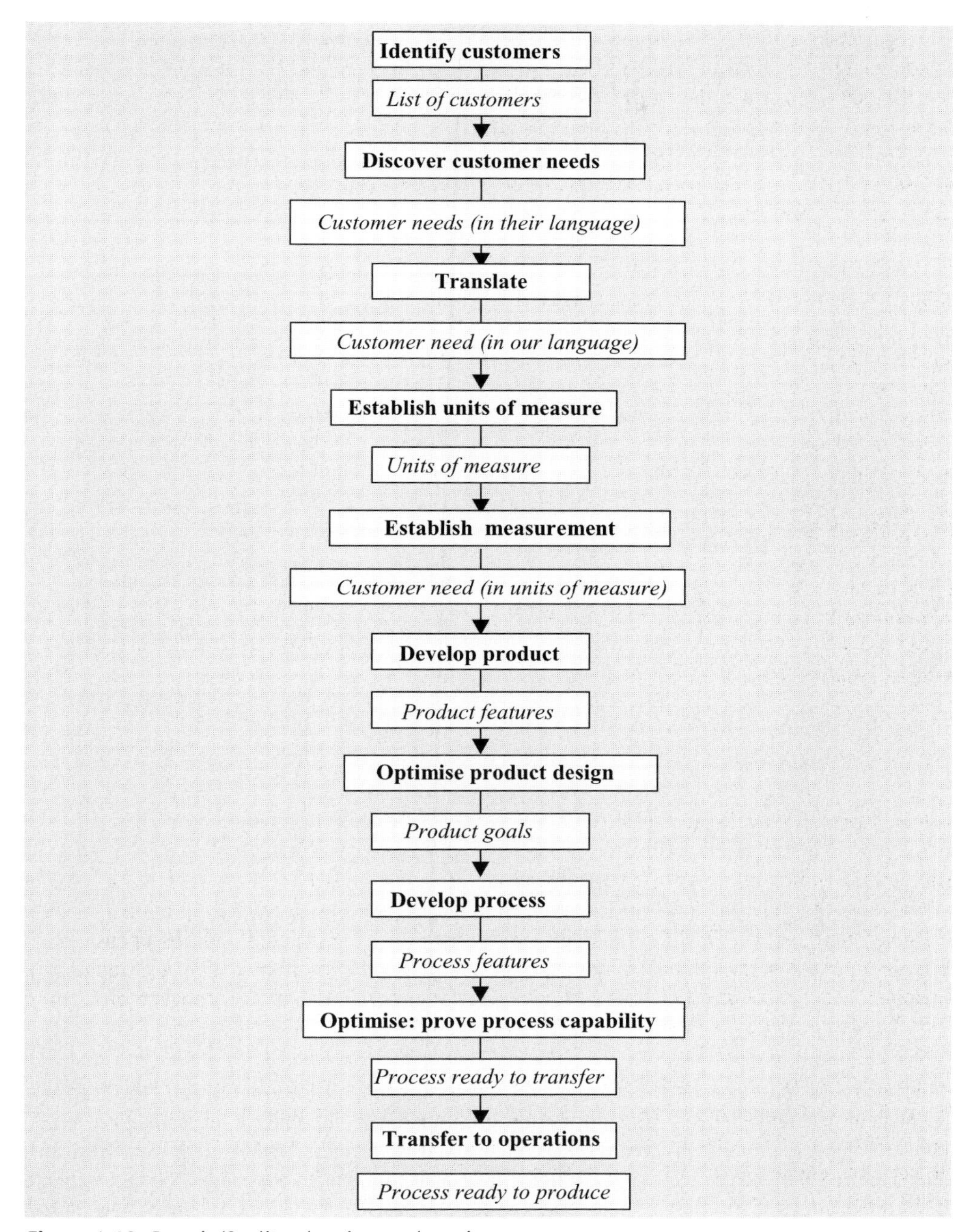

Figure 4.10. Juran's 'Quality planning road map'.

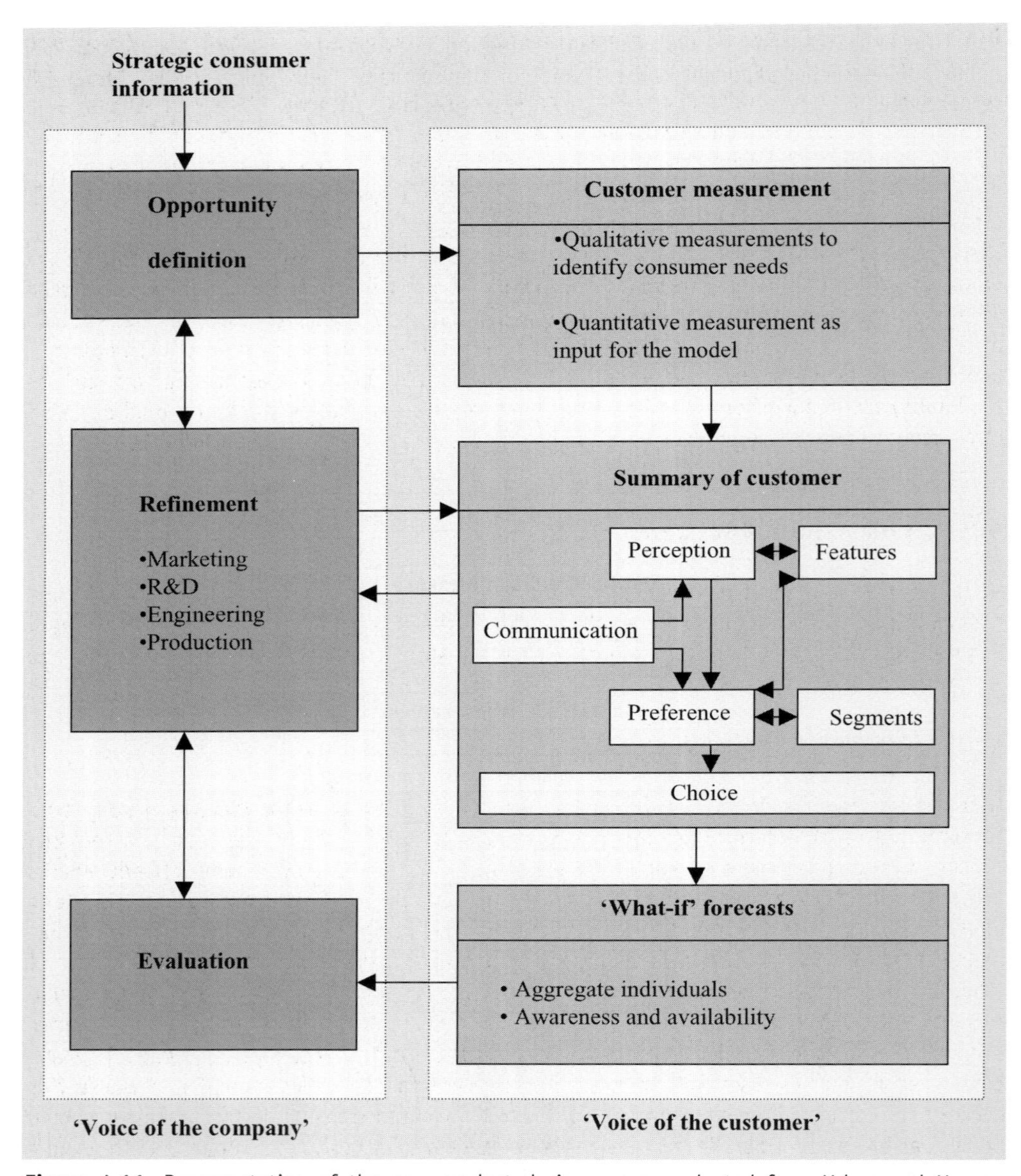

Figure 4.11. Representation of the new product design process adapted from Urban and Hauser (1993).

The *managerial part* concerns decisions, with respect to opportunity definition (the market), refinement (benefit delivery and optimisation through the integration of marketing, R&D, engineering and production) and evaluation, to assess business opportunity. Van Trijp and Steenkamp (1998) added pro-active strategic consumer research to the model to make explicit that opportunity definition should find a strong basis in continuous and deep-rooted understanding of the consumer in the relevant product categories.

The *consumer part* concerns customer measurement (to make an assessment of market potential before substantial funds are invested in product development), modelling consumer behaviour (to judge and verify product or service alternatives) and "what-if" forecasts to make an assessment of strengths and weaknesses of the new product concept/prototype. The consumer behaviour model of Van Trijp and Steenkamp (1998) assumed that consumer-oriented behaviour finds its basis in consumers' evaluative reactions to the product (product preference). Product preferences find their basis in product attributes as perceived by the consumer (perception). These attribute perceptions, in turn, may be generated through manipulation of the physical product features or through communication (marketing). It is assumed that consumers attach different weights to their attribute perceptions in preference formation. Therefore, recognition of the differential weighing of benefits may provide a relevant segmentation basis. Finally, it is important to recognise that choice does not automatically follow preference. Factors such as availability and point-of-sale promotion may interfere in this relationship and should be taken into account early in the product development process (Van Trijp and Steenkamp, 1998).

4.5 Cross-functional design

Modern design management includes collaboration between different disciplines inside and outside the company. In this section collaboration is discussed with the concept of concurrent engineering. Separately cross-functional teams are described being a basic condition for effective exchange of knowledge.

4.5.1 Concurrent engineering

Transforming an idea for a new product or service into an actual product is usually time consuming and expensive. The costs, associated with making major changes to a product concept or even rejecting a concept, increase dramatically as the product development process progresses. In the near past, product life cycles were so long that mass production and a long product life cycle easily returned the costs of developing. However, current trends like fragmenting of markets, mass customisation, rapid technological advancements, shortening of product life cycles and intense global competition have changed the mass-production approach. In fact, these challenges call for a drastic change in the way in which products are designed and in the role of production department and experts in the development process. In many companies design processes were performed according to a sequential approach. However, this concept is not suitable in a dynamic market where time and costs of design processes should be kept as low as possible. For this purpose the concept of concurrent engineering offers more opportunities. Both concepts are explained below.

In a *sequential approach* the information flow is predominantly downstream (Figure 4.12). The product concept is transformed into functional product and process specifications are turned over to production and purchasing. At this stage the decisions are made and the available options are usually dependent on earlier decisions. As a consequence, by the time that production is involved in product and process designs, it might be too late to make changes without initiating seemingly endless engineering changes. This can lead to the looping effect, which rapidly lengthens the product development process as well. The design process becomes less flexible and costs of incorporating engineering changes can increase by a factor of 10 as the product moves from concept stage to market launch. Studies have shown that the design of a product accounts only for about 5% of total costs. However, the cost impact of additional quality changes, manufacturing and equipment modification and technology acquisition can be at least 70% of the total product. The sequential approach is also known as the *over-the-wall approach* (Noori and Radford, 1995).

An alternative to this approach is called *concurrent or simultaneous engineering* (Figure 4.12, Table 4.5). It is an organisational tool that can facilitate integration by creating cross-functional teams. Functional product and process specifications are developed concurrently. Ideally, concurrent engineering involves the firm's customers and suppliers along with a broad range of representatives of different functional areas. Inclusion of customer representatives in the design team might help to ensure that a product really meets their needs. Suppliers may have a large influence on product costs and quality, therefore, their involvement from the beginning can reduce costs and improve quality. Collaboration with suppliers is even more crucial when a firm transfers partial or even full design responsibilities for specific components to its suppliers (Noori and Radford, 1995).

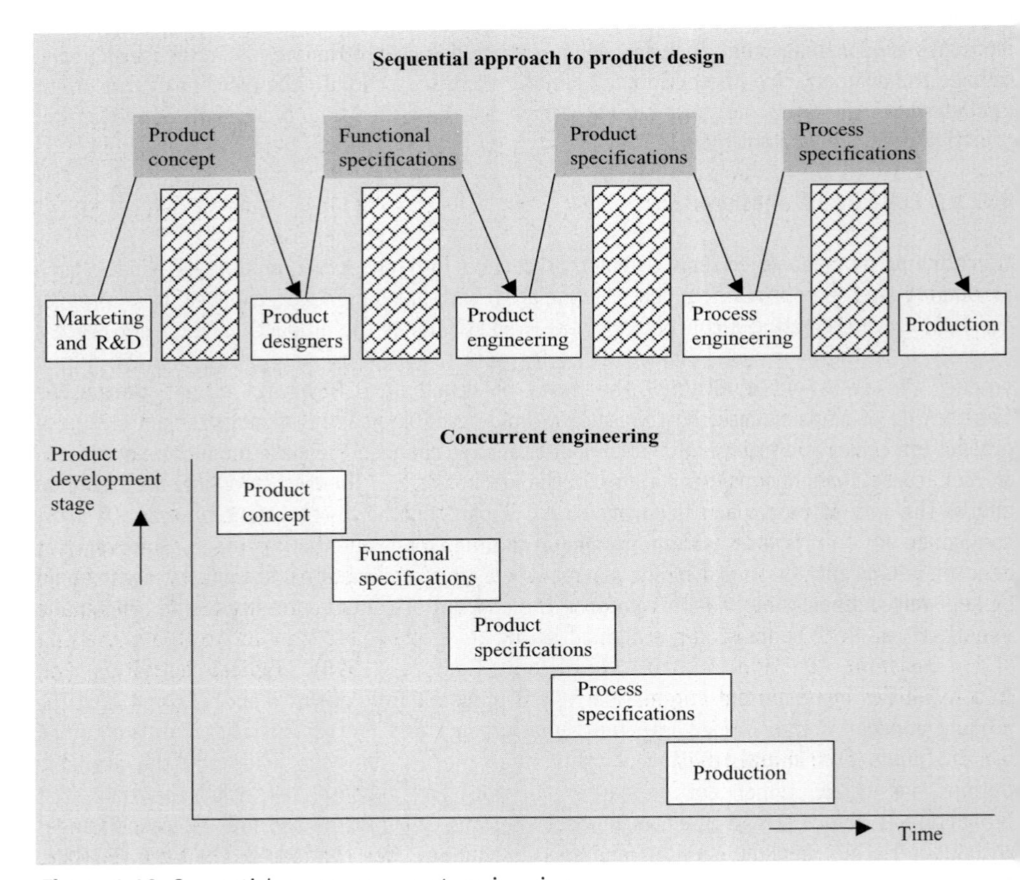

Figure 4.12. Sequential versus concurrent engineering.

Table 4.5. Activities and responsibilities in concurrent engineering.

Conceptual design
Marketing: Proposes and investigates product concepts
Engineering: Proposes new technologies and stimulates performances
Operations: Proposes and investigates manufacturing and delivery processes
Product design
Marketing: Defines markets and specifies objectives
Engineering: Chooses components and key suppliers
Operations: Defines process architecture and estimates costs
Product and process engineering
Marketing: Conducts customer tests on prototypes
Engineering: Builds full-scale prototypes for evaluation and refinement
Operations: Builds system to manufacture prototype, plans full-scale system, tests tooling and new procedures
Pilot development and testing
Marketing: Prepares for market roll out, trains sales force
Engineering: Evaluates and tests pilot unit
Operations: Builds pilot units in commercial processes, refines the system, trains personnel, checks suppliers
Volume production and launch
Marketing: Fills distribution channel, sells and promotes, gains feedback from target customers
Engineering: Evaluates customers' experience with product
Operations: Builds up plant to volume targets, refines quality, yield, and cost performance
Post-sale service
Marketing: Gains customer feedback
Engineering and Operations: Study warranty data

In order to make concurrent engineering successful, it is necessary to change the manner of operating and communicating in the functional areas. Functional managers must delegate decision-making authority to their representative in the team so that the team can review, modify and approve designs quickly. As a consequence, current lines of authority and reporting relationships must be reviewed and realigned. Many decision-making authorities might feel threatened and reluctant to pass on power to others. Conflicting relationships that might exist between functional areas must be overcome. Career paths of functional specialists might be no longer well defined. Therefore, changes in the reward structure are needed to encourage teamwork, rather than individual efforts. Changes are also needed to minimise the possibility of inconsistent functional goals (Noori and Radford, 1995). Typical activities and responsibilities in concurrent engineering are shown in Table 4.5.

4.5.2 Cross-functional teams

To make proper decisions in the multidisciplinary process of product and process development, knowledge of each relevant aspect should be included. Nowadays companies use more and more cross-functional teams, in which relevant experts and/or departments are involved. As illustrated in Figure 4.13, a development team should ideally consist of the following departments or experts (Fuller, 1994; Cohen, 1995):

- *Senior management* to assure that selected ideas fit the corporate image and comply with company's objectives.
- *Financial department* should monitor development costs and must keep it within budgetary limits.
- *Legal adviser* in order to consider legal implications, like allowed amount of additives or labelling, to prevent the team from wasting time and money pursuing activities against the law.
- *Marketing and sales* department(s) are responsible for monitoring the marketplace and getting information on consumer demands. Marketing must also consider the impact of new products on company's current brand image, whereas sales should think about the impact on the shelves at the retailer.
- *Warehousing and distribution* department(s) should indicate opportunities and restrictions for new products in the existing distribution and storage system. Special requirements, such as cooled and/or modified atmosphere storage require additional facilities, thus additional costs.
- *Engineering personnel* must evaluate the processes needed for manufacturing of the new product(s). If new or innovative processes are necessary than additional costs and time must be established.
- *Production or manufacturing* screens the impact of new product(s) on current plant production systems. They indicate if required skills, personnel and physical plant are available or must be attracted.
- *Research and development* department plays a vital role in the screening procedure. They control the technology of the development process and indicate technological possibilities and restrictions of ideas and prototypes.
- *Purchasing department* must balance between availability of proper materials (like ingredients, packaging), costs (*e.g.* of materials, of transportation to get the materials), reliability of supply and quality (*i.e.* complying with specifications of design). Product costs are a major screening tool.
- *Quality control department* should consider potential hazards and indicate relevant control points.

In addition also *packaging expertise* is required to advice proper packaging materials with suitable properties for the new product at an early stage of product development. In practice, packaging considerations commonly enter the process at the end, often resulting in (unnecessary) shelf life problems.

This cross-functional approach is required for optimal screening throughout the development process. For example, marketing, product development, process engineering and quality control should closely collaborate to assure product quality and safety. Product design and manufacturing must know what is feasible with respect to purchasing of raw materials from suppliers. Finally it must be remarked that in smaller companies one person often fulfils more skills, whereas large companies have distinct departments.

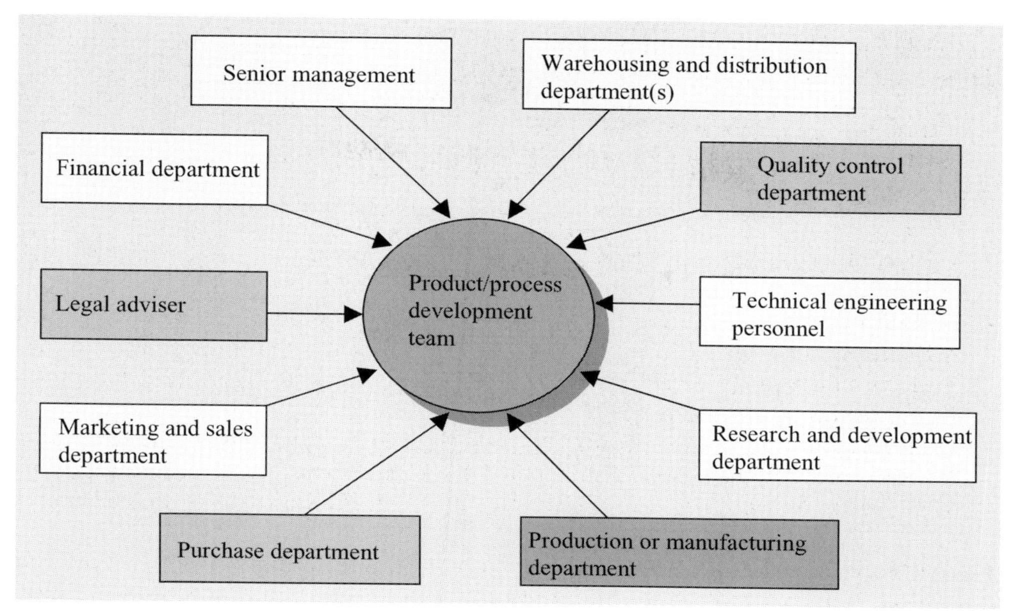

Figure 4.13. Composition of a product/process development team.

4.6 Managing the design process

In previous sections of this chapter the steps in the design process have been described, including different techniques and concepts for effective design performance. Nevertheless, design processes will not succeed if they are not properly managed. In this section two important management activities will be discussed. The product development strategy provides the basis for embedding product development in the organisation and long term strategy of the company. Project management provides techniques and organisation principles for effectively and efficiently realising design goals and objectives.

4.6.1 Product development strategy

The design process has to be managed actively in order to reach design goals both in qualitative terms and in delivery time. In Figure 4.14 design management is described as product development strategy (Wheelwright and Clark, 1992).

Wheelwright and Clark (1992) invoked the image of a funnel, illustrated by the dotted lines, to depict activities that support a fast-cycle product development activity. At the front end of the funnel, two ongoing business processes must initiate the process. Technology assessment and forecasting predict technological developments that a business should anticipate. This function evaluates new process technologies that will emerge from research and development actions, in order to find practical applications in the market place. This technology assessment then continues by weighing the benefits of acquiring essential capabilities from outside suppliers, for example by strategic alliance, by simple purchase, or by developing them internally (Melnyk and Denzler, 1996).

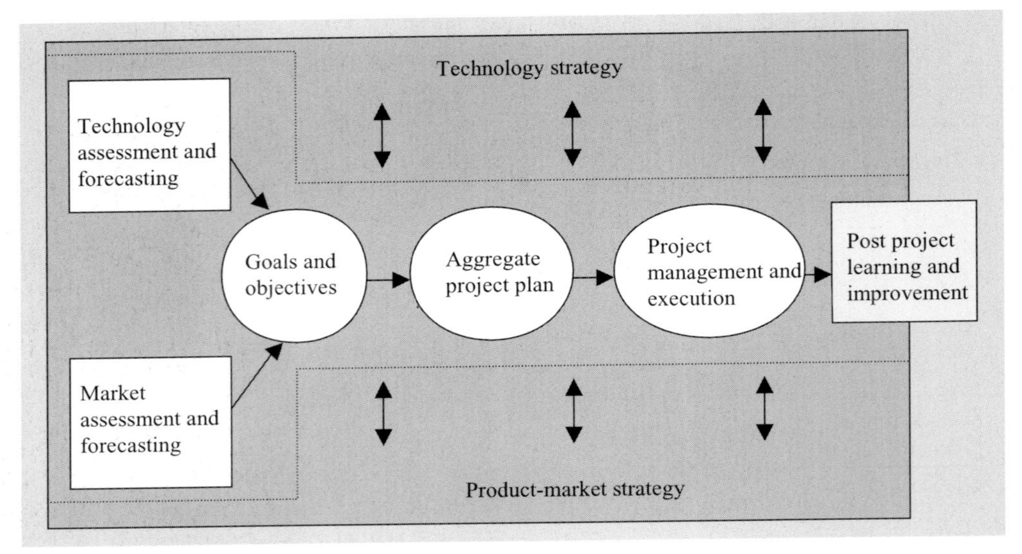

Figure 4.14. Product-development strategy.

Roussel and co-authors (1991) emphasised the importance of creating strategic technology plans. In a competitive environment there should be an ongoing effort to estimate the strengths of competitors in technologies, which are important to the business and which are different in respect to the successful conduct of R&D. They described three types of R&D:

- Incremental R&D, *i.e.* clever exploitation of existing scientific and engineering knowledge in new ways; characterised by low risk and modest reward,
- Radical R&D, *i.e.* creation of knowledge new to the company and possible new to the world for a specified business objective; characterised by higher risk and high reward,
- Fundamental R&D, *i.e.* creation of knowledge new to the company and probably new to the world, to broaden and deepen understanding of a scientific or engineering arena by the company; characterised by high risk and uncertain applicability to business needs.

A product development strategy continuously has to consider the contribution of R&D projects to the creation of unique knowledge.

In addition to technology information , the product development strategy needs an assessment of the firm's current markets and those that it would like to enter. First the firm must complete these assessment processes, because their outputs become major inputs to the firm's technology strategy and product/market strategy. Both of these component strategies specify directions for introductions or innovations into the marketplace. The technology strategy determines whether the company will introduce new technology in small increments or whether the nature of the technology permits a major breakthrough strategy. Of course, these decisions depend on the firm's overall willingness to take risks. Similarly, the marketing strategy determines if the firm introduces large-scale innovations or sequences of incrementally enhanced models. The technology and product/market strategy must also comply with the firm's general business strategy.

From general business strategy, product development goals and objectives must be translated to project levels. Statement of goals and objectives also specify process development because each activity competes with the firm's innovation-related funds. Planning is the next step and here is estimated how much and when each product-innovation project will contribute to the firm's achievement of its overall goals and objectives. Performance measurement must have a place as an integral part of this planning process. Beyond routine sales and profit objectives, the product and process innovations should be evaluated on three additional criteria: costs, time to market and quality. After setting and qualifying goals and objectives, the innovation infrastructure continues with the development of an aggregate project plan. Then it must be checked if the firm plans its deployment of resources to direct the right amount of resources to the right type and to each project at the right time, to maximise likelihood of success. To create an aggregate plan, projects are roughly planned by means of project planning techniques. However these project plans are subject to considerable uncertainty. Quantifying, at least approximately, is necessary to avoid the risk of the projects running out of control (Melnyk and Denzler, 1996).

4.6.2 Project management

A project is an interrelated set of activities that has a defined starting and ending point. It should result in a unique product or service. The steps in a design project are described in Figure 4.2. Projects often cross organisational lines because they need various skills of multiple professions and departments. Figure 4.15 shows how different departments in the organisation contribute to a variety of project activities, which are logically organised for project purposes (Kampfraath and Marcelis, 1981). Furthermore, each project is unique even if it is a routine. Uncertainties, such as development of new technologies or timing of certain events, can change the character of the project. Finally, projects are temporary activities because personnel, materials and facilities are assembled to accomplish a goal within a specified time frame. After that time the project team is disbanded (Krajewski, 1999).

Successful project management involves co-ordination of tasks, people, organisations and other resources. Three major elements are the project manager, the project team and the project management system.
The project manager has the responsibility to integrate efforts of people from various functional areas to achieve specified project goals. Traditional organisational hierarchies tend to slow progress on project due to lack of communication, co-ordination and sometimes motivation. A project manager can overcome these roadblocks to project completion. The manager is responsible for establishing project goals and providing the means to achieve them. The project manager must also specify how the work will be done and ensure that appropriate hiring is done and any necessary training is conducted. The project manager must demonstrate leadership and provide motivation necessary to accomplish the tasks required. Finally, the project manager evaluates progress and takes appropriate action when necessary. In product development projects, the project manager will most times be recruited from the marketing department or the R&D department.
The project team in product development will consist of members described in section 4.5, *i.e.* cross-functional teams. Sometimes there are representatives of outside firms involved in the project-team. Size and composition of the team may fluctuate during the life of the project. The team will be disbanded when the project has been completed.
The project management system consists of an organisational structure and an information system. The organisational structure defines the relationship of the project team members to

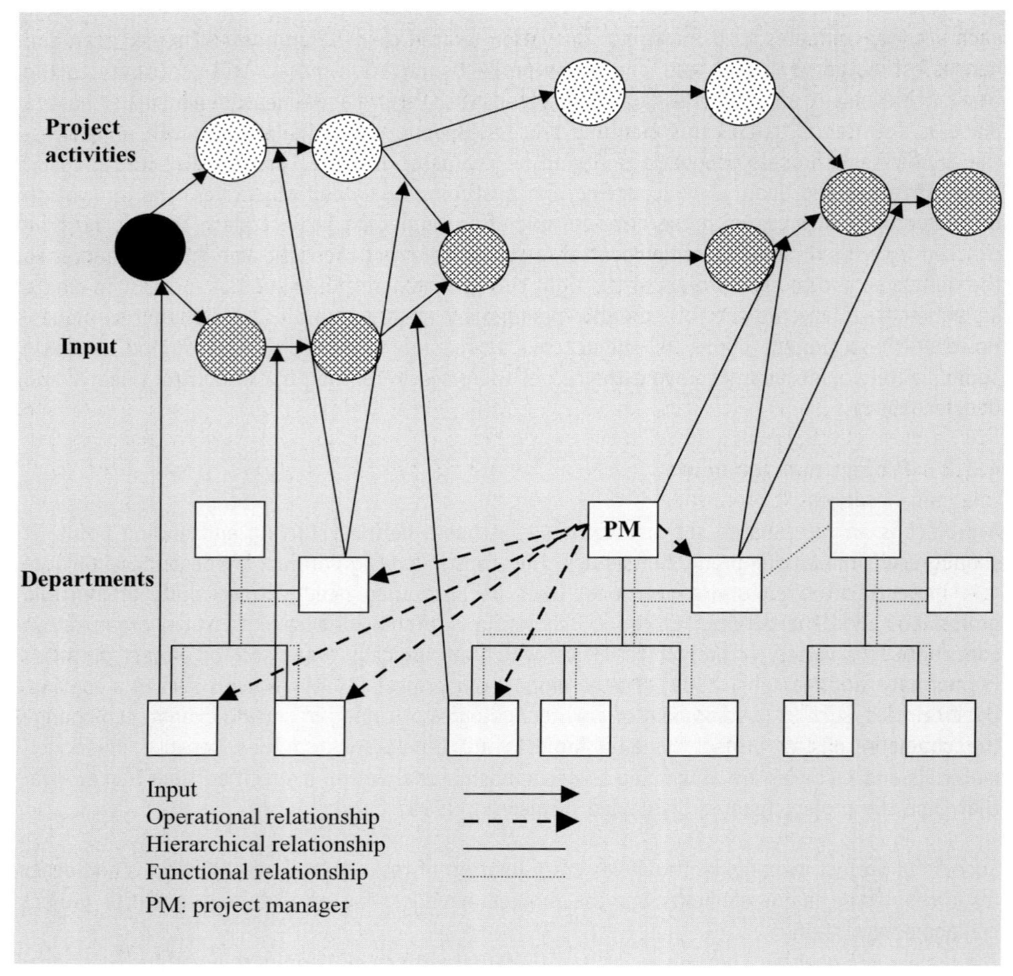

Figure 4.15. Project from a basis organisation view.

the project manager. The traditional structure is the functional organisation, whereby the project is housed in a specific functional area, presumably the one with the most interest in the project. The project manager must negotiate assistance from personnel in other functional areas. Under this structure, the project manager has minimal control over the timing of the project, but resource duplication across functional areas is minimised. At the other extreme is the pure project structure, whereby team-members work exclusively for the project manager on a particular project. Although this structure simplifies the lines of authority for the project manager, it could result in significant duplication of resources. A compromise is the matrix structure depicted in Figure 4.15. Each functional area maintains authority over who will work on the project and the technology to be used. In most cases the hierarchical manager is also the one with a functional relationship. The project manager has an operational relationship and has control over the team-members for the period they work in the project. Because team-

members have more than one boss, co-ordination between bosses is an important condition for success of matrix organisation.
The project management system also provides an integrative planning and control of the project, the accumulation of information related to performance, costs, schedule changes, and the projected time and costs of project completion (Krajewski and Ritzman, 1999).

Project planning is mostly based on network planning methods. A well-known technique is PERT, program evaluation and review technique, which identifies and controls many separate events required to complete a project. The use of PERT diagrams, such as the one shown in the upper part of Figure 4.15, is common in project management. The diagram shows the relationship between various phases of the project, with key activities (the arrows) being identified according to time requirements. The model helps to control a project by clearly identifying all required activities, the paths between them and the dates for their completion. It ensures that all activities on all paths are executed in proper sequence and in time. The path requiring the longest time is designated as the critical path and it attains special attention in the control process (Schermerhorn, 1999).

Besides project planning, project control is very important because of the high probability of changes in both market condition and engineering. Changes increase product design costs, especially when they are made late in the design process. As illustrated in Figure 4.16, the design becomes less flexible and the cost of incorporating engineering changes tends to increase by a factor of 10 as the product moves from concept stage to market launch (Noori and Radford, 1995). Therefore in design projects, project control based on go/no-go decisions accompanying each phase is very important (Figure 4.17).

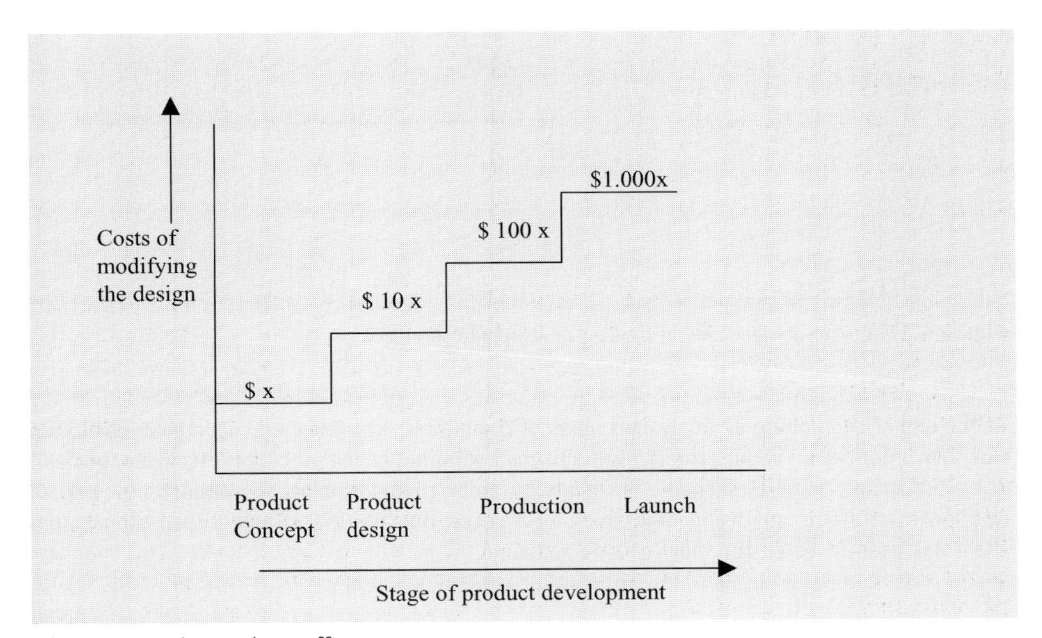

Figure 4.16. The escalator effect.

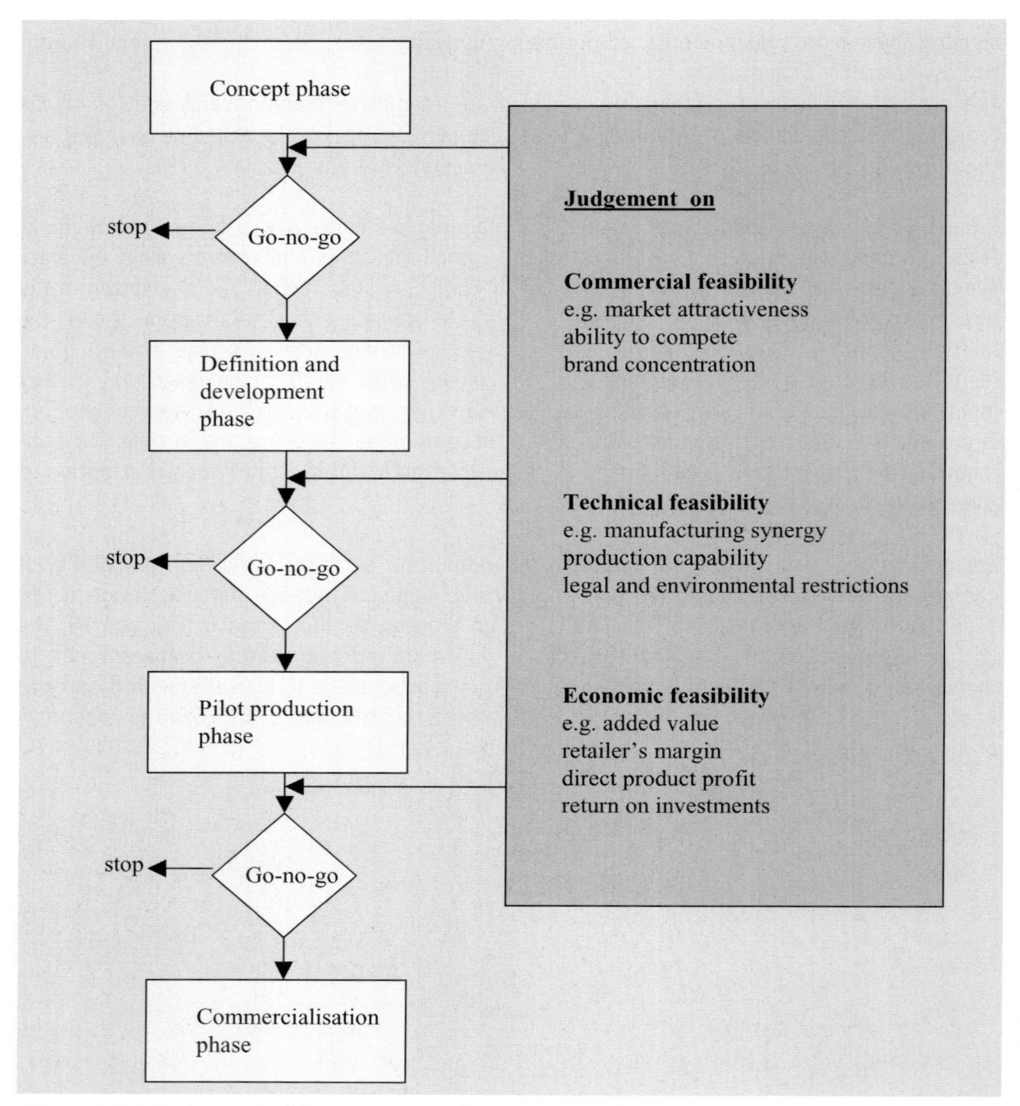

Figure 4.17. Go/no-go decisions in product development projects.

At the end of each phase an analysis is made of commercial, technical and economic feasibility. However, information about project results in the beginning of the process is often very unclear. Therefore proper estimates should be made in order to judge expected project outcome in relation to strategic goals and objectives. Less successful projects can be ended then before there has been invested too much money and time.

4.7 Quality design in the food industry

Product design can be characterised on two dimensions, *i.e.* product origin and difficulty. Product origin refers to where the product idea was created, which can be divided in originated from Research and Development (R&D) or from marketing. Difficulty reflects complexity of the product and of the technology used. Based on these two dimensions, pharmaceuticals are often characterised as originated from R&D and highly difficult, whereas foods are typified as originated from marketing and not difficult. Although these characterisations tend to overemphasise extremes, it clearly shows how food product design is generally considered. As a matter of fact a part of the (new) food products can be classified as originated from marketing and not difficult. However, there is a trend towards more complex food products like ready-to-eat meals and nutraceuticals. In these food concepts the technological degree is high and therefore must be considered as difficult. Moreover, the drive from the market has been intensified in the last years by an increased critical attitude of consumers and enlarged competition. Thus food design has become more difficult, whereas the consumer driven potential has increased. As a consequence, the food industry faces a big challenge, which might be too difficult for many companies in this sector. The reasons for this can be due to the following aspects that are typical for the food industry:

- *Product design as major activity of Research and Development*. In the food industry usually R&D proceeds the full design process and considers it mainly from a technological viewpoint. Sometimes even hobbyism is decisive instead of strategic objectives set by top management. People can be biased by the technology they have developed themselves and to which they can not renounce.
- *A restricted market orientation*. Many companies do not have much contact with their customers or consumers and market research is often limited.
- *The functional organisation structure*. The functional structure is still dominant in the food industry and often results in discipline focus and lack of co-operation between departments. Cultural differences between the departments of marketing and sales versus technical research departments hinder proper collaboration. For example, difference in language and communication between marketing and R&D results in misunderstanding and frustration.
- *Process design activities are initiated too late*. Modifications in process design are often carried out just after problems occur during pilot plant production. At this stage of the design, process costs are already rather high. Positive experiences were obtained from the food industry when applying simultaneous and integrated product and process design (Traill and Grunert, 1997).
- *Lack of proper project organisation*. Product design is commonly considered as a task of one department. Instead it should be considered as a joint project that is properly managed by using, amongst others, appropriate project planning methods.
- *Restricted collaboration* with other members from the production chain. Although intensive mutual dependencies exist between members of food productions chains, there is few collaboration between suppliers and customers in the area of design. One of the factors is knowledge protection, but also lack of proper chain management plays a major role.
- *Feedback of design process to the strategy of the organisation*. It regularly occurs that designs must be modified or design projects must be cancelled because strategic goals were changed.

This list of potential reasons for bottlenecks in food design processes is not complete but shows that there are many improvement opportunities.

From the techno-managerial approach, solutions for this type of problems should combine the unique technological aspects with appropriate management attention. This should occur as well in supply, production like distribution. Especially three points in the design process require special attention *i.e.*,

1. At the initiation phase, where consumer demands and needs must be translated to technical and technological specifications.
2. During product and process design (concept and prototype phases), where technological and technical knowledge must be adequately transferred to prototypes.
3. In the process where technical and technological knowledge and experience of production and distribution must be perfectly translated to production and distribution concepts.

According to the techno-managerial approach, Quality Function Deployment (QFD) can be a strong tool at the initiation phase. QFD is a tool that enables a structured collaboration between people from different disciplines (or departments) in order to design a new product based upon consumer demands. The house of quality and other planning matrices of QFD provide a systematic approach to translate consumer (marketing) language into technological language, therefore supporting communication between different departments like marketing, sales, engineering and production. Moreover, QFD forces discussions about prioritising quality attributes in terms of consumer demands, future production and costs. In this way it enables pro-active planning and control of the design process so preventing the escalator effect as shown in Figure 4.16.

Phases two and three can be strongly supported by modelling techniques. These techniques can predict behaviour of *e.g.* microbial, enzymatic, chemical or physical processes in fresh and processed food. Also more complex food-packaging interactions can be modelled. These technological models provide knowledge and information at an early stage (even before prototype production), which enable managers to consider consequences of different choices. Based upon these considerations, for example, better choices can be made with respect to packaging concepts where a proper fit between barrier properties and product requirements is necessary. This type of information can also enable a better tuning of time-temperature requirements of sensitive products on the one hand, and logistic planning on the other hand. So, although the existing models are often simplified reflections of much more complex situations, they can support in making decisions in an early stage of product development.

Another method that requires attention is the use of Taguchi principles. It enables a proper focus to those technological aspects in design that are really relevant with respect to quality and costs. In this way, the design process can be effectively judged and modified by management. This principle can be applied at all stages of the design process.

Above mentioned methods and techniques are only valuable and effective when they are properly embedded in the organisation. Consequently, attention should be paid to principles of concurrent engineering and cross-functional collaboration, not only with other departments within one organisation but also between partners of the production chain (like suppliers and customers). Moreover, a continuous checking of project results against the strategy of the company is required. This can be realised by consistent project organisation coupled to the information flow as reflected in Figure 4.18.

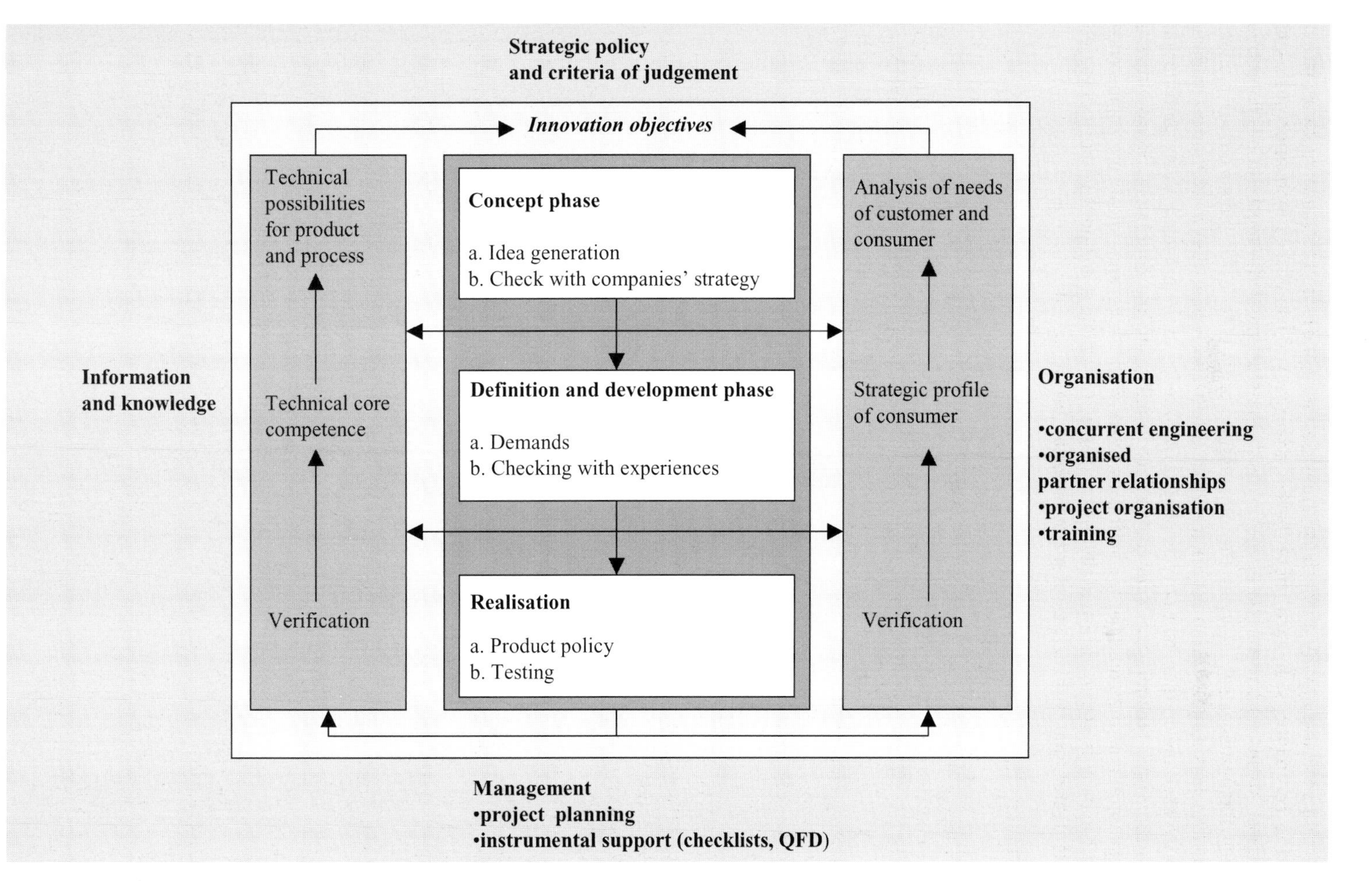

Figure 4.18. The product design process (modification of Urban and Hauser model, 1993).

Figure 4.18 shows a summary of the suggestions described above. In this model much attention is paid to 'control loops'. In each phase there is a feedback to established innovation goals and objectives. In this way, there is a direct link with the strategy of the organisation. From this strategy top management provides the organisation with criteria for judgement of product designs. An important addition to the model of Urban and Hauser (1993) is that a company needs to check the development of technological knowledge in the project with its contribution to the company's technological core competence. Similarly to the continuous check with the strategic profile of the consumer, a continuous check with the technological core competence of the company is proposed. Especially for companies that build a competitive position based on their technology is this check of major importance.

5. Quality control

Once product and production processes have been designed and developed, the system must be controlled during manufacturing of agri and food products. Controlling not only means inspection, but also taking corrective actions when the performance is not in compliance with specifications. The major aim of quality control is to produce a product that complies with targets within set tolerances. Therefore, a profound understanding of the sources of variation is required. Another reason for quality control is quality improvement. The effect of any improvement cannot be measured if the process is unstable and shows great variation. Quality control can be considered as a major process to realise quality of products and productions.

Quality control has been described as the ongoing process of evaluating performance and taking corrective action when necessary (Evans and Lindsay, 1996). It is generally considered as that part of the quality management system which is focused on operational techniques, and the processes applied to fulfil quality requirements (ISO, 1998). In this context, the manufacturing-based definition for quality is dominant, *i.e.* conformance to targets within specified tolerances (*e.g.* specified by designers or legislative regulations). Quality control involves both technological and management elements. Typical technological elements include *e.g.* the statistical and instrumental methods used. Typical management aspects refer to responsibility for quality control, relationships with suppliers and distribution, but also education and instruction of personnel to enable them to control quality. Both elements are described in this chapter. Firstly, the general process of quality control and typical aspects with respect to agri-food production are explained. Next, the technological tools and methods that can be used to perform quality control are described in detail, including acceptance sampling, statistical process control and analysis and measuring. The latter sections are focused on the managerial aspects of quality control, including control and business performance and managing of the control process. Furthermore, the control activities are considered from a product and resource perspective at supply, production and distribution. The last section includes a description of the current situation in the food production chain with respect to control activities, and the situation is considered from the techno-managerial perspective.

5.1 Quality control process in the agri-food production

Any control system consists of the following components: a measuring or inspection unit, comparison of actual results with a target value (*i.e.* norm, standard, goal or specification) with tolerances, and if required corrective actions; this is also called the 'control circle'. The general components of the control circle are explained in detail in the next section. Furthermore are described characteristics of agri and food production that may have consequences on quality control.

5.1.1 General principle of quality control

In order to control a product, process or system, it is important to understand the sources of variation.
Shewart (1931) proposed that causes of variation could be distinguished in *common* and *special* causes of variation, which has been later adopted in the foundations of the Deming Philosophy

(1993). Variations in a production process or system can be due to different sources including people, materials, machines and tools, methods, measurements, and environment. Common causes of variation are inherent to the product, process or system and involve the combined effect of all individual sources. Common causes generally account for 80-90% of the observed variation. The extension of common causes can only be reduced by structural improvements in people (*e.g.* by education), materials, machines, tools, methods, measurement and/or environment.
Specific causes of variation are derived from sources not inherent to the product, process or system and account for 10-20% of the variation. Examples are a bad batch from a material supplier, an inexperienced operator, or improper calibration of measuring instruments (Evans and Lindsay, 1996). Specific causes can be detected by the use of (graphical) control charts.

The control circle can be considered as a major principle of quality control. Control circles are not only applied on the level of production but also on management levels. However, this section is focused on aspects of control circles used in the operational manufacturing process. Control circles generally include four basic elements, *i.e.*

1. *Measuring* of the process parameter, *e.g.* temperature [T];
2. *Testing*, which means comparison of the measured value with the norm with specified tolerances, *e.g.* temperature must be 25 ± 2°C;
3. *Regulator* determines which kind and how much action must be carried out, *e.g.* the thermostat;
4. *Corrective action* involves the actual action that is carried out, *e.g.* the heating system that is turned up or down.

It should be noted that the regulator and corrective action are often embodied in one handling, person or equipment.

Ad 1. Measuring

In the measuring step the process or product is analysed or measured. Important characteristics of the measuring unit involve the signal/noise ratio, the reaction velocity to changes in the process and the correspondence between the signal determined and the actual situation of the process. The results obtained from analyses or measuring must properly reflect the actual situation of the process.
The measuring unit can be automatic or a manual handling performed by the operator. The measurement can be visual or instrumental like pH, time, temperature or flow rate. It can also involve analyses like microbial or sensory analyses. These analyses usually require more time than direct measurements, which can have consequences for the time span before corrective action can be carried out. Measuring units can be distinguished in five typical variants (Callis, 1987).

- Off-line, *i.e.* manual sampling followed by transport to laboratory for analysis or measurement.
- At-line, *i.e.* manual sampling followed by analysis or measurement at location.
- On-line, *i.e.* automatic sampling followed by automatic analysis.
- In-line, *i.e.* the signal is observed by a sensor (thus no sampling) in the product stream and translated to an external signal.
- Non-invasive, *i.e.* measuring of a signal in a product stream without physical contact with the product.

Ad 2. Testing

Testing is comparing the outcome of measurement or analysis results with the established target and tolerances. The outcome can be a number (*e.g.* amount of colony forming units of a certain pathogen) or a visual result (*e.g.* colour, appearance). For this part of the control circle, control charts are often used, in which the actual results, the target and tolerances are graphically reflected. With respect to the causes of variation, the combined effect of common causes is reflected in the tolerance levels, which can be statistically established. Specific causes can be noticed in the control charts as the out-of-control situations.

There are several types of control charts. The choice of a certain control chart depends on *what* must be monitored and the *type* of data. For example, different charts are used for variable (*i.e.* measured along a continuous scale) and attribute data (*e.g.* pass or fail). More details about control charts are described in the section on statistical process control.

Ad 3. Regulator

The regulator determines which corrective action is required based upon the result of the comparison with the target value. The extent (many or few) and direction (positive or negative) of the corrective action are thus set by regulator. The regulator must be selected on the basis of the character of the process and the required accuracy. A wide variety of regulators are used in practice. They vary from very simple regulators to complex systems. Some common regulators include:

- the *open-close regulator*, it is the simplest regulator, which has only two fixed positions
- the *proportional regulator*, in this regulator the magnitude of the action is proportional to the deviation of the process parameter
- an *optimising regulating system*, in such a system several process parameters are compared to various targets in order to obtain an optimal regulation
- an *expert system*, this is a more sophisticated regulator, in which expertise and knowledge of the system is incorporated For example, in the washing process of re-usable PET bottles a pattern recognition system is used. Headspaces of bottles are quickly analysed (within seconds) and the patterns of volatile compounds are identified and compared with the information in the expert system. The bottles are removed if the patterns do not fit, *e.g.* because there are still contaminants in the bottles after washing.

Ad 4. Corrective action

The corrective action is the actual handling that is carried out due to exceeding the target tolerances, *i.e.* out of control situation. The corrective action can be an instrumental (*e.g.* temperature increase) or a human handling (*e.g.* removal of the non-conformance products). The accuracy of the corrective action is essential for a successful execution of the control circle.

There are various forms of control circles, but the basic principle is the same. Two common control circles are the feedback and the feedforward control circle (Van den Berg, 1993), as illustrated in Figure 5.1. In the *feedback* control circle, the corrective action is performed *after* the trouble has occurred in the process. In the *feedforward* control circle, the process trouble is recognised *before* it occurs. The forward circle anticipates the expected situation based on earlier measurements in the process. For example, the composition (*e.g.* sugar content) of incoming raw tomatoes can be analysed in order to modify the recipe prior to production to obtain ketchup that complies with the specifications.

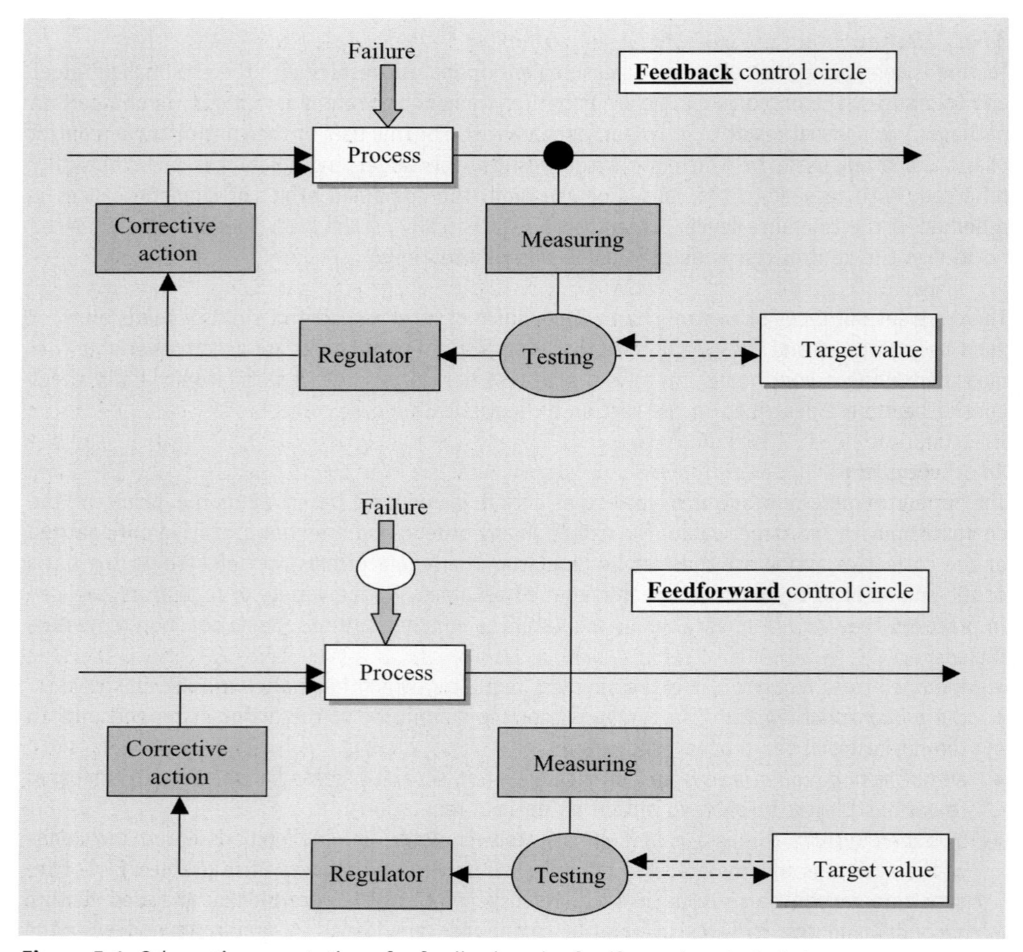

Figure 5.1. Schematic presentation of a feedback and a feedforward control circle.

For well functioning of the control circle it is important that the basic elements are properly tuned. A profound knowledge of the process is required to assess the appropriate measuring parameters that correspond with the quality attributes that must be controlled. Moreover, the control circle must consist of all basic elements and must be tested *in situ*, prior to use.
Besides tuning, another important aspect is the run-time of the control circle. The process stability must be such that during the run-time (*i.e.* the time between measuring the deviation and performing the corrective action) no unacceptable troubles occur. In other words, the run-time must be short enough to get no troubles in the mean time. As a matter of fact, the corrective action cannot always be carried out in the same process. Where the action is performed depends, among others, on

- time between measuring or testing *and* the corrective action;
- type of process: is it a batch or continues process.

5.1.2 Technological variables in control of agri-food production

Variations in a production process or system can be due to different sources including materials, environment, methods and measures, machines, equipment, tools and people as shown in Figure 5.2. For quality control it is important to understand these sources of variation. With respect to the control of quality of agri-food production, a number of typical variables are important to mention.

- (Raw) *materials* in agri-food processing often form a large source of variation. This can be assigned to the following aspects. Raw materials from plant or animal origin have a natural variation of 10%. For example, a large amount of variation can occur due to seasonal and harvesting effects. Moreover, raw materials are often unstable due to a variety of reactions that occur immediately after *e.g.* harvesting, disruption or processing (see Chapter 2). Furthermore, batches of materials or raw products are often inhomogeneous, which can complicate quality control. Low concentrations of agents with a high impact (*e.g.* off-flavours or pathogenic bacteria) can put high demands on quality control. Finally, the origin of raw materials is often unknown, which renders quality control more difficult. As a consequence of these sources of variation, the tolerances should not be specified as too narrow unless it will result in unsafe products.
- Moreover, the large variation in raw materials put high demands on the *methods and measurements* used. Sampling especially (taking the right number of samples at the appropriate location) and sample preparation are crucial for the control process. The time span of some sampling methods (*e.g.* microbial analyses can take more than three days) can make it difficult to react quickly to disturbances in the process. Furthermore, the methods are often destructive, which means that the products that are actually consumed have not been controlled. The methods and measurements, especially the way of sampling, can thus result in a large source of variation.
- Another important source of variation is the *people*. On the one hand, in agri-food production processes often people with rather low education levels are employed . On the other hand, hidden safety risks (like contamination by growth of pathogens due to unhygienic conditions or improper processing) require a rather profound knowledge of microbial processes in food. Shift work can also form a great source of variation. Mistakes due to misunderstanding or too low levels of knowledge can result in quality problems. Another people-related aspect is the fact that in agri-food processing, many control processes include visual inspections. This type of measurement is very sensitive towards variation due to the objectivity of the measure and the low accuracy.
- Finally, the *machines and equipment* used in agri-food production are often very specific for a certain application. As a consequence, equipment is not easily exchangeable. For each type of equipment other types of controls are required, which can form a potential source of variation. In addition, the design of equipment and machines for agri-food processing is essential with respect to food safety. Microbial contamination due to poor design and/or improper cleaning put high demands on the control process and is major sources of variation.

As shown in Figure 5.2, quality control is usually carried out at three major points in agri-food production processes, *i.e.* at receipt of the (raw or semi-processed) materials, during processing and prior to distribution of the (end) product.

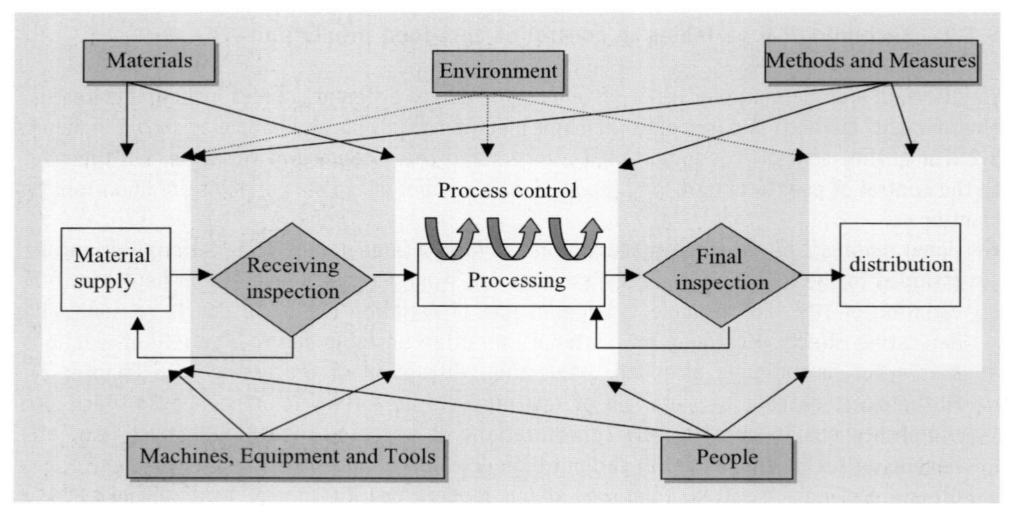

Figure 5.2. Typical control points in agri-food production processes.

At receipt of the incoming materials, quality control is carried out by product inspection. At this stage it is decided whether batches are rejected or accepted for processing. For this purpose different methods are available, like spot-check procedure, hundred-percentage inspection or acceptance sampling.
During processing or manufacturing, quality control is executed by measuring quality attributes at critical points in the process. For a proper functioning of the process, the principles of statistical process should be applied, using control charts to monitor the process.
Before distribution and storage, final inspections are carried out to check the actual quality of the manufactured or processed product. For this purpose similar techniques can be used, like acceptance sampling.

Finally, it has to be mentioned that shifts occur in quality control processes. In the last decade, a shift can be observed from quality control by acceptance sampling towards controlling the supplier by auditing or certification. The supplier is now responsible for the quality of the raw materials as delivered to the customer.
Moreover, a shift can be noticed from final inspection to quality control by (statistical) process control. The quality system HACCP (Hazard Analysis Critical Control Points) is a typical illustration of this shift. The HACCP system is developed to control critical points in the process instead of controlling quality by inspection of the end products.

5.2 Technological tools and methods used in quality control

As previously mentioned, the typical aspects of agri-food production require knowledge of appropriate sampling techniques. Important aspects that must be considered before sampling or measuring are:

- How and where should samples be taken; thus how to take a representative sample from, for example, an inhomogeneous product, which changes rapidly after harvesting or processing

- Which analysis or measuring method should be applied; is it destructive or not, what is the reaction time, does it change the content of the quality attribute etc.
- On what basis are batches or products rejected or accepted; what are the selection criteria

Statistical methods can support in taking correct decisions with respect to, for example, acceptance or rejection of a batch of raw materials. Some regular statistics used in quality control are explained in 5.2.1 (acceptance sampling) and 5.2.2 (statistical process control). Furthermore, an overview and considerations are given on major measurements and analyses used in quality control of agri-food production.

5.2.1 Acceptance sampling

Usually a quality control activity occurs at delivery of incoming raw materials, *i.e.* delivery control. The aim of delivery control is to decide whether a lot or batch will be accepted or rejected based upon its compliance with set specifications. Several inspection methods are commonly used in the agri-food industry, such as spot-check procedure, 100% inspection and acceptance sampling.
In the *spot-check procedure*, a fixed percentage of the batch is selected for inspection. For example, 10% of each lot or a periodical selection of every 10th product are selected for inspection. This method has no statistical basis. It gives no information about the risk of making an incorrect decision. The spot-check procedure is often used to check quantities (*e.g.* number of packaging materials delivered) and not as a decision tool for quality verification (Evans and Lindsay, 1996).
Hundred-percentage inspection is in fact a sorting method and will in theory eliminate all non-conforming products from a batch or lot. This inspection method is often used for grading of size and/or shape of agricultural products. In addition, 100% inspection is also applied for products with very critical safety requirements or with high failure costs. The limitations of 100% inspection are:

- The accuracy is typically 85%, because the monotony and repetition can create boredom and fatigue of inspectors;
- It is only applicable for non-destructive testing;
- It is often costly and impractical (Evans and Lindsay, 1996; Kehoe, 1995).

An alternative to 100% inspection is acceptance sampling. To decide which inspection method should be applied, the cost of inspection and the cost of failure should be considered. The cost of failure depends on the implications of non-conforming products to customers and the likelihood of failure occurring.
Acceptance sampling is commonly used at delivery of incoming materials, but can also be applied before materials are submitted to an expensive operation, handling or processing step. The method is based on statistical principles and provides a judgement of the risk of a decision that is taken. The risk refers to: on the one hand the *producer's risk* (α), *i.e.* the probability that a batch is rejected whereas the actual batch quality was acceptable and on the other hand, *consumer's risk* (β), *i.e.* the probability that the batch is accepted whereas the batch quality level is not sufficient (Kehoe, 1995).

Acceptance sampling includes the following steps.

- Inspectors take a statistically determined random sample (n) from a batch or lot (N), *i.e.* *sampling*.

- Then the number (or amount, or level of, etc.) of non-conforming items is determined by visual inspection, instrumental measurements or analyses.
- Subsequently, the results will be compared with acceptance criteria (c).
- Finally, it will be determined if the batch or lot is accepted or rejected, *i.e. batch or lot sentencing*.

Application of acceptance sampling involves several aspects that must be considered, as described below (Kehoe, 1995).

Location and type of data

First, the location of acceptance sampling should be determined. Besides at delivery control, it can also be used *e.g.* after changes in responsibility or before high added-value processes. Furthermore, the type of data that will be used as acceptance criteria should be assessed. Criteria can be either attribute data, whereby the product is inspected on a pass (*i.e.* conforming) or fail (*i.e.* non-conforming) basis, or variable data, whereby the measured value of the inspected parameter is recorded and used as basis of evaluation.

Sampling design and Operating Characteristic (OC) curve

Design of a sampling plan is another important aspect of acceptance sampling. It includes two key questions *i.e.* how many should be sampled (n) and how many defectives are allowed in the sample (c). Sampling plans can be designed by using distribution tables (*e.g.* modified Thorndike chart) or by national or international (ISO 2859) standards tables. The modified Thorndike chart reflects the relationship between the average number of defectives (x-axis) and the probability of occurrence of 'c' or less defectives (y-axis) for different acceptance numbers (c). The design of a sampling plan can influence the probability of acceptance. For a specific sample size (n) and acceptance number (c), the modified Thorndike chart can be used to determine the probability of accepting the batch for a range of values of percentage defectives. Plotting this relationship generates the typical Operating Characteristic (OC) curve as illustrated in Figure 5.3. The OC-curves show that the probability of accepting a batch (*e.g.* with 30% defectives) is much higher in a sampling plan with a small sample size (n=5) and high acceptance criteria (c=2), than in a plan with a larger sample size (n=20) and more severe

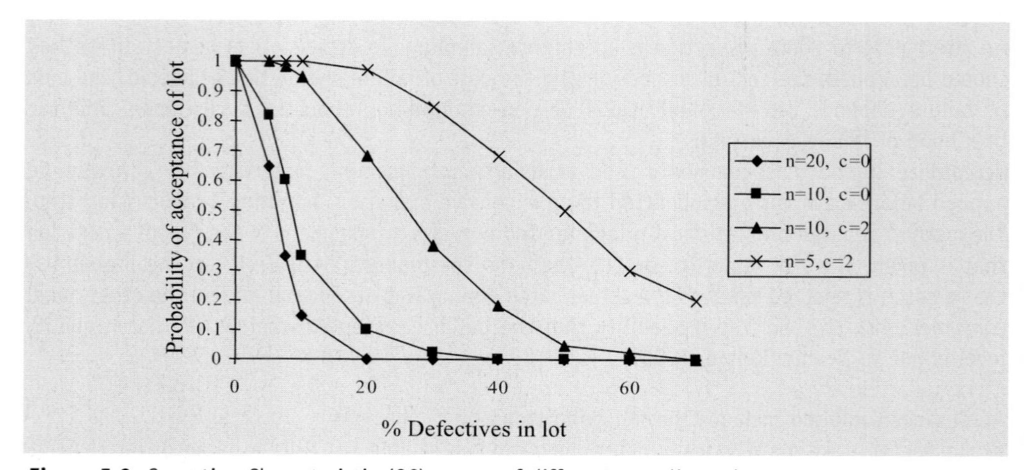

Figure 5.3. Operating Characteristic (OC) curves of different sampling schemes.

acceptance criteria (c=0). In other words, the chance that the batch is unjustly accepted is higher in the first than in the second sampling plan.

Quality levels

Another aspect is the quality level against which conclusions are drawn with respect to acceptance or rejection of the lot. Two common types of quality levels are:

- The Acceptable Quality Level (AQL) is the maximum defect level in the complete batch that is still acceptable to the customer. The government can assess the AQL (*i.e.* legal levels), but usually in practice, both the supplier of raw materials and the purchaser reach an agreement on the AQL level.
- The Limiting Quality Level (LQ) or Lot Tolerance Percent Defective (LTPD) is the defect level in the complete batch that is considered to be unacceptable to the customer. This is the level, assessed by the government or purchaser, at which the batch should always be rejected. The chance that a batch with defects at the LQ level is unjustly accepted must be below 0.1.

If the Accepted Quality level (AQL), the limiting quality level (LQ), the producer's risk (α) and consumer's risk (β) are agreed, then a final OC-curve can be developed using the Thorndike chart. Figure 5.4 gives an illustration of the AQL and LQ level in an OC-curve. For statistical details on OC curves one is referred to literature (Kehoe, 1995).

The OC-curves described above are suitable only for single sampling by using attribute data (*i.e.* inspection on pass or fail basis) and require rather many samples. However, in many practical situations it is desirable to reduce the costs of acceptance sampling. For example, when the process under inspection had previously shown that it is capable of producing materials well within the acceptable quality level (AQL) and outside the limiting quality level (QL). Moreover, the OC curve is only suitable for attribute and not for variable data. For such situations other types of sampling plans are available, which require smaller sample sizes or sampling plans that can be used for variable data. Some commonly used sampling plans are briefly described below. For details one is referred to literature (Kehoe, 1995; Otten and Verdooren, 1996):

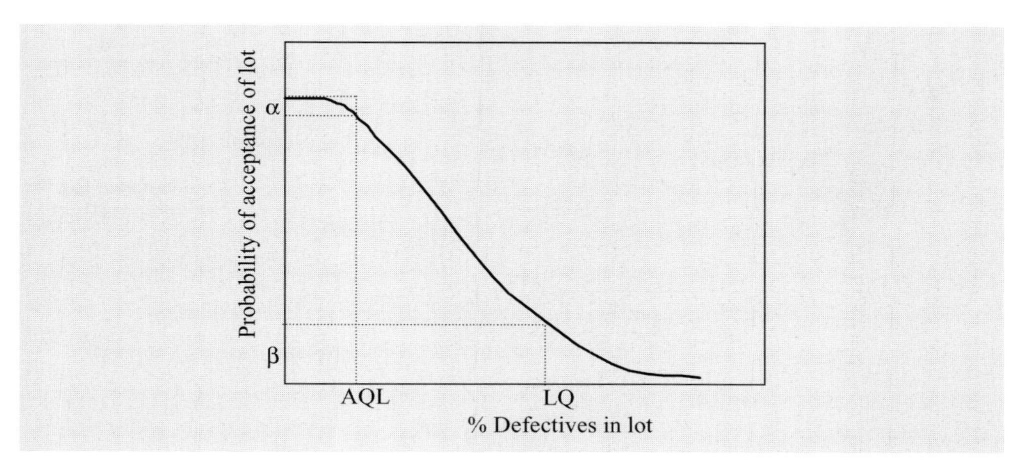

Figure 5.4. Operating Characteristic (OC) curve with acceptable quality level (AQL) and limiting quality (LQ) level.

A common type of sampling that requires less samples is called multiple sampling, where up to a maximum of seven samples can be taken. In this sampling plan:

- More but smaller sample sizes (n) are selected and analysed.
- The number of defectives in the initial sample is determined and compared with the acceptance number (Ac).
- If it is less, then the whole batch is directly accepted (using fewer samples than at simple sampling, thus less product loss).
- If the number of defectives exceeds the rejection number Re, than the batch is immediately rejected.
- In the intermediate range (Ac< number of defects < Re), a new sample is selected and analysed.
- The number of defects in the first and second sample are summarised and this total number must be compared with the Ac and Re of the second sample.
- Based on this comparison, the batch is directly accepted or rejected, or a further sample is taken.
- At the final (maximal 7th) sample, the decision to accept or reject the sample is based on the cumulative number of defects detected in all samples.

A second sampling plan that can be applied to reduce costs in sampling is sequential sampling. This is an important technique often used to monitor ongoing quality of production. The principle of this method is as follows:

- A sample is taken (n=1) and immediately analysed.
- Subsequently a new sample is taken and analysed.
- The results of both samplings are cumulated and judged against the acceptance and rejection lines as illustrated in Figure 5.5.
- A batch is *accepted* if the cumulative number of defects passes the line of the accept region. A minimum number of samples are required to pass this line, *i.e.* minimum sample size for acceptance.

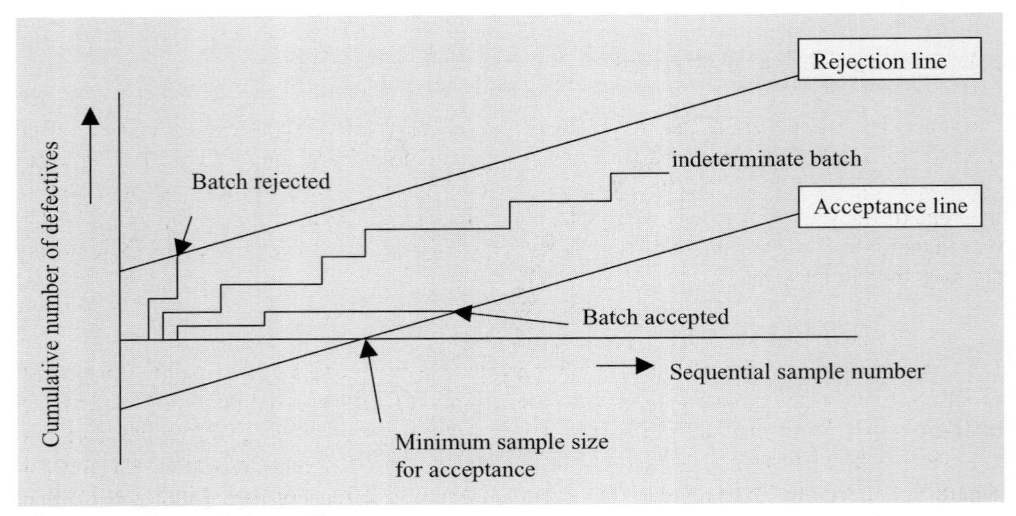

Figure 5.5. Sequential sampling.

- A batch is *rejected* if the cumulative number of defects passes the line of the rejection region.
- If the value of the cumulative number of defects remained between both lines of acceptance and rejection, then no decision can be made and a multiple sampling plan must be selected.
- The graphical acceptance and rejection lines (sampling design) can be constructed by using the values of (1-α, AQL) and (β, LQ) or by using national or international standards such as BS 6001.

The sampling schemes discussed above are suitable for attribute data only, which means for pass-failure decisions. However, in case of variable data (assuming normal distribution of data), other statistical techniques should be applied with
- x as the normally distributed variable
- μ as the mean of the random variable x
- σ^2 as the (known) variance of x
- $x_1,......x_n$ as the a-select sampling with $\bar{x}$ as sample mean

Similarly to the attribute data, also for variable data single, multiple and sequential sampling plans can be constructed. For example, in the case of a single sampling plan, it is characterised by sample size (n), acceptance criteria (c) and sample mean $\bar{x}$, which is compared to acceptance criteria. If a lower μ is associated with a better quality (*e.g.* the lower the bacterial count the better) then the batch will be accepted if sample mean $\bar{x} < c$ (acceptance criteria), it will be rejected if $\bar{x} \geq c$.
If a higher μ is associated with a better quality (*e.g.* protein content in milk) then the batch will be accepted if sample mean $\bar{x} > c$ (acceptance criteria). It will be rejected if $\bar{x} \leq c$.
The values for (n) and (c) of a simple sampling plan can be statistically calculated, using 1-α (α= producer's risk) and (β=consumer's risk) as boundary conditions. For more statistical details and construction of multiple and sequential sampling plans for variable data, one is referred to Otten and Verdooren, (1996) and Evans and Lindsay (1996).

An overview of different sampling methods used for acceptance sampling is given in Table 5.1.

5.2.2 Statistical process control

Statistics are also applied to control manufacturing processes. Statistical process control (SPC) is a specific approach to monitor the operational process in order to identify special variations. Specific variations are out-layers, which exceed the common variation. Common variation is inherent to the process methods, materials, environment and people used. In this section, the procedure for SPC is described and the different types of process control charts that can be used are explained in more detail.

A. Procedure of statistical process control

The aim of statistical process control (SPC) is to *monitor* the process and to *distinguish* normal variation from special variation. Normal variation is due to natural variation, which is inherent to the process. It includes all those factors that are not pertinently controlled, such as the relative humidity or the temperature of the environment. Special variations represent unusual variability in the process *e.g.* due to occasional extreme large seasonal differences in raw materials. Prior to the use of statistical process control, a number of aspects must be

Table 5.1. Overview of sampling plans used for acceptance sampling.

Sampling	Characteristics
Single sampling	• One attribute data • Relatively large sample size (n) • Acceptance criteria (c) • Critical level (m) • Use of OC-curve to assess probability of accepting batch at certain (n, c)
3-Class single sampling	• One attribute data • Relatively large sample size (n) • Critical level divided in 3 classes: $x<m$, $m<x>M$, and $x>M$ • If the analysed data $x<m$ then the batch is accepted • A restricted number c (*i.e.* acceptance criteria) is allowed in the middle class $m<x>M$, whereas at $x>M$ the batch is rejected • A 3-dimensional OC curve can be used to assess probability of accepting batch at certain (n, c, $P_{acceptance}$, $P_{rejection}$)
Multiple sampling	• One attribute data • More but smaller sample sizes, n1, n2, n3 ...n7 • More acceptance criteria 0($c1<c2<c3.... <c7$ • Ac: acceptance number • Re: rejection number
Sequential sampling	• Attribute data • Samples taken one by one (n=1) • Number of defects are cumulated and compared with rejection and acceptance line • If cumulated values remain in intermediate area then other sampling plan
Simple sampling with variable data	• Normally distributed variable data • Known variance σ^2, expectation μ and sample mean $\bar{x}$ • For *e.g.* $\mu_0>\mu_1$ (*i.e.* higher μ if better quality) if $\bar{x} > c$ than batch accepted if $\bar{x} \leq c$ than batch rejected • Calculation (n) and (c) for $\mu_0>\mu_1$ with boundary restrictions if $\mu \geq \mu_0$ then chance of acceptance $P_{acc} \geq 1-\alpha$ if $\mu \leq \mu_1$ then $P_{acc} \leq \beta$ with $P_{acc} = P[\bar{x} > c]$

considered, as illustrated in Figure 5.6. The procedure for SPC application includes the following steps (Kehoe, 1995; Evans and Lindsay, 1996):

Understanding and defining of the process. Firstly, the process must be described (*e.g.* by developing flowcharts); subsequently the relevant (or critical) process parameters and the available control measures must be identified. Several techniques can be used for this purpose such as:

- *Failure Mode Effects and Analysis* (FMEA) for processes, *i.e.* a systematic procedure identifies failures in terms of cause, effects and controls. Moreover, an indication of the risk is

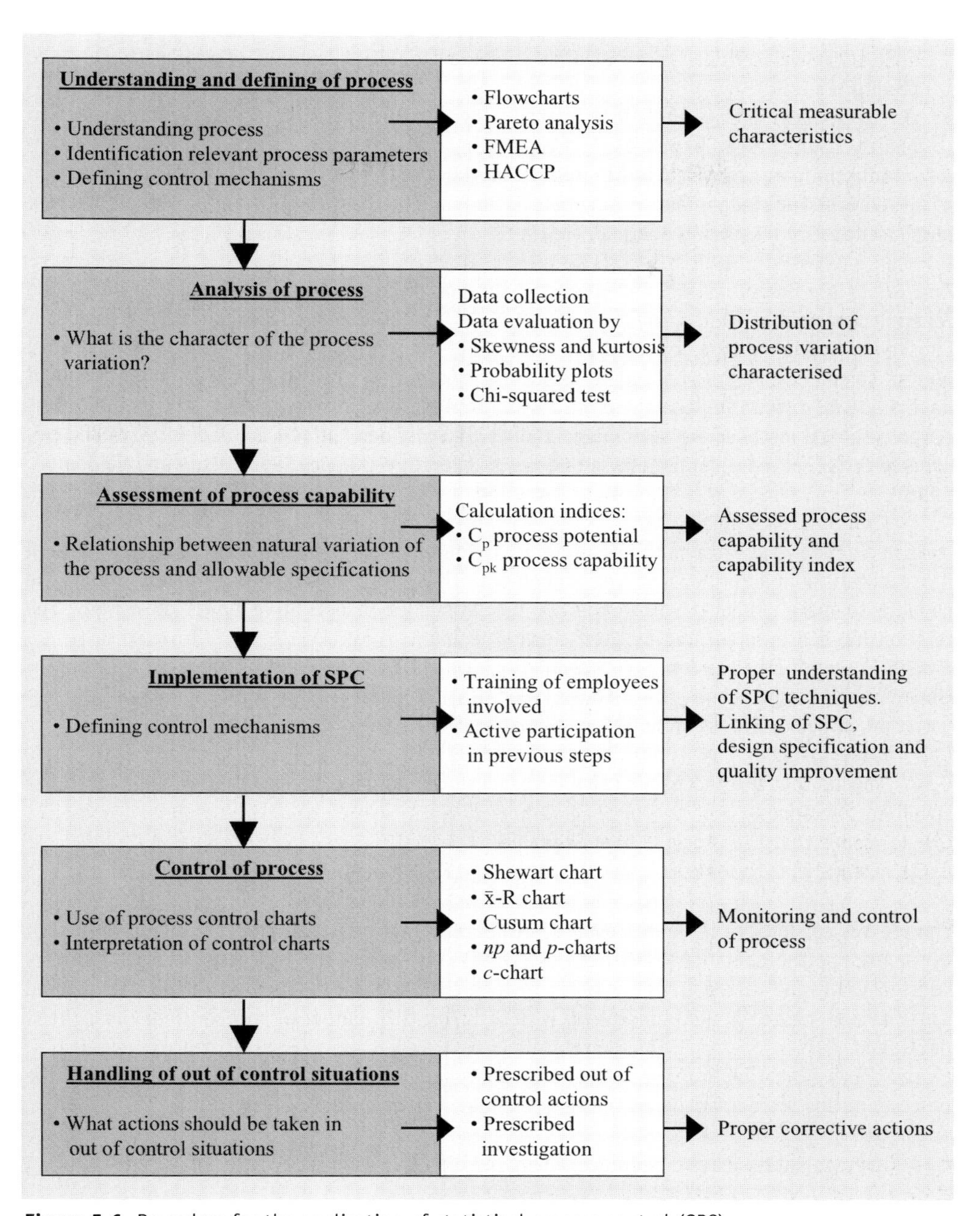

Figure 5.6. Procedure for the application of statistical process control (SPC).

obtained by calculation of the risk priority number (RPN) of the particular failure mode (or critical point).

- *Hazard Analysis Critical Control Points, i.e.* a similar approach to FMEA, but it focuses on safety failures (*i.e.* hazards) with respect to the production of agri-food products. The HACCP analysis reveals critical control points and identifies the required control measures (see Chapter 7).

- *Pareto analysis, i.e.* a method to identify the most important processes with respect to quality failures (see Chapter 6).

An interdisciplinary team should be used to carry out this part of SPC to assure incorporation of all relevant knowledge. The team should include people from various areas, like production, purchase, quality inspection and maintenance.

Analysis of the process is required to assess the normal variation of the process; what is the inherent variability? Firstly, the type of data, *i.e.* attribute or variable, should be assessed. Subsequently, the data are collected and the level of the process parameter (*i.e.* mean or centre) and its spread (*i.e.* range or standard deviation) must be determined. For SPC, the data should have a *normal distribution* and therefore distribution of variation has to be established. For this purpose several methods are available including *skewness and kurtosis, probability plots* and *chi-squared tests*. For more details one is referred to literature (Kehoe, 1995; Otten and Verdooren, 1996).

Assessment of process capability is required to determine the relationship between natural process variation and the specified tolerances. Thus the manufacturing process should be capable to make products within the set limitations (*e.g.* legislative limits or norms). The relationship can be presented by the following index:
Process Potential Index C_p (or process capability), which is the quotient of tolerated variation (= Upper Specification Limit (USL)-Lower Specification Limit (LSL)) and the actual process variation, which is 6σ assuming a normal distribution (Banens *et al.,* 1994; Kehoe, 1995).
$C_p = (USL-LSL)/6\sigma$
$C_p = 1$ means that the process is capable, because the actual process variation is equal to the specified tolerances.
$C_p > 1$ means that the process is very capable
$C_p < 1$ means that the process is not capable (see Figure 5.7)

The process capability can be improved by either reducing the actual process variability (6σ) or by increasing the specified tolerances (USL-LSL) . However, the process potential index C_p considers only the variation. The extent to which the variability is centred within the required specification range is not considered with C_p. For this purpose, another index has been developed, *i.e.*
Process Performance Index C_{pk} (or process capability index), which indicates the deviation of the process, means in comparison to the target value; it is given by:

C_{pk} = (average value-nearest tolerance)/ half the effective range
$C_{pk} = (\bar{x} - LSL)/3\sigma$ or $(USL-\bar{x})/3\sigma$

When the process is correctly centred, then the process potential index (C_p) and process performance index are equal ($C_{pk} = C_p$). However, in practice the C_{pk} is smaller that the C_p, because the process is not operating in the centre of the specification range (see Figure 5.7). The indices often used in practice are: C_p is 1.67 or even 2.00 to be sure that the process variation fits within the specifications. A common value for C_{pk} is 1.33.

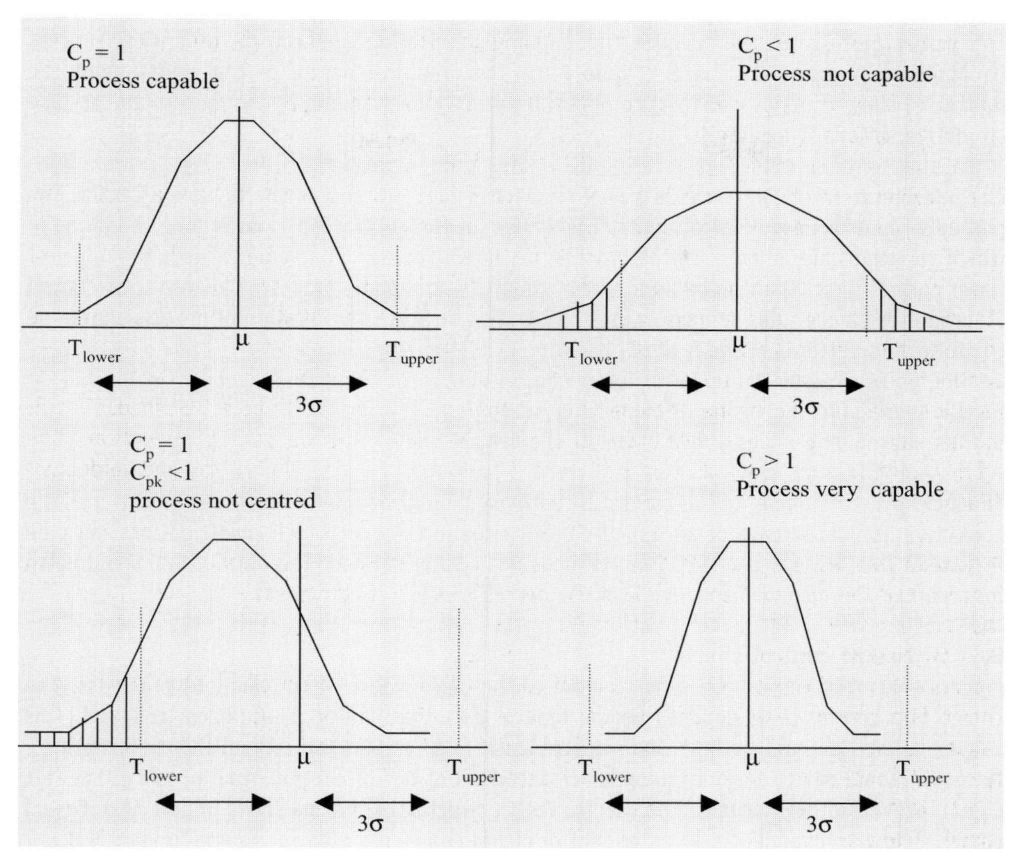

Figure 5.7. Illustration of process potential index (or process capability C_p) and process performance index (or capability index C_{pk}).

The process capability indices can be used to select raw material suppliers, or to select and accept a new process, and/or to design product tolerances (Kehoe, 1995). Although these indices are commonly applied in (capital) good industries, the indices are not yet frequently applied in the food industry. This might be due to the fact that the use of statistics to control processes is not yet very common in the food industry. Moreover, large (often natural) variations in raw materials and ingredients may complicate the assessment of process capability.

Implementation of SCP is the next step. It includes proper training of the people involved in statistical process control. SPC software is widely available but is sometimes used before people fully understand the principles of statistical process control. Therefore, the employees should preferably be involved from the beginning of the SCP preparations, *i.e.* starting with understanding the process, the analysis stage and assessment of process capability indices. By the time the SPC group reaches the implementation stage, they are familiar with the techniques and aware of the background (Evans and Lindsay, 1996; Dale *et al.*, 1998). Furthermore, the application of SPC should not only be just the application of a technique, but people should know the philosophy of understanding and improving process variation (Kehoe, 1995).

The actual *process control* includes the ongoing monitoring and control of critical process parameters using process charts (Figure 5.8). The graphical nature of the charts helps the operators, supervisors or managers to obtain a better understanding of the process and to take corrective actions if required.

The interpretation of the charts is a major aspect of SCP. All process charts have a central line and upper and/or lower control limits. The values for the control limits determine the range in which variations are inherent to the process, *i.e. normal causes*. If a value exceeds the upper or lower control limit, than probably a *special cause* disturbed the process, *i.e.* an out-of-control situation. The three rules commonly used in practice for the interpretation of process charts are the following patterns (Banens *et al.*, 1994):
1. One point exceeds the upper or lower control limit
2. Six subsequent increasing or decreasing values
3. Ten subsequent values above or below the central line

Finally, how to *handle out-of-control situations* should be clearly defined. What are the corrective actions in case of an out-of-control situation and/or which investigations must be initiated? This last stage of SPC is sometimes not clearly defined. The data of SPC should also form part of the management review of the overall process effectiveness.

B. Use of control charts

Process control charts can be used to monitor the performance of (production) processes. The choice of a control chart depends on the type of data that can be measured or analysed. This can be attribute (pass or fail) and variable data (*e.g.* amount of component, temperature). Moreover, some charts are only suitable for detection of out-of-control situations (*e.g.* Shewart chart), whereas other charts can be used to detect structural but small shifts in data (*e.g.* Cusum chart). Below an overview is given of control charts commonly used in industry (Figure 5.8).

Charts for variable data

Control charts often used for variable data include the Shewart chart, the $\bar{x}$-R chart and the Cusum-chart.

In the Shewart chart the control variable is plotted against the time (t) or the sequence sampling number (n), as illustrated Figure 5.8. The control variable can be *e.g.* the sample mean ($\bar{x}$) or the median (m) or the sample range (R). The Shewart chart consist of the following lines:

- The *Central Line* (CL), which is located at the target value. The central line can be the estimated sampling mean ($\bar{x}$), or the median of the sampling (M), or the estimated process variation (2).
- The Lower Warning Limit (LWL) and the Upper Warning Limit (UWL) which are often calculated as follows CL -/+ 2σ (σ = standard deviation of the control variable).
- The Lower Control Limit (LCL) and the Upper Control Limit (UCL) which are often calculated as follows CL -/+ 3σ.

Generally, action is taken if one datum point exceeds the UCL or LCL, or when two subsequent datum points exceed the respective upper or lower limit line. The Shewart chart described is a two-sided control chart. However, in some situations there is only an upper or a lower control limit. For example, often legislative regulations put demands on the maximum level of *e.g.* pesticide residues or residues of veterinary drugs. In such cases the Shewart chart has only an

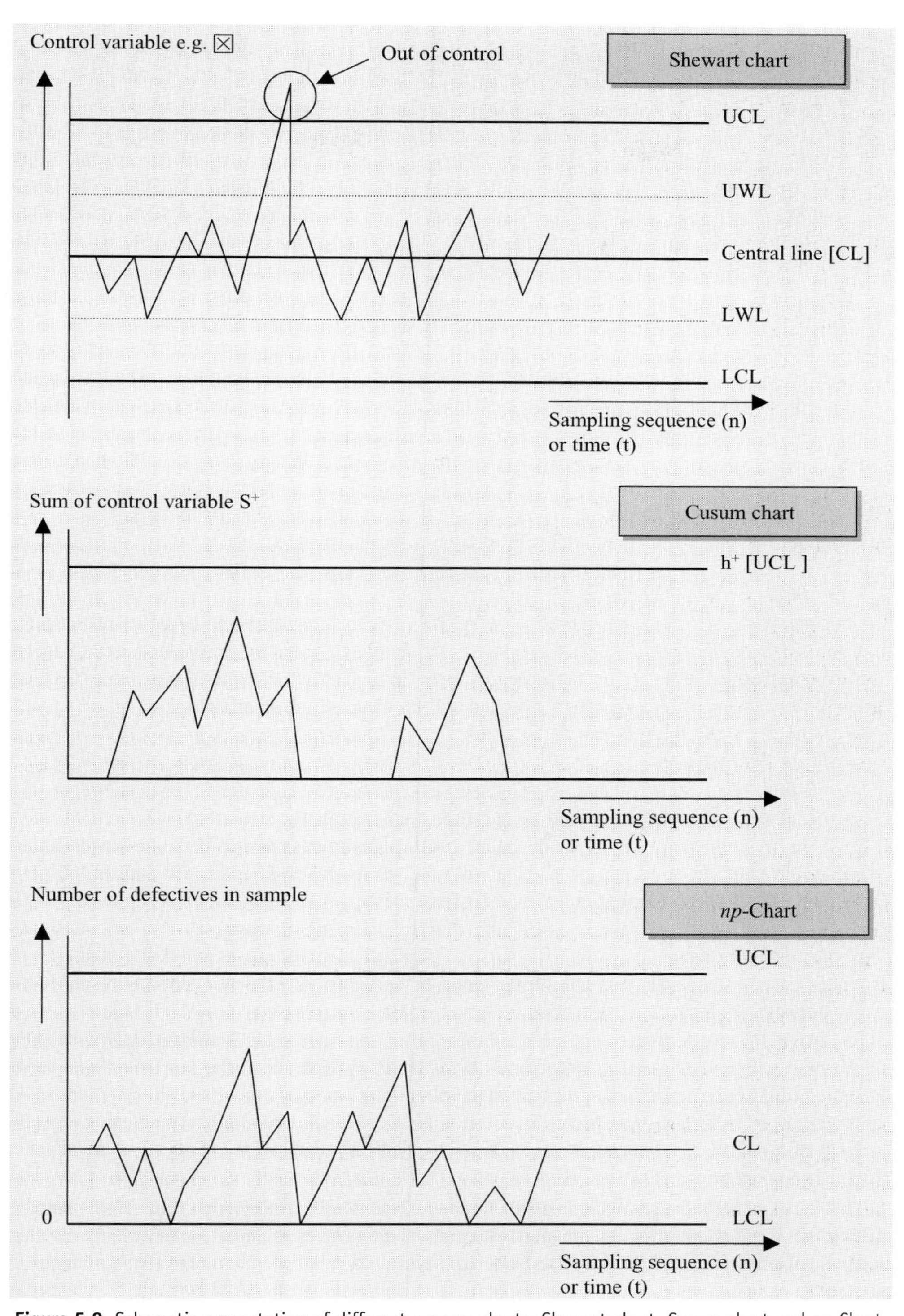

Figure 5.8. Schematic presentation of different process charts: Shewart chart, Cusum chart and *np*-Chart.

Upper Control Limit, *i.e.* one-sided. A one-side lower control limit can be applied when a certain composition is claimed, for example the amount of Vitamin C; it may not be lower than the LCL.

The $\bar{x}$-R charts are in fact combined Shewart charts with control variables $\bar{x}$ and R (not shown). The $\bar{x}$-chart (*i.e.* x-bar chart) is used to monitor the centring of the process, whereas the R-chart (*i.e.* Range chart) is used to monitor the variation in the process. The control limits in the $\bar{x}$-charts depend on the average range. If special causes occur in the R-charts, they may produce unusual patterns in the $\bar{x}$-chart. Therefore, the R-charts should be analysed first to determine if the process is in statistical control.

The Cusum chart (*cu*mulative *sum*) was designed to identify small but sustained shifts in a process much faster than common $\bar{x}$-charts. In contrast to the Shewart chart the decision to take action is not only based on the last sampling but include all sampling data. The Cusum chart incorporates all previous data by plotting cumulative sums of the deviations of sample values from a target value, *i.e.*
$S_0 = 0$
$S_t = S_{t-1} + (\bar{x}_t - \bar{x}_0)$
Where S_0 is the starting value,
S_t is the sum of deviations of sample values from a target value
$\bar{x}_t$ = the average of the tth sampling
$\bar{x}_0$ = the target or reference value
The Cusum charts can be constructed for two-sided (*i.e.* both positive and negative deviations) and one-sided deviations; the latter are relatively simple to use. The one-sided charts can be used to monitor positive (S^+) or negative (S^-) deviations. Figure 5.8 shows a one-sided positive Cusum chart. Only positive values for the cumulative sum are used as follows:
$S^+_0 = 0$ and $\bar{x}^+_0$ = the reference or target value
$S^+_t = S^+_{t-1} + (\bar{x}t - \bar{x}^+_0)$
If $S^+_t < 0$ then reset $S^+_t = 0$
If $S^+_t > h^+$ then stop the process and investigate the cause of the deviation
If $0 < S^+_t < h^+$ then continue; h^+ is the upper control line in case of positive one-sided charts
Likewise, the negative sums can be plotted, with $S^-_t > 0$ reset $S^-_t = 0$, and lower control limit h^-.
The values for h^+ and h^- depend on the process variation.

Charts for attribute data
Attribute data assume only two values *e.g.* good or bad, pass or failure. A distinction must be made between defects and defectives. Defect refers to a single non-conforming quality characteristic, whereas defective refers to an item that can have several defects. Attribute data are often visually evaluated or counted and have a broad application in the (agri-food) industry.
A common attribute chart is the *p*-chart, also called the fraction defective chart. It monitors the fraction of defectives produced in a lot. Like the Shewart and $\bar{x}$-R charts, the *p*-chart consists of a central line, an upper and a lower control limit (*i.e.* zero defects).
The *np*-chart is a control for the absolute number of defectives in the sample (Figure 5.8). The *np*-chart can only be used if the sample number is constant. The *np*-chart is often used as alternative for the *p*-chart. The absolute number of defectives is more meaningful than the fraction of non-conforming products and therefore easier to understand for production personnel.

In some situations it is required to monitor the *number of defects* in the sampling. For this type of control the c-charts are suitable. The *c*-charts control the total number of defects per unit when sampling size is constant. It consists of a central line, an upper and a lower limit (*i.e.* zero defects), similar to the Shewart chart. For more details about the statistical background and construction of charts one is referred to literature (Montgomery, 1985; Banens *et al.*, 1994; Evans and Lindsay, 1996; Otten and Verdooren, 1996).

5.2.3 Quality analyses and measuring

Quality analysis and measurements are an important facet of quality control processes. Analyses or measurements are applied to evaluate or quantify relevant quality attributes or to control processes.
Different sample types can be distinguished (Pomeranz and Meloan, 1994). Incoming *raw materials samples* are analysed to check if they are in compliance with the specifications set by the suppliers. New suppliers submit *buying-samples* of raw materials to test them for practical use. *Process control samples* must often be rapidly measured or analysed (*e.g.* temperature, pressure, pH) to adjust the process in order to obtain uniform quality products. *Finished products* are analysed to check if the agri-food products meet legal requirements, comply with product specifications, are acceptable to their customers and/or have the desired shelf life. *Complaint samples* are submitted by customers or consumers and are analysed to detect process mistakes. Finally, *competitor's samples* are analysed to obtain information for product development.

Food samples can be controlled by direct measurements, *e.g.* measuring of pH or visual inspection of colour. However, if samples cannot be directly measured, then an analysis procedure is needed before measuring including (1) sampling (2) sample preparation and (3) actual measurement or analysis (Figure 5.9). The different aspects of the analysis procedure are described below because they are often a major source of variance.

Sampling
Sampling often contributes the most to total error, whereas the actual analysis or measurement comprises least. Ideally, the sample should be identical to the bulk and must reflect the same intrinsic properties, *e.g.* same textural properties, or equal concentration of toxins, or same taste. According to Pomeranz and Meloan (1994) typical causes of variation in sampling of agri-food products include:
- Irregular shapes, *e.g.*, round particles flow more easily into the sampler compartments than angular particles of similar size.
- During or after sampling changes in composition may occur, such as water loss, evaporation of volatile compounds or enhanced enzymatic reactions by mechanical injury.
- The non-homogeneity of many agri-foods both between products and within the product. For example, compounds can be located in specific compartments or unevenly distributed.

For information about statistical sampling plans, one is referred to the previous section. Which sampling plan should be chosen depends on:
- The purpose of inspection, *e.g.* for acceptance sampling or process control.
- Nature of the material to be tested, *e.g.* homogeneity, what is the history, costs of the raw material?

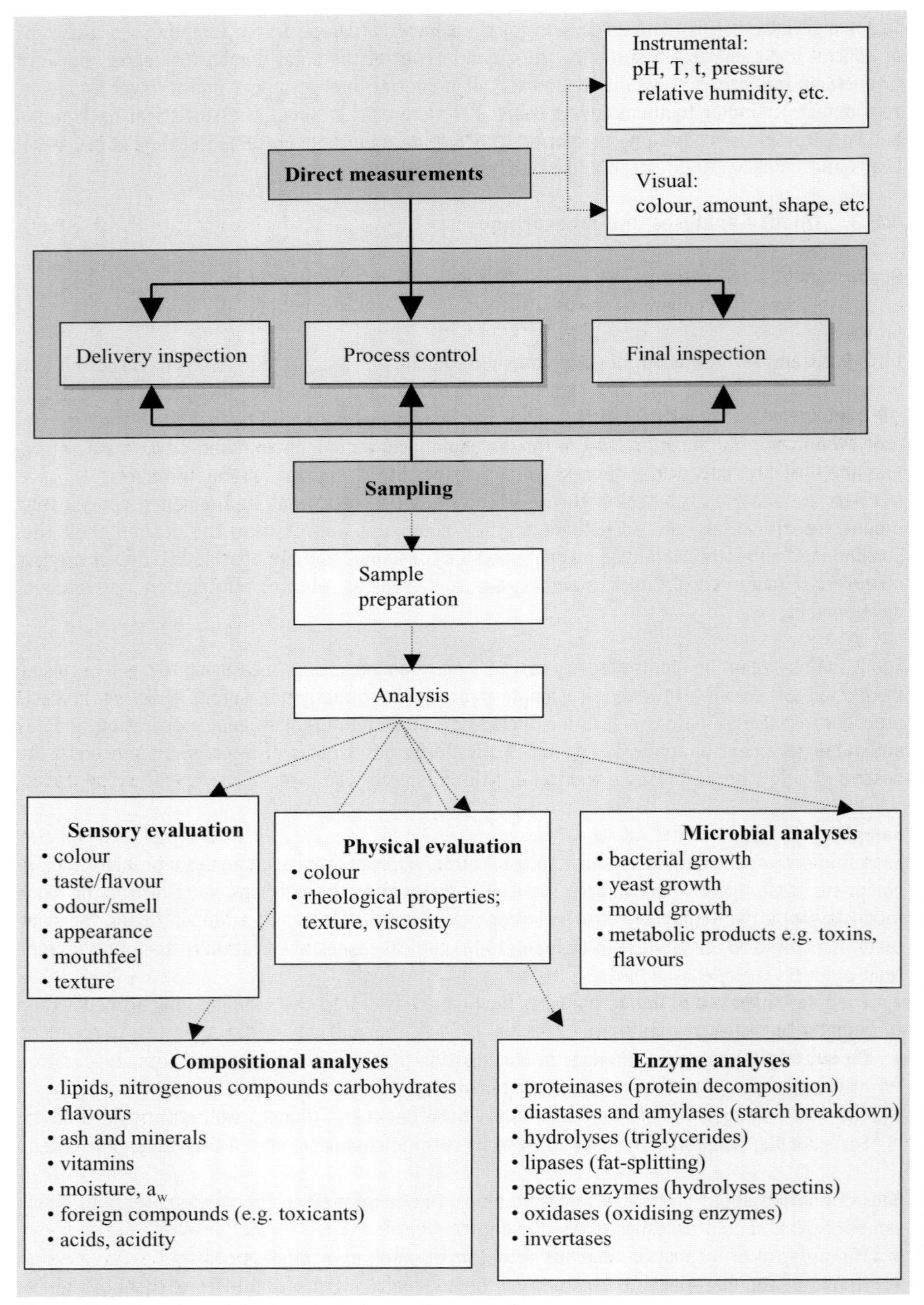

Figure 5.9. Overview of food analysis for quality control.

- Character of the test/lab procedure, *e.g.* (non)destructive, importance of test.
- Nature of lot/batch/population, *i.e.* sizes and number of units involved, how are sub-lots treated?
- Level of assurance required, costs versus information obtained.
- Characteristic of sampling parameter, *i.e.* attribute or variable (Gould and Gould, 1993).

Sample preparation

Sample preparation is a critical step in agri-food products analysis. Major aims of sample preparation are:

- Minimising of undesired reactions *e.g.* enzymatic and oxidising reactions
- Preparation of homogeneous samples
- Prevention of microbial spoilage of samples
- Extraction of the relevant compounds

Different preparation methods can be distinguished (Pomeranz and Meloan, 1994), *i.e.*

- Mechanical grinding of dry and moist materials in order to obtain homogeneous samples;
- Enzymatic and chemical treatment are also used to disintegrate various materials;
- Enzyme inactivation by heat treatment or by inorganic compounds that cause irreversible enzyme poisoning;
- Controlling oxidative and microbial spoilage by storage at low temperatures under nitrogen and/or by addition of preservatives.

Analysis

Finally the analysis or measurement is performed. The reliability depends on several aspects including specificity, accuracy, precision and sensitivity of the analysis (Horwitz, 1988; Pomeranz and Meloan)

- Specificity is the ability to measure what actually should be measured. Specificity is affected by interfering substances, which react similarly to the actual compound to be measured.
- Accuracy is the degree to which a mean estimate approaches a true estimate of the measured substance. Deviation may be due to inaccuracy inherent to the method, effects of foreign compounds and alterations in the compound during analyses.
- Precision is the degree to which a determination of a substance yields an analytically true measurement of that substance. As a rule, analyses should not be made with a precision greater than required. *Repeatability* is the within-laboratory precision where *reproducibility* is the total between-laboratory precision.
- Sensitivity is the ratio between the magnitude of instrumental response and the amount of the substance.

An overview of major methods and measurements applied in the control of agri-food products is presented in Table 5.2. For more detailed information one is referred to literature (Gould and Gould, 1993; Pomeranz and Meloan, 1994).

Sensory evaluation

The sensory properties of a product are importing influencing factors in the selection and acceptance of food. Sensory evaluation can be applied at final inspection, during product development, to determine the effects of process changes or to characterise and compare own

Table 5.2. Overview of typical methods and measurements for analyses of agri-food products.

Quality aspects	Methods and measurements	
Sensory evaluation		
Colour, appearance Flavour, odour, taste Mouthfeel, texture Off-flavours	• Consumer acceptance (preference) tests • Panel difference methods like: • paired comparison • triangle taste panel, triangular flavour score panel • ranking test • numerical scoring • descriptive terms and quantitative descriptive analysis • visual inspection using *e.g.* reference colour charts	
Physical evaluation		
Colour	CIE tristimulus system (X,Y,Z); Simplified tristimulus colour systems: • Munsell system: hue, value, chroma • Hunter colorimeter: L = lightness, a = red, green, b = yellow, blue • Spectrophotometric analyses of extracts • Chemical analysis of colour pigments	
Rheological properties: • Viscosity, elasticity • Texture *e.g.* mealiness, crispiness	• Major rheological devices for analysis of viscosity and elasticity • Rotational viscosity meter • Capillary viscosity meter • Shear press device • Wide variety of empirical devices • Methods and devices used for textural properties • Tenderometer • Shear press device • Texture meter • Penetrating measurements	
Microbial analyses		
• Growth of bacteria, yeast and moulds • Metabolic products	• Physical methods *e.g.* impedance, micro-calorimetric, cytometric • Chemical methods *e.g.* heat stable nuclease, ATP-measurement • Characterising and finger printing, *e.g.* serotyping, DNA probes, DNA amplification • Immunologic methods, *e.g.* fluorescent antibody, serology, ELISA	
Compositional analyses		
• Lipids, nitrogenous compounds, carbohydrates • Flavours • Ash and minerals • Vitamins • Moisture, a_w • Foreign compounds etc	• Gas-liquid chromatography • Mass spectrometry • High-performance liquid chromatography • Thin-layer chromatography • Column, size exclusion and ion exchange chromatography • Infrared spectroscopy	• Fluorimetrics • Visible/ultraviolet spectroscopy • Flame photometry, atomic absorption • Coulometrics • Conductivity • Electrophoresis • Nuclear magnetic resonance • Radio-immunoassay, -activity
Enzyme analyses		
• Enzyme activity • Metabolic products	• Determination of substrate • Assays for determination of (specific) enzyme activity • Inhibition studies	

products with competitor's products. Two major types of sensory evaluation can be distinguished (Gould and Gould, 1993) *i.e.*

- Consumer acceptance (preference) tests. Consumer panels ideally include a true cross section of the population market for which the product is meant. The consumer panellists evaluate the product by indicating their preferences. Consumer panels consist of relatively large number of people to obtain reliable and meaningful results.
- Difference-methods. For these methods well-trained analytical panels are used. The panellist are selected on their ability to detect the four taste senses (sweet, sour, bitter and salt), their ability to repeat themselves at any particular threshold level (*i.e.* repeatability) and their ability to detect the relevant flavours at low thresholds (*i.e.* sensitivity). A wide range of 'difference-methods' are available, some are shown in Table 5.2; for details one is referred to literature (Stone and Sidel, 1985; Piggott, 1988; Meilgaard *et al.*, 1991; Lyon *et al.*, 1992; Koeferli *et al.*, 1998).

In addition, in practice *product experts* are often used to control the quality of the specific products. They are trained to profile or describe sensory characteristics of specific products, using complex descriptive language. The descriptive quality attributes are specifically developed for the specific product groups. For example, wine experts have their own flavour terms to describe the quality of wines. Sometimes instead of product experts, *plant panellists* are used for quality control during and after production. They are trained in recognising off-notes, but they assess if products are 'in spec' or out of specification (Moskowitz, 1994).

Sensory tests are often supported with instrumental analysis to identify the chemical and/or physical parameters involved, to establish sensory observed differences and/or to correlate instrumental and sensory measurements.

Physical evaluation

Physical evaluations, with respect to agri-food quality control, include analyses of colour and rheological properties. The latter refers to viscosity and elasticity but also texture (*e.g.* meatiness, crispiness) can be considered as a result of rheological properties (Pomeranz and Meloan, 1994).

Rheology is concerned with the relationship between stress (*i.e.* compressive, tensile or shear) and strain, which is measured by deformation, as function of time. The rheological properties of agri-foods can be analysed by either an analytical or an integral approach. In the analytical approach the material properties are related to basic rheological parameters (deformation-time curves). In the integral approach, empirical relationships between stress, strain and time are determined. Empirical tests are commonly based on experienced relationships between (poorly) defined properties and textural quality. In imitative tests various properties are measured under conditions similar to practice *e.g.* during processing, handling or consumption (Pomeranz and Meloan, 1994).

Examples of rheological measurements are

- Measuring of the kernel hardness of wheat to estimate milling characteristics
- Viscosity measurement of dough to obtain information on changes in the bread-making process
- Measuring meat tenderness to evaluate consumer acceptance of meat
- Texture evaluation of fruits and vegetables to determine ripeness

Common techniques used in practice are shown in Table 5.2. For detailed information one is referred to the literature (Gould and Gould, 1993; Pomeranz and Meloan, 1994; Prins *et al.*, 1994).

Colour is an appearance property that is attributed to the spectral distribution of light; whereas gloss, transparency, haziness and turbidity are material properties that contribute to the manner in which the light is reflected and transmitted (Pomeranz and Meloan, 1994).
Colour analyses are applied *e.g.* to control the amount of synthetic colorants added, to determine the ripeness of fruit and vegetables, or to determine colour changes during processing and/or storage.
Commonly used methods for colour measurements are mentioned in Table 5.2. For detailed information one is referred to the literature (Gould and Gould, 1993; Pomeranz and Meloan, 1994).

Microbial analyses
Most methods for detecting and characterising micro-organisms are based on their metabolic activity for certain substrates, measurements of growth response, analysis of some parts of cells, or combinations of these. Jay (1996) distinguished different methods *i.e.*
- *Physical methods* that are based on measuring *e.g.* conductivity, enthalpy changes, or flow sorting, which correspond to the micro-organisms metabolic activities.
- *Chemicals methods* to determine metabolic products, or endotoxins or typical bacterial enzymes.
- *Characterisation* and fingerprinting methods to identify the micro-organisms.
- *Immunology* methods are used for detection and quantification of micro-organisms and/or their metabolic products in foods.

Currently, DNA techniques are used in microbial analyses as well. For detailed information one is referred to the literature (Jay, 1996).

Compositional analyses
A wide range of analytical techniques is available to determine composition or specific compounds. Major techniques used are mentioned in Table 5.2 and for details one is referred to the literature (Gould and Gould, 1993; Pomeranz and Meloan, 1994)

Enzyme analyses
Enzymes are heat-labile proteins, which act as specific catalysts in a wide range of, desirable and undesirable, chemical reactions in agri-food products. Enzymes occur naturally in living plants, animal tissues and microbial cells, *i.e.* endogenous enzymes. Especially in raw (unheated) materials, these endogenous enzymes can influence *e.g.* textural properties, flavour and off-flavour formation, and de-coloration processes. Moreover, enzymes are used as an aid in agri-food processing (*i.e.* exogenous enzymes) for a wide range of applications, such as accelerating fermentation processes, improving cheese flavour, tenderising of meat texture and bleaching of natural colour pigments (Fox, 1991; Gould and Gould, 1993; Pomeranz and Meloan, 1994).

Major applications of enzyme analyses as part of quality control include:
- Determination of the *quality status and/or the history* of the food product. For example, in milk the presence of bacterial dehydrogenase is an indicator for unhygienic conditions,

whereas high catalase levels suggest that the milk might be derived from sick cows (mastitis). Not properly stored grains (high humidity and elevated temperature) show increased fat acidity, which is measured by determining the lipase activity.
- *Monitoring* or *determination of efficiency* of the *heat treatment*. For example, peroxidase activity is determined in fruit and vegetables to establish the efficiency of the blanching process. Likewise, phosphatase activity is measured to establish the pasteurisation process of milk.

Enzyme activity can be determined in several ways including,
- Measuring of rheological changes in substrate. For example, (-amylase (*i.e.* starch liquefying enzyme) activity can be measured by monitoring changes in viscosity of the starch.
- Analysis of the degradation products after enzyme action. For example, peptidase activity can be monitored by measuring the free amino acids released from peptidase activity.
- Monitoring specific effects of enzyme activity, such as coagulation of milk caused by proteolytic enzymes, which indicate the quality of milk for cheese manufacturing (Pomeranz and Meloan, 1994).

Common techniques for quantitative monitoring of enzyme reactions include spectrophotometric, manometric, electrometric, polarimetric, chromatographic and chemical methods. For details one is referred to the literature (Bergmeyer, 1983; Fox, 1991).

Both direct measurement and analysis procedures can be applied at acceptance of incoming materials, during processing and at the final inspection of products. The choice on the type of control tests is dependent on the availability of appropriate measurements or analyses, which reflect the actual quality attribute (*i.e.* accuracy of the test). The time span between sampling and the outcome of the analysis or measurements can also be important. If a quick response is required to adjust a process, then fast methods are necessary, which may be not as specific as a more time consuming analysis. The costs can also play a decisive role in the selection of test methods.

5.3 Quality control and business performance

A management control system is a planned and ordered scheme of management control. It enables the firm to assess where the firm actually 'is' at a point in time relatively to where it wants or expects to be. Specifically, control helps a firm to adapt to changing conditions, it limits the magnification of errors, assists in dealing with increased complexity and helps to minimise costs. For effective execution of control activities it must be integrated with planning, so that actual results can be compared with planned projections. Moreover, the ability to control activities must be considered during the planning process. Control is not only focused on objective measurable targets; it also includes control on performance in terms of more subjective and difficult to quantify criteria, such as the firm's image, workers' morale and animal welfare. The relationship between planning and controlling continues as a long-term cycle. Plans are made and control is used to evaluate the effectiveness or organisational activities relative to those plans. If the control system indicates that things are proceeding as they should, the current plan can be maintained. If progress is inadequate or the control system may indicate that the situation has changed (*e.g.* the competitive or legal environment), new plans have to be developed.

In the chapter about quality management a distinction is made between product and resource decisions (section 3.3.2). It has been hypothesised that both product decisions and resource decisions influence business performance. However, a strong relationship exists between them. Following the input-output model shown in Figure 3.1, the relationship between control and business performance will be analysed with respect to the three basic control processes, *i.e.*

- Input, the supply control
- Transformation, the production control
- Output, the distribution control

All these processes will be described from both the product and resource point of view, as shown in Figure 5.10.

	Supply	Production	Distribution
Product control	•Incoming material control •In storage control	•Product control •In storage control	•Distribution control •Service control •Complaint service
Resource control	•Supplier evaluation •Supplier relationships	•Process control •process capability analysis •Equipment control	•Distribution channel control •Customer relationships •Market analysis

Figure 5.10. Quality control activities from a product and resource perspective.

5.3.1 Supply control

In a company, the purchasing department is responsible for obtaining supplies (*i.e.* materials, parts and services) needed to produce a product or to provide a service. The importance of purchasing is clear, considering that in manufacturing up to 60 percent of the costs of finished goods comes from purchased parts and materials. In retail and wholesale companies it sometimes can exceed 90 percent. However, not only costs are important, especially in food chains, where quality and timing of deliveries of goods have a significant impact on operations. In fresh-product business, quality of end products is almost completely dependent on supply-quality. Control activities in supply management can be summarised in the purchasing cycle as sown in Figure 5.11.

The purchasing cycle begins with a request from within the organisation to purchase material, equipment, services or other items from outside the organisation. The cycle ends when the purchasing department is informed that a shipment is received in satisfactory condition and that it is properly stored so that quality continuously will meet production requirements. The main steps in this cycle are (Stevenson, 1999):

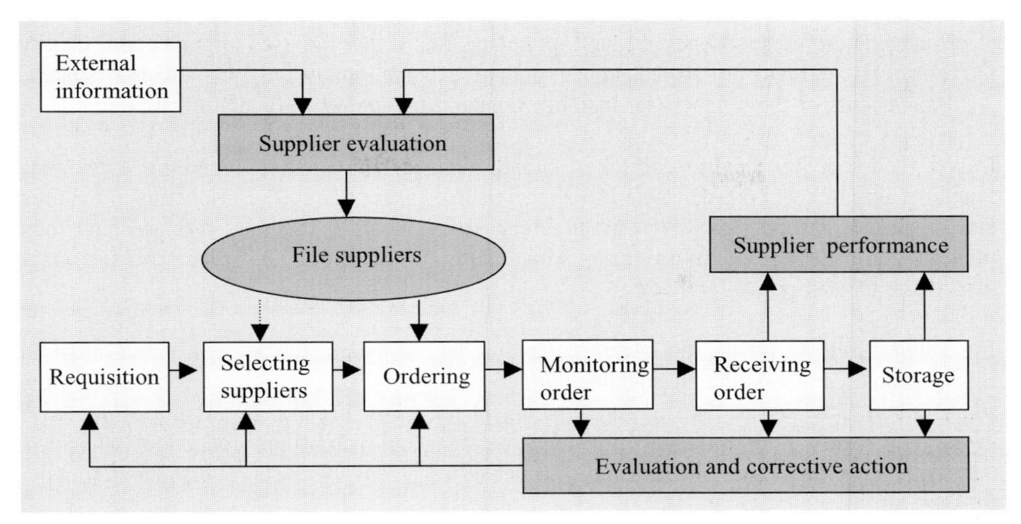

Figure 5.11. Supply control.

1. *Receiving the request (requisition)*. The request includes:
   - a description of the item or material desired
   - the quality and quantity necessary and
   - desired delivery dates
2. *Selecting suppliers.* The purchasing department must identify suppliers who have the capability of supplying the desired goods. If no suppliers are currently listed in the files, new ones must be identified. Vendor ratings may be used to choose appropriate vendors. Rating information can be communicated to the vendor with the idea of upgrading future performance.
3. *Placing order.* If the order involves a large expenditure, particularly for a one-time purchase for example of equipment, vendors will usually be asked to bid on the job. Operating and design personnel may be asked to assist in negotiations with a vendor. Large-volume, continuous-usage items may be covered by blanket purchase orders, which often involve annual negotiation of prices with deliveries subject to request throughout the year. Moderate-volume items may also have blanket purchase orders, or may be handled on an individual basis. Small purchases may be handled directly between the operating unit requesting the items and the supplier, although some control must be carried out to keep it in hand.
4. *Monitoring orders.* Routine follow-up on orders, particularly large orders or those with lengthy delivery schedules, allows the purchasing department to predict delays and communicate this information to the appropriate operating units. Likewise, the department must inform suppliers about changes in quantity and delivery needs of the operating units, so they have time to adjust their plan.
5. *Receiving orders*. Receiving must check incoming shipments from vendors for quality and quantity. It must report to purchasing, accounting and the operating unit that requested the goods. If the goods are not received in satisfactory condition, they may have to be returned to the supplier for credit or replacement, or they must be subjected to detailed inspection. Again, purchasing, accounting and the operating unit must be notified of this information. In either case, vendor evaluation records must be updated.

6. *Storage.* Received goods must be kept in storage before they are used in production. Quality control is needed on stored goods and/or storage conditions due to the specific requirements of the product (for instance temperature control in food storage).

In the food industry, quality control in supply management is often based on acceptance sampling, due to cost consideration. However, in some cases full (100%) inspection is required due to specific customer requirements or safety issues. Figure 5.11 shows the feed back loop from quality inspection points to phases where corrective action can be taken. This feed back loop is the *product-control* activity.

The *resource-control* activity can be summarised as measuring supplier performance, evaluating supply sources and developing supplier relationships. Good suppliers are a vital link in the supply chain (Stevenson, 1999). Late deliveries of materials or parts, or missing or defective items, can cause large problems for manufacturers by disrupting production schedules, increasing inventory costs and causing late deliveries of end products. Substandard supplier services can have similar results.

Choosing a supplier involves taking into account many factors. A company should consider price, quality, the supplier's reputation, past experience with the supplier and service after sale; this process is called vendor analysis. Very important is that a company provides suppliers with detailed specifications of the materials or parts that are needed, instead of buying items of the shelf. In this way the suppliers can comply with the ordered quantities and specific production requirements. The relationship with the supplier depends on the situation. For instance, in the food industry intensive communication is often required to design packaging materials that comply with the specific requirements of the manufacturer regarding product demands, packaging and distribution process characteristics.

The main factors a company should take into account when evaluating a supplier are:
1. *Price.* This is the most obvious factor, along with any discounts offered, although often it is not the most important one.
2. *Quality.* What is the quality reputation of the supplier, what quality assurance system does the supplier have, is the supplier able to collaborate in future design processes.
3. *Services.* Special services can sometimes be very important in selecting a supplier, like replacement of defective items, sharing purchase-market information, fast response on complaints, which can be key factors in selecting one supplier over another.
4. *Location.* Location has impact on shipping time, transportation cost and response time.
5. *Emergent cases.* Sometimes quality claims due to origin can play a role (food and drinks).
6. *Inventory policy of supplier.* When customer demands are fluctuating (for instance seasonally) guarantees are needed about in-time deliveries.
7. *Flexibility.* The willingness and ability of a supplier to respond to changes in demand, and to accept design changes, can be important considerations.

Periodic *supplier audits* are a mean of controlling supplier's production capabilities, quality and delivery problems and resolutions, as well as suppliers' performance on other buying criteria. If an audit reveals problem areas, a buyer can address these before they will result in serious problems. Among the factors typically covered by a supplier audit are management style, quality assurance, material management, used design process, process improvement policies, procedures for corrective actions and follow-up. Supplier audits are also an important first step in supplier certification programs.

Supplier certification is a detailed examination of the policies and capabilities of a supplier. The certification process verifies that a supplier meets or exceeds the requirements of a buyer. This is generally important in supplier relationships, but it is particularly important when buyers are attempting to establish long-term relationships with suppliers. One advantage of using certified suppliers is that the buyer can eliminate much or all of the inspection and testing of delivered goods. Although problems with supplier goods or services might not be totally eliminated, there is much less risk than with non-certified suppliers. Many companies rely on standard industry certifications such as ISO 9000, perhaps the most widely used international certification (see also Chapter 7).

Nowadays, *supplier partnership* has become increasingly important. In the past, too many firms regarded their suppliers as rivals and dealt with them on that basis. The relationship between customer and supplier has been seen an arms' length relationship. It was necessary to keep at a distance, neither becoming too dependent upon a supplier nor allowing a supplier to become too dependent upon any customer. This view led naturally to multiple source as a norm because single source, *i.e.* having one supplier for each good, placed to much power in the hands of the supplier.
The concept of managing the total supply chain, however, stressed the benefits of mutual dependence between supplier and customer. This is totally different from the traditional approach. In the mainstream activity, the survival of each partner is correctly perceived as being dependent on the other. This implies a long-term commitment, not simply to do business with each other, but also to find new business for each other. Suppliers support their customers even at the cost of short-term difficulties for themselves. Benefits in terms of profitability or of quality improvement are there to be shared, based upon an open relationship. This is therefore the notion of partnership based upon a perception of mutual advantage, or the so-called win-win synergy (Syson, 1992).
Juran (1989) has developed a useful framework to distinguish between adversarial (rivals) and teamwork relationships with suppliers (Figure 5.12)

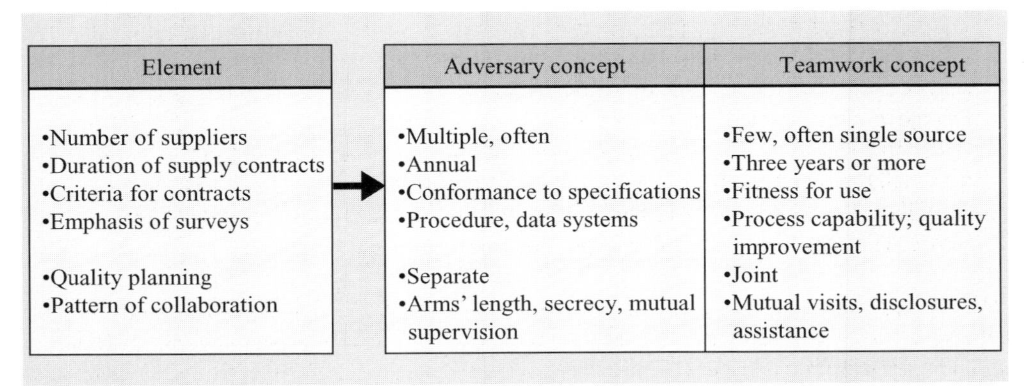

Element	Adversary concept	Teamwork concept
•Number of suppliers	•Multiple, often	•Few, often single source
•Duration of supply contracts	•Annual	•Three years or more
•Criteria for contracts	•Conformance to specifications	•Fitness for use
•Emphasis of surveys	•Procedure, data systems	•Process capability; quality improvement
•Quality planning	•Separate	•Joint
•Pattern of collaboration	•Arms' length, secrecy, mutual supervision	•Mutual visits, disclosures, assistance

Figure 5.12. Juran's trends in supplier relationships.

Dean and Evans (1994) discussed that in a teamwork concept suppliers must be recognised as crucial to organisational success, because they make it possible to create customer satisfaction. Neither the quality nor the cost of the organisation's product can reach competitive levels and continuous improvement without the contributions of suppliers. Customer-supplier relations, therefore, must be built on trust rather than suspicion. Besides, from the obvious teamwork implications, monitoring supplier or customer behaviour does not add any value to the product. Trust is developed over time through a pattern of success by all parties to fully and faithfully deliver what was promised.

5.3.2 Production control

The purpose of quality control of production activities is to assure that processes are running in an acceptable manner. Companies accomplish this by measuring process outputs using statistical techniques (see also 5.2.1 and 5.2.2). If the results are acceptable, no further action is required; unacceptable results call for corrective action. Controlling products involves *acceptance-sampling* procedures. In addition, more and more companies emphasise on designing quality into the process, thereby greatly reducing the need for inspection of final products (Stevenson, 1999). In fact, attention is nowadays shifted towards controlling the process, which is referred to as (statistical) *process control.* In process control, feedforward mechanisms are preferred rather than feedback control (see also 5.1.1). A feedforward system will prevent defects and variation. As a matter of fact, product inspections will still be necessary when applying process control, but inspection results are then intended to verify that the process is running conform to standard.

Production control activities are summarised in Figure 5.13, wherein both product and resource control are described. The main steps are (steps 1 to 4 are product control and steps 5 and 6 are resource control):

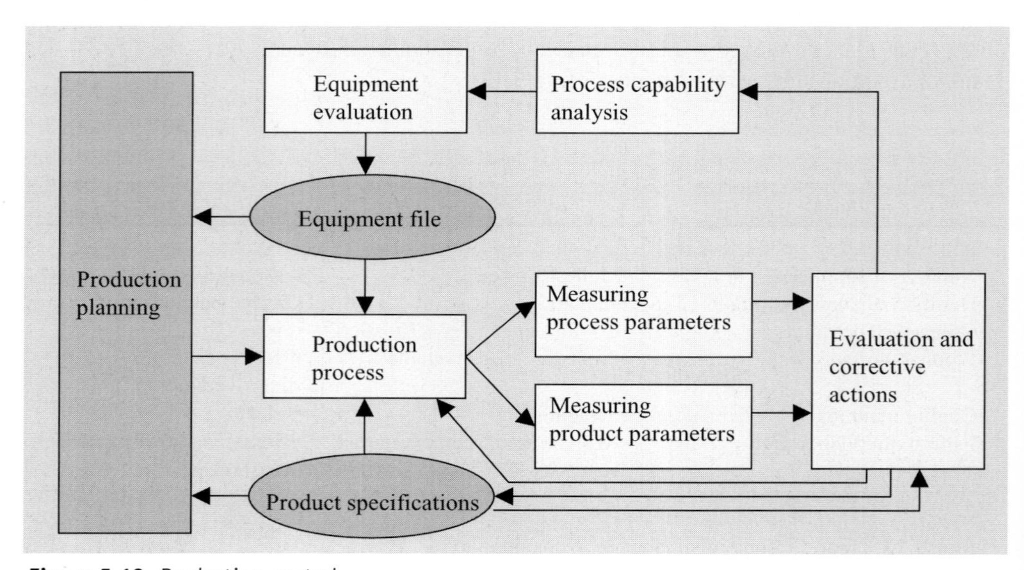

Figure 5.13. Production control.

1. *Production planning.* Production planning is usually not recognised as a quality control activity. However, it should be noted that it has great influence on quality performance. Production planning combines information about equipment (availability, quality performance, and cost) and customer orders, including product specifications, into a production schedule. Many problems with product quality can be introduced here, when costs and delivery time result in production schemes wherein time pressure makes it impossible to meet quality standards.

2. *Production process*. Quality activities differ with the type of process-organisation. Four types of process-organisation can be classified (Noori and Radford, 1995):
   a. *Job-shop production*: a highly skilled workforce using general-purpose equipment makes a wide variety of customised products. These processes are referred to as a jumbled-flow process because there are many possible routings through the process.
   b. *Intermittent flow*: a mixture of general-purpose and special-purpose equipment is used to produce small to large batches of products. Often specifications of the product are new, or especially for one batch (example: a catering firm).
   c. *Repetitive flow*: several standardised products follow a predetermined flow through sequentially-dependant work centres. Workers typically are assigned to a narrow range of tasks and work with highly specialised equipment. Food manufacturers often can be typified as repetitive flow producers.
   d. *Continuous flow*: commodity like products flow continuously through a linear process (example: a sugar refinery).

   Quality activities are clearly different in each process-organisation. In the job-shop production, professional skills are very important. During production, the workers carry out quality control by checking their own work. Occasionally, quality is not clearly defined because unique products are made in an experimental stage. On the other hand, in continuous flow production, products are specified in detail and there is much knowledge about the product and the production process. Quality control is for a great part incorporated in the process and automatically worked out. Nevertheless, operators control the process and bring with their experience and insights the finishing touch in the process, being responsible for the important last percentages of quality.

3. *Measuring.* Measuring in the production process can occur at three points: before, during and after production. The logic of checking conformance before production is to assure that inputs are acceptable. The logic of checking conformance during production is to make sure that the conversion of inputs to outputs is proceeding properly. The logic of checking conformance of output is to make a final verification of conformance before passing goods to customers (Stevenson, 1999).
   Questions that arise when developing measuring procedures are:
   - how much to inspect
   - how often
   - and at what points in the process inspection should occur

   Low-cost, high volume items (continuous flow production) often require little inspection, because cost associated with defectives are quite low and the processes that produce these items are usually highly reliable, defectives are thus rare. Of course when there is a safety risk like in food production, process conditions are controlled very intensively, *e.g.* the

intensive cleaning and disinfection to prevent microbiological problems. When a clear relationship between product requirements and process parameters exists, then the production processes can be controlled by measuring these process parameters. For instance, control of temperature and pressure in food processes are often closely related to product requirements.
In high-cost, low-volume production, more intensive inspection is necessary due to large costs associated with passing defectives. In general, inspection costs must not exceed the decrease in failure costs. Concerning the question of where to inspect, some typical inspection points are:
- raw materials and purchased parts
- finished products
- before a costly operation (for instance before adding expensive ingredients)
- before an irreversible process (for instance before filling and labelling food products in cans)

4. *Evaluation and corrective action.* It is very important that specifications of the product and process parameters consist of both standards and tolerances. Management must establish a definition of out-of-control situation. Even a process that is functioning as it should, will not yield output that conforms exactly to a standard, simply because of natural variations inherent to all processes (especially for food processes with biological materials). The main task of quality control is to distinguish random from non-random variability, because non-random variability means that the process is out of control. When a process is judged out of control, corrective action must be taken. This involves identifying the cause of non-random variability (for instance work equipment, incorrect methods or materials, failure to follow specified procedures) and correcting it (Stevenson, 1999).

5. *Process capability analysis.* When evaluation indicates an out-of-control process, an analysis of process capability should be made. Capability analysis means determining whether the inherent variability of the process output falls within the acceptable range of variability allowed by the design specifications for the process output.
The technique of a process capability study is already described in par. 5.2.2, so the focus here is on possible solutions for out-of-control processes (situations). Stevenson (1999) mentioned a range of solutions:
- Redesign the process so that it can achieve the desired output,
- Use an alternative process that can achieve the desired output,
- Retain the current process but attempt to eliminate unacceptable output using 100% inspection, and
- Examine the specifications to determine if they are necessary, or could be relaxed without negatively affecting customer satisfaction.

6. *Equipment evaluation.* Equipment is evaluated on the possibilities of improvement through redesign or modification of the equipment and on changing maintenance (and cleaning) programs. Redesign and modifications are mainly intended to improve reliability and, in food industry, hygiene. Here the connection between quality control and quality design is made. In Chapter 4, quality design techniques for reliability analysis (FMEA) and hygienic design have been described in more detail.

Maintenance includes all activities involved in keeping a system's equipment in working order. Maintenance is concerned with avoiding undesirable results due to system failure. Even when the results are not catastrophic, they can be disruptive, inconvenient, wasteful and expensive.

Maintenance falls into two categories: preventive maintenance and breakdown maintenance (Heizer and Render, 1993). *Preventive maintenance* involves performing routine inspection, servicing and keeping facilities in good repair. Preventive maintenance activities are intended to build a system that will find potential failures. It makes changes or repairs that will prevent failure. Preventive maintenance is much more than just keeping machinery and equipment running. It also involves designing technical and human systems that will keep the productive process working within tolerance; it allows the system to perform. The emphasis is on understanding the process and allowing it to worn without interruption.
Breakdown or corrective maintenance is remedial; it occurs when equipment fails and must be repaired on an emergency or priority basis. Figure 5.14 indicates that failure occurs at different rates during the lifetime of equipment; it may follow different statistical distributions.

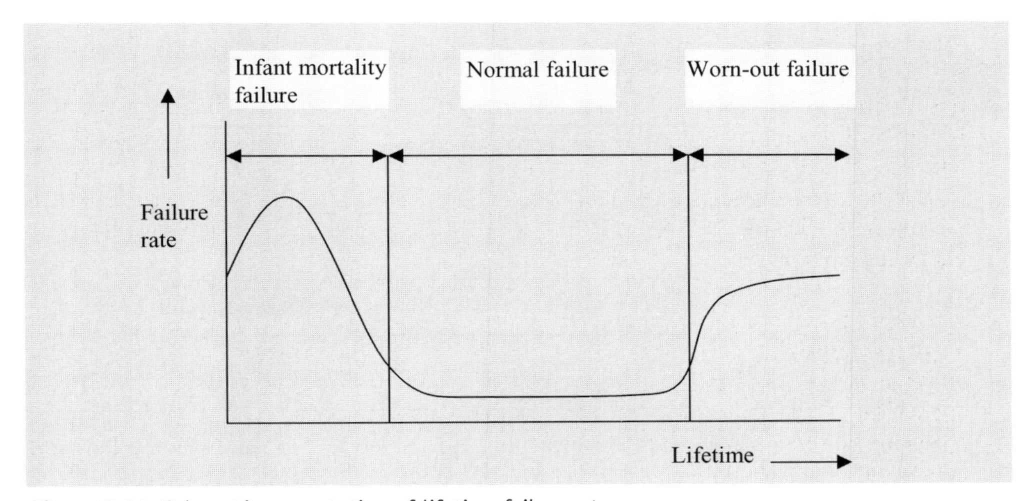

Figure 5.14. Schematic presentation of lifetime failure rates.

A high failure rate, known as infant mortality, exists for most equipment in the initial (starting) phase. Once the machine or process is settled, a study can be made of the MTBF (mean time between failure) distribution. When the distributions have a small standard deviation, then it is a candidate for preventive maintenance. Whether preventive maintenance is applicable depends on the relationship between preventive maintenance costs and the costs of breakdowns. With modern technical instruments for diagnosis of wear-out (measuring metal parts of oil, ultrasonic measuring of vibrations, infrared thermograph) prevention costs can be kept low, whereas breakdown costs tend to become higher due to safety risk and losses in goodwill and image. When reliability cannot be achieved and preventive maintenance is not appropriate or does not work, management can enlarge or improve the repair facilities. Remedial maintenance can then be performed in order to put the system back into operation.

Marcelis (1984) stressed the importance of management processes in maintenance. The analysis was based upon a study amongst 150 companies, leading to the conclusion that the results of maintenance activities highly depend on management factors. Equipment design processes and co-operation between the production departments and technical departments appeared to be important for high quality in maintenance.

5.3.3 Distribution control

Whereas purchasing deals with the incoming flow of materials, distribution deals with the outgoing flow of materials (Krajewski and Ritzman, 1999). Distribution control involves management of the flow of materials from manufacturers to customers and from warehouses to retailers. It also includes the storage and transportation of products. Distribution broadens the marketplace for a firm, adding time and place value to its products. Quality control in distribution management concerns decisions around transportation and storage of products and measuring the whole process, including consumer usage of the products (Figure 5.15).

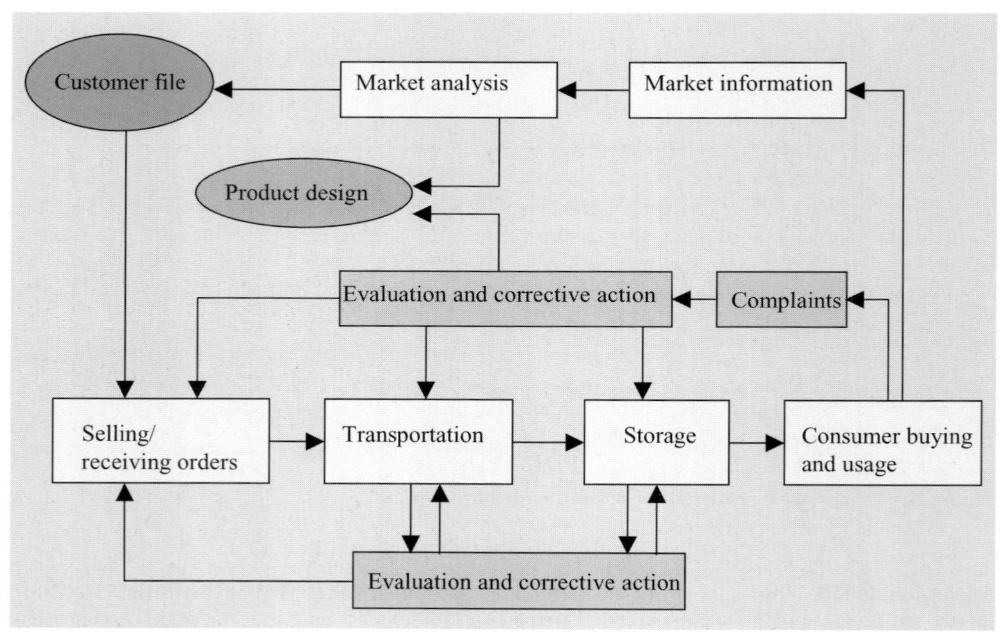

Figure 5.15. Distribution control.

The main steps in distribution control include product and resource decisions. They are:

1. *Selling, receiving orders*. In food industry, selling is commonly established in year contracts and within one contract many delivery-orders are agreed. Quality in selling means that no deals are made that cannot be sustained in the course of the year. Detailed agreements about delivery time, price and quality specifications are a part of this process (Van den Berg and Delsing, 1999).

2. *Transportation and storage*. In food distribution, transportation and storage can be complex processes within a long supply chain, often requiring international transport movements. The major amount of food products is following a "common" routing from manufacturers to retailers' distribution centres and from there in combined freights to retail stores.
   Product control during transportation and storage concerns maintaining quality of the products by monitoring the product and taking corrective action when necessary.
   Resource control involves transportation equipment, distributor organisation and storage conditions, both in distribution centres and in retail stores. In food industry temperature and humidity conditions are very important control parameters, as well as hygienic conditions. Distributors often are obliged to introduce quality assurance systems in their organisation to ensure acceptable quality levels.
3. *Buying and consumer usage*. In Figure 5.15, the common situation in the food business is described where retailers (storage) are considered as customers and the individual person buying from the retail store is regarded as the consumer of food products. Quality control in the stage of buying and consumer usage is often restricted to, for instance, information on package, in leaflets and magazines.
4. *Complaints, market information and analysis*. Complaints and market research provide lots of information about quality and quality performance. This information can be used for quality improvement. Firstly on a short-time basis, when consumer problems have to be solved immediately and secondly on a long-term, where this information can be a starting-point for product redesign. In Chapter 4 (customer-oriented product design) the process of turning market information into product design is already described.

A complaint procedure is described in Figure 5.16 (Korteweg, 1991).

The advantages of a controlled complaint procedure are having more insights in complaints per product, more insights in cause and effect relationships, and faster solutions of complaints. The essential elements in a complaint procedure are:
- Central receipt of complaints, where complaints are registered, controlled and reported. This is also the central point for external information.
- A procedure for analysis and research, to ensure that complaints are analysed from a technical, commercial and juridical point of view.
- A procedure for adequate action taking. Firstly, complaints should be settled in a customer/consumer friendly way. Secondly, corrective action could be necessary, for instance delivering new products, or obtaining back products and sorting out bad materials, making a deal that is satisfying for the customer. Thirdly, preventive action should be taken to improve quality, so making quality complaints a basis for organisational learning. Of course a complaint procedure must have the linkage to recall procedures, which are an element of the HACCP system of the company (see also Chapter 7).

Considering supplier relationships automatically addresses a discussion about *customer relations*. In the food business chain relationships are getting more important, in particularly the relationship between manufacturer and retailer. Lancaster and Massingham (1999) referred to the emergence of the concept of relationship marketing, which is about developing and maintaining mutually satisfying long-term relationships with customers, which has, of course, major implications for product strategies, distribution strategies, etc.

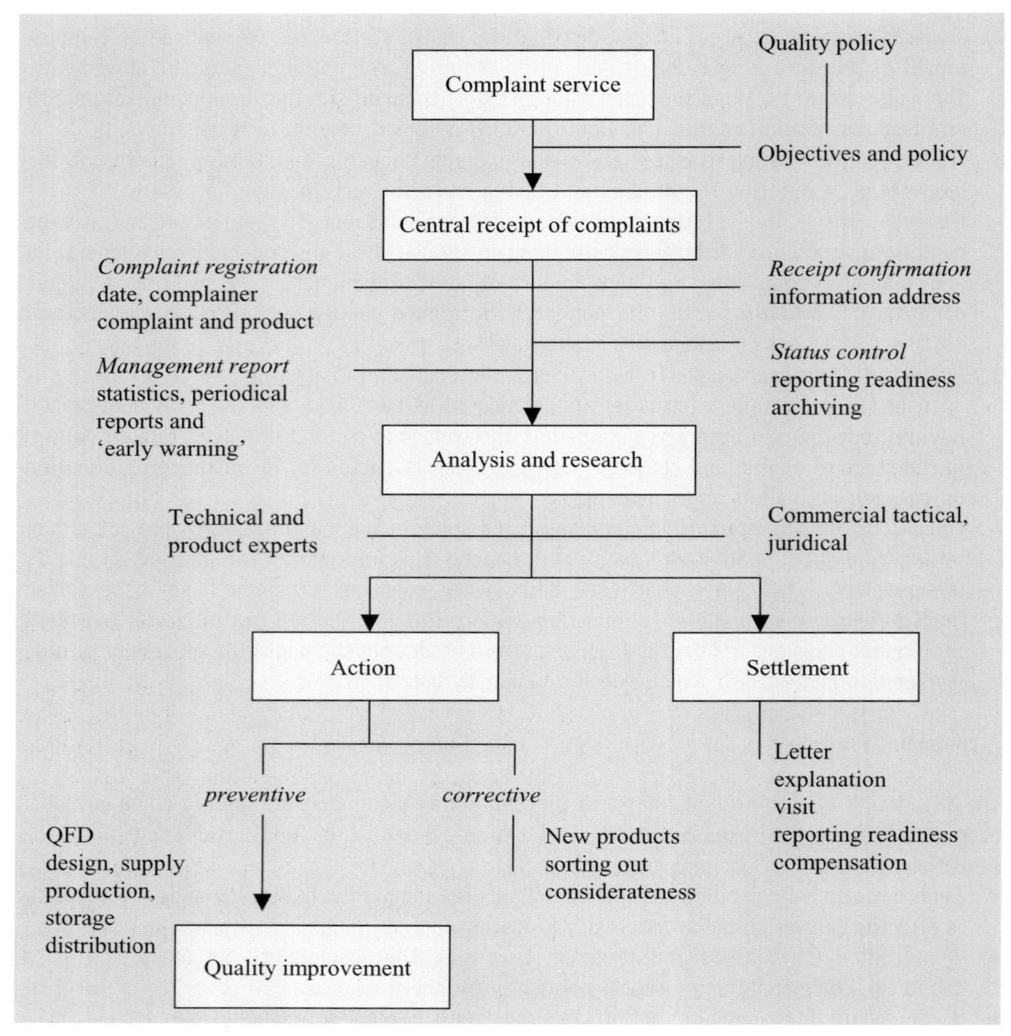

Figure 5.16. Complaint procedure.

Relationship marketing is the opposite of transaction marketing. The first one is more oriented on product benefits than on product features. It has a higher customer service emphasis with an extensive contact and high commitment to the customer. With respect to quality, relationship marketing demonstrates quality concern in its all functions, whereas transaction marketing accepts quality to be restricted to the production function. Close relationships are on the long-term a steady basis for quality and quality improvement, leading to benefits for both parties.

5.4 Managing the control process

Surely it is not evident that broadening control activities implies a growth in effectiveness of quality performance. In this section, requirements for effective controls will be discussed. Subsequently, embedding control in the organisation will be analysed as well as the costs and benefits of control.

5.4.1 Effective controls

Control must be carefully managed to be successful. Effective control systems are typically well integrated with planning and are flexible, accurate, timely and objective (Gatewood *et al.*, 1995).

- *Integration with planning*
  Control should be integrated with the planning process. Specifically, objectives have to be set that may readily be converted into performance standards that will reflect how well plans are being carried out. A close link with planning ensures that efforts at increased control may be easily and accurately evaluated in terms of meeting organisational objectives.
- *Flexibility*
  Flexibility is an important factor in the development of an effective management control system to allow the firm to respond to changes in the business environment. The more dynamic or complex the business environment, the more flexible the control system should be. If changes occur in either the production process or the required quantity of resources needed, for example, due to new technologies or changing consumer demand, control must be flexible enough to readily accommodate modifications while remaining effective.
- *Accuracy*
  Control systems are useful only to the extent that the information on which they rely (and therefore produce) is accurate. If a quality control system somehow leaves workers the opportunity to hide product defects, the potential exists for errors that may render the control system useless, because it accurately measures or reports on what is supposed to be.
- *Timeliness*
  An effective control system provides performance information when it is needed. In general, the more uncertain and unstable the situation the more often information will be needed.
- *Objectivity*
  To be effective, the control system must provide unbiased information. Moreover, objective, control related information demands for assessing the information received. Instead of simply reporting how many defects occurred, one should analyse how these defects were developed.

In practice, a resistance towards control activities can exist. Common reasons for resistance to control include over-control, inappropriately focused control (control systems that do not focus on relevant issues that make sense), control that rewards inefficiency or control that results in enhanced accountability. To overcome resistance to control, one should create effective control by careful planning, encouraging employee participation and involvement of both employees and management by objectives (converting organisational objectives to personal objectives). In addition, there must be a system of check and balances (*i.e.* provide information and

documentation for control decisions). Often workers must be educated about the purpose and function of control, as well as about how their activities relate to control objectives.
Basic elements of a control system that is not operating effectively are (Gatewood *et al.,* 1995):

- A high incidence of employee resistance to control;
- A department meets control standards but fails to achieve its overall objectives;
- Increased control does not lead to increased or adequate performance;
- The existence of control standards that have been in place for an extended period of time;
- Organisational losses in terms of sales, profits or market share.

5.4.2 Forms of organisational control

Managers have in general two options with respect to control, *i.e.* internal and external control (Schermerhorn, 1999). Managers can rely on people who exercise self-control over their own behaviour. This strategy of internal control allows motivated individuals and groups to exercise self-discipline in fulfilling job expectation. Alternatively, managers can take direct action to control the behaviour of others. This is a strategy of external control that occurs through personal supervision and the use of formal administrative systems. Organisations with effective control typically use both strategies to good advantage. However, the trend today is to increase the emphasis on internal or self-control. This is consistent with the renewed emphasis on participation, empowerment and involvement in the work place.

Figure 5.17 illustrates the typical differences between the two control models, addressed as bureaucratic (external) and organic control (internal, self-control) (Gatewood *et al.*, 1995).
Bureaucratic control attempts to control the firm's overall functioning through formal, mechanistic, structural arrangements. It attempts to gain employee conformance through strict administration of rigid, straightforward policies and procedures, and its reward system focuses on individual employee compliance with either an implied or formal written code of behavioural standards. As such, bureaucratic control allows limited employee input into organisational activities.
Organic control attempts to regulate overall organisational functioning through reliance on informal organic structural arrangements. It tries to stimulate strong employee commitment by vigorously encouraging employee input and group participation, rather than setting strict behavioural standards as in bureaucratic control systems. Organic control relies on self-control

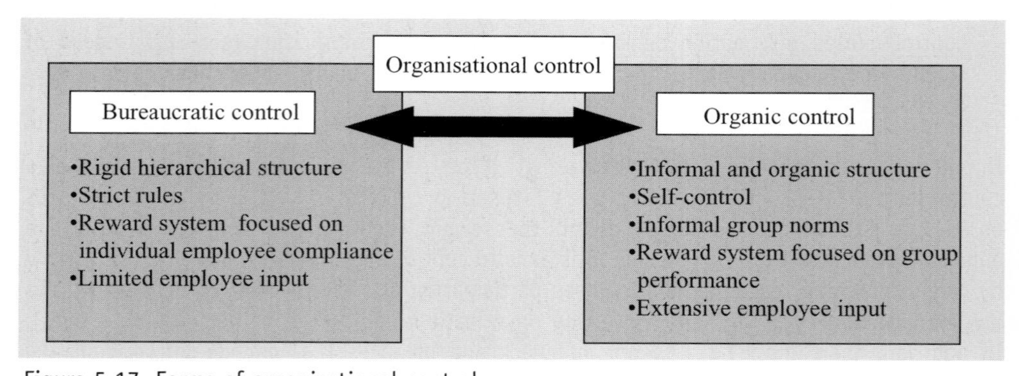

Figure 5.17. Forms of organisational control.

and informal group norms to effectively create a relaxed, but sharply focused, working environment.

It should be kept in mind that large organisations consist of a number of departments, which can differ widely in their emphasis on bureaucratic or organic controls. For example, if the production department faces a relatively stable environment and the marketing department faces a changing one, managers of the two departments are likely to choose different ways to divide and co-ordinate the work.
The contrast between bureaucratic and organic control also exists in quality management in the traditional and quality-based models of control, the latter being based on the principles of the organic model (Ivancevich, 1994).

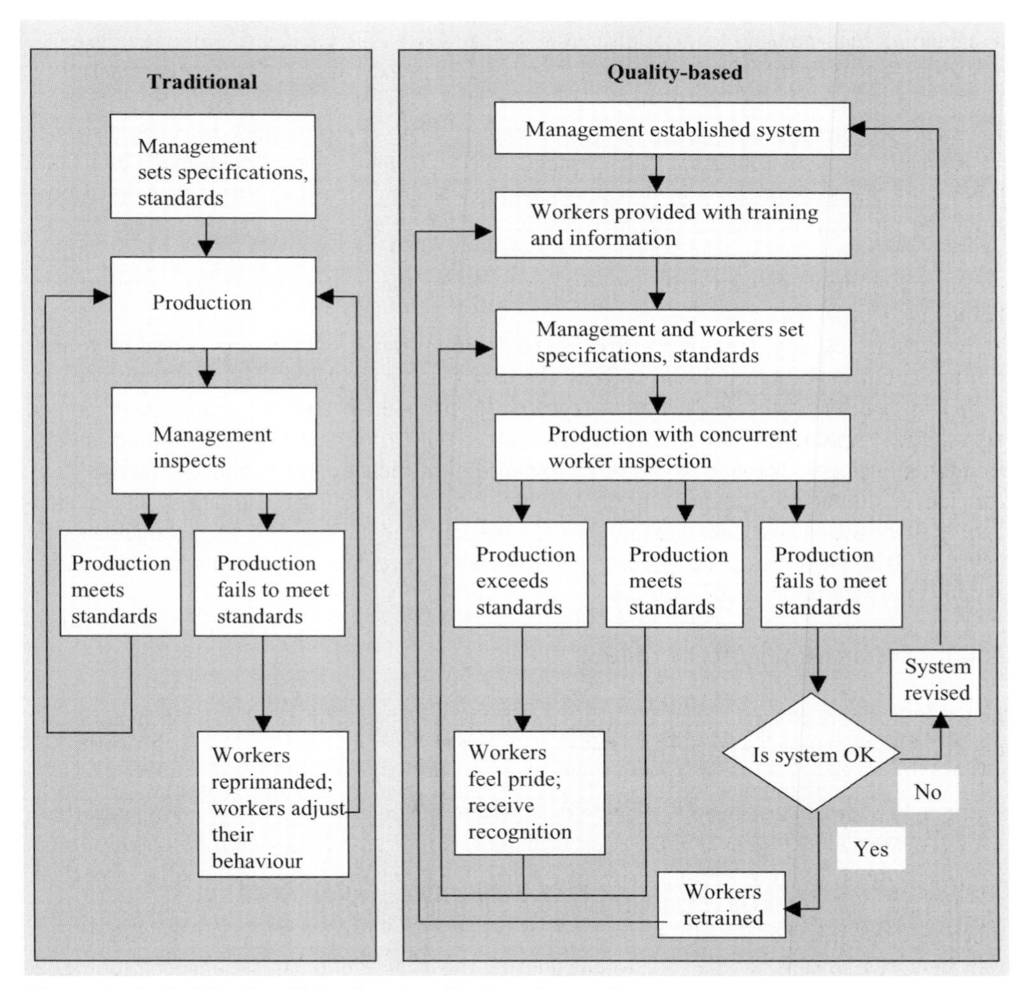

Figure 5.18. Models of traditional and quality-based control.

Figure 5.18 shows the contrast between both models. The traditional approach does not include training for workers; management inspects the results of production, and failure to meet production specification results in worker reprimands. In contrast, the quality-based control model includes worker training; workers inspect the results of production during the production process, and failure to meet production specification results in revision of the system.
In quality management, employee involvement can thus be considered as the essential ingredient (Heizer and Render, 1993). Employee involvement means including the employee in every step of the process from product design to final packaging. Building communication networks that include employees encourages employee involvement. This can be achieved through:

- Open and supportive supervisors;
- Moving quality responsibility from the quality control department and inspectors to production employees;
- Building high-morale organisations;
- Using formal techniques, such as quality circles (see next chapter).

All these measures are consistent with the concept of empowerment, which has been discussed in Chapter 3. Also consistent with empowerment is employees maintaining their own equipment. As a matter of fact, it requires training of employees and introducing knowledge from the technical department to the operators, and here are indeed the limitations for this concept. Whatever maintenance policies and techniques are decided upon, they must include an emphasis on employees accepting responsibility for the maintenance they are capable of doing.
In food manufacturing, there is a restriction to decentralisation because specialised tests must be carried out in the lab, thus not on line. The central issue in the decision concerning on-site or lab inspection is whether the advantages of specialised lab tests are worth the time and interruption needed to obtain the results (Stevenson, 1999). Reasons favouring on-site inspection include quicker decisions and avoidance of introduction of extraneous factors (*e.g.*, damage or alteration of samples). On the other hand, specialised equipment and a more conditioned test environment (less noise and confusion, reliability) offer strong arguments for using a lab. Many food companies rely on self-inspections by operators. This places responsibility for errors at their origin.

5.4.3 Costs and benefits of control

As with all organisational activities, controls should be pursued if the expected benefits are greater than the costs of performing them.

Figure 5.19 shows a cost-benefit model for judgement of the effectiveness of an organisation's control system (Hellriegel, 1992). The horizontal axis indicates the amount of organisational control, ranging from low to high. The vertical axis indicates the costs and benefits of control, ranging from zero to high. For simplicity, the curve of cost of control is shown as a direct function of organisational control.
Managers have to consider tradeoffs when choosing the amount of organisational control to use. With too little control, costs exceed benefits and the organisation controls are ineffective. As the amount of control increases, effectiveness also increases up to a certain point. Beyond this point, further increases in the amount of control result in decreased effectiveness. For example,

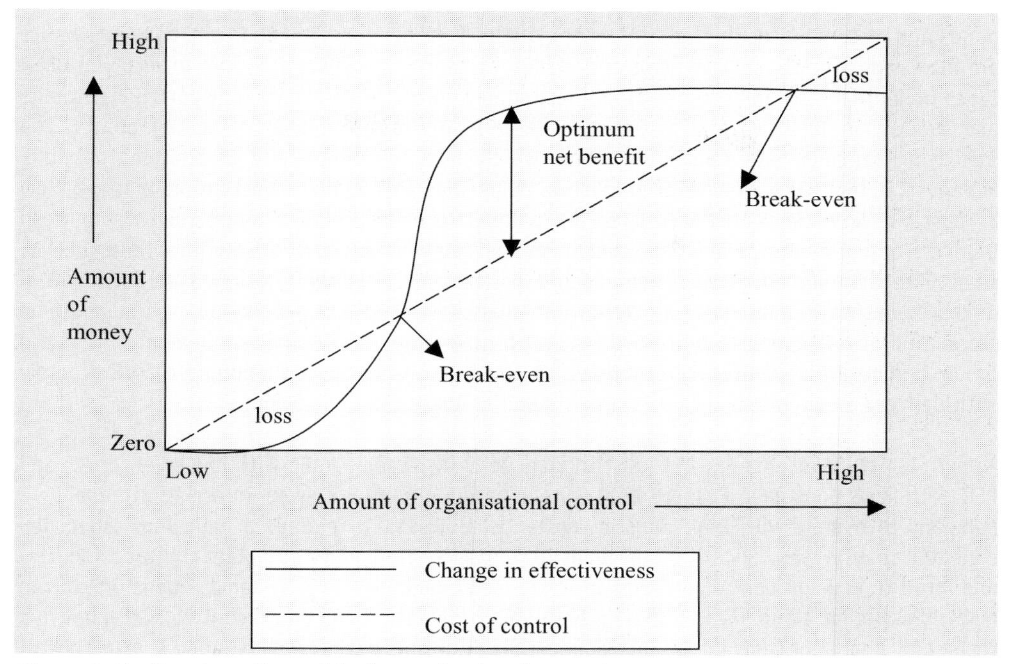

Figure 5.19. Cost-benefit model of organisational control.

an organisation might benefit from intensifying inspection of finished goods, reducing the number of defects in shipped goods. However, when by good sampling all batches having defect products are detected, more inspection only cause losses due to the elimination of inspected items. Figure 5.19 shows two break-even points. They indicate where the amount of organisational control moves from a loss to a net benefit and then return to a net loss. The optimal point is difficult to calculate. Effective managers probably come closer to achieving it than ineffective managers do.

5.5 Quality control in the food industry

Variation is a typical characteristic of food products because they consist of biological materials. Especially raw materials directly delivered from primary production (fruit, vegetables, milk, meat etc.) to processing companies show considerable variation. The large variation in incoming materials is often just a fact and can be scarcely influenced by primary production. Standardisation is partly realised by modification of recipe and/or processing conditions. However, for fresh fruits and vegetables, the variation remains during the whole production chain, including the moment of consumption. A consequence of the large variation for agri-food production chains is that there is considerable uncertainty about actual properties of raw materials and products. This uncertainty can only be partly overcome by control processes.

Another complicating factor typical for the food industry is logistics. Logistics refer to the movement of materials within a production facility, and to incoming and outgoing shipments of goods and materials. For the food industry logistics is rather complex, and demands on logistics are increasing. The complexity involves the high number of transport activities for many product types from different supply points. From these supply points, products are transported via storage and production facilities and distribution centres to a large number of selling points. The increasing demands refer to delivery time, which must be reliable and as short as possible. Moreover, in home stock levels of production companies and retailers must be preferably as low as possible. Sometimes these demands are parallel to quality requirements, such as for fresh products, which require small stock levels because they have a restricted shelf life. Nevertheless, sometimes there are conflicting requirements. For example, availability of raw materials is often linked to optimal harvest time. However, once it is harvested it is a bulk volume that must be stored and often cannot be processed immediately. Another conflicting requirement is that the production must be fast at low costs. However, sometimes too fast production, in order to comply with delivery requirements, results in high material loss during operation.

Modern logistic concepts like JIT (Just In Time delivery, accompanied with low inventories) and ECR (Efficient Consumer Response, a system based on point-of-sale information) enable a company to comply with high logistics demands. However, it requires a high and, especially, constant quality level. Such high demands on logistics and quality stimulate the food industry and retail to develop control systems, which exceed companies' own systems. This is one of the major reasons for companies to change to chain co-operation and chain control. Well-known is the supply chain concept, *i.e.* an interconnected set of linkages between suppliers, warehouses, operations and retail outlets. An essential aspect of this concept is that not only the direct supplier or customer is controlled, but also previous or next partners in the production chain; for example, when a producer and distributor of orange juice is involved in cultivation of the oranges in order to guarantee a high quality standard. Although the supply chain concept seems very attractive on paper, in practice only a few initiatives succeeded in developing chain co-operation structures. This is assumed to be due to lack of mutual trust between partners.

Another typical problem for the food industry, which complicates control, is the location of specialised engineering knowledge. This problem originates from the typical process character of food industry, which can be typified by repetitive flow processes. Control in such situations is mainly focussed on process parameters, like temperature and pressure during operation. Moreover, many control systems are automatic. As a matter of fact, most knowledge about the production process is located in a specialised department, like the engineering department. Often the knowledge is not even available in the company itself, but is located at the companies supplying the equipment. This results in reduction of possibilities to influence and control the production process. This problem becomes even clearer when, due to higher demands on quality and flexibility, more influence and control on processes is required.

The functional organisation form is very popular in the food industry, which forms a basis for the type of problems described above. In functional organisations, specialisation and expertise are clustered in functional departments (*e.g.* purchase, engineering and production department). With respect to quality, specialised departments like R&D and quality assurance often feel responsible for it. They provide other departments with quality services and knowledge, thus making them dependent on the quality and/or R&D department. A possible consequence of this situation is that control is focused on controlling each other instead of controlling the process to comply with consumers' demands.

Besides these organisational hurdles, control as such is rather difficult. Difficulty in control refers, on the one hand to large variation in product and process conditions, and on the other hand to the high number of product attributes that are commonly required (like colour, taste, nutritional value, but also low costs). Based upon the complexity of control, one would expect statistically supported control systems in food industry. In practice, however, the use of statistics (*e.g.* statistical process control approach) is very much restricted in the food industry and sometimes control frequency depends only on the number of available inspectors.

In this type of situations. the techno-managerial approach would choose for management activities that are focused on major points in the process that can be technologically or technically influenced. The basic idea beyond it is that only a restricted number of control points are determining the quality as requested by consumers. In other words, control measures should focus on those points that are relevant for influencing final product quality. An advantage of this approach is that control activities can be restricted and attention of employees and operators is now focused on only those points in the process that are relevant for quality (like sensory properties, safety etc).

Concepts such as HACCP (Chapter 7), FMEA (Chapter 4) and the Taguchi principles (Chapter 4) enable companies to perform a profound analysis of the process and to assess points which are critical with respect to safety and quality. The use of these 'tools' should be combined with management actions like decentralisation and training, as described in the model of quality-based control (Figure 5.18). This combination creates opportunities to control and take corrective actions when required.

Furthermore, it is of great importance to develop proper measuring and information systems. Quick measuring units, which can be used by operators themselves, can be very supportive. However, only those data should be collected that are actually used for process control, whereby a process can also refer to purchase or supply and not only operational processes. An important technical measure to support control processes is introduction of automation. Automation provides operators with the possibility to influence and control processes by themselves, and it is not necessarily aimed at reducing the number of operators. Another advantage of automation is the reduction of set-up times, which increases flexibility of the company.

6. Quality improvement

In food and agribusiness quality, improvement of product and processes is of great importance. Competition demands for innovation and improvement of products and processes and reduction of costs. Juran distinguished quality control and quality improvement as two different activities, besides quality planning. They are all part of the Juran's trilogy: planning, control and improvement (Juran, 1990, see also Figure 3.26). Quality planning is the continuous process of the development of products or services, which meet customer demands. Quality control includes activities undertaken to ensure that actions lead to achievement of objectives. It involves evaluation of actual quality performance, comparing with quality objectives and acting on differences. Quality improvement actually embodies the need for change, breaking through the status quo. It implies that the firm has to "learn". It requires facilitating structures, such as means of communication, procedures and reward systems, focused on identification of changes in internal and external business environment and on avoiding routine and rigid structures in the firm.

There is a widely held belief that an organisation would have a few, if any, problems if workers would do their jobs correctly (Edge, 1990). In fact, the potential to eliminate mistakes and errors lies mostly in improving the systems through which the work is done, not in changing the workers. Juran and Deming have proposed since the early 1950s that at least 85% of organisation failures are due to flaws of management-controlled systems. Workers control fewer than 15% of the problems. This observation has evolved in the rule of thumb that at least 85% of problems can only be corrected by changing the systems (which are very largely determined by management). For example, a worker in the production line cannot do a top quality job when working with defect tools or parts. Even when it does appear that an individual is doing something wrong, often the trouble lies in how that worker was trained, which is a system problem.

An important base for quality improvement is thus an approach with both management and employee involvement. Management should initiate the driving force for this process, because improving means changing, getting away from routine structures. In many cases within firms, routine-structures exist that cause rigidity. They hamper the improvement process aimed at maintaining gained methods, hierarchy, etc. Improvement will only be accepted if there is sufficient commitment.

In this chapter, firstly the improvement process will be analysed, followed by a summary of the quality guru's ideas about improvement. After a short description of the well-known improvement tools, attention is focused on working with improvement teams and -due to the importance of management and employee involvement-on strategies of organisational change.

6.1 The improvement process

Quality improvement is a systematic approach to improving a system. It involves documentation, measurement and analysis for the purpose of improving the functioning of a system. Typical goals of quality improvement include increasing customer satisfaction, achieving higher quality levels, reducing cost, increasing productivity and accelerating the process.

Kampfraath and Marcelis (1981) perceived the improvement of administrative systems as a double looped cybernetic process (Figure 6.1).

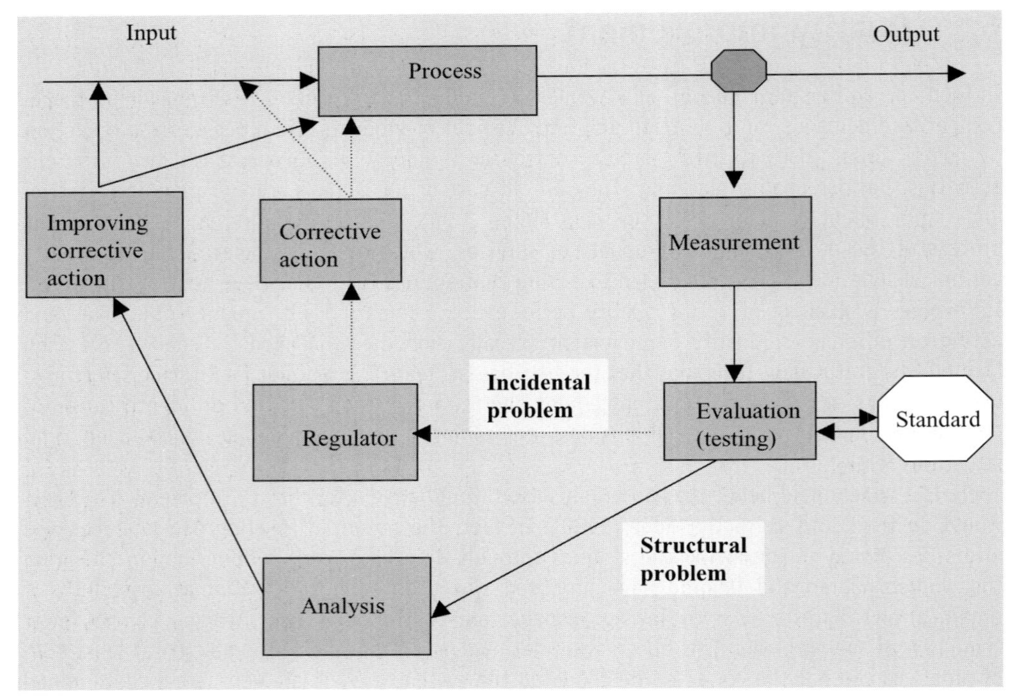

Figure 6.1. Schematic reproduction of the quality control and quality improvement circle.

The first loop is identical to normal control processes and deals with incidental problems, bringing the process back between its control limits. The second loop arises when problems have a structural nature, for instance quality problems always occurring at the end of the week during production peak, or quality problems that only arise at incoming materials from one specific supplier. This kind of problem first has to be analysed before corrective action could prevent these problems in the future. After analysis, the next step in the loop is improvement as a corrective action, and bringing the system on a higher level of quality control.

Stevenson (1999) provided three crucial steps in the improvement of processes within a system:

1. Map the process.
   a) Collect information about the process; identify each step in the process; for each step, determine inputs and outputs, people involved and decisions that are made.
   b) Document measurements like time, cost, space used, waste, employee morale and employee turnover, accidents and/or safety hazards, working conditions, revenues and/or profits, quality and customer satisfaction, as appropriate.
   c) Prepare a flow chart that accurately depicts the process; note that too little detail will not allow for meaningful analysis, and too much detail will overwhelm analysts and will be contra productive. Make sure that key activities and decisions are represented.
2. Analyse the process.
   a) Ask the following questions about the process: is the flow logical, are any steps or activities missing, are there any duplications?

b) Ask the following questions about each step: is the step necessary, could it be eliminated, does the step add value, does any waste occur at this step?
c) Analyse for improvement by asking: what causes arise for known problems, could they be eliminated, can the process be improved by shortening time or by reducing costs, can the process be improved by providing better conditions for quality?

3. Redesign the process.
   Use results of analyses for redesign of the process. Document improvements; potential measures include reductions in time, cost, space, waste, employee turnover, accidents, safety hazards, increases in employee morale working conditions, revenues/profits, quality, and customer satisfaction.

Quality improvement is in a way contradictory to the belief in optimisation, which was very strong in the 1960s and the 1970s. Optimisation had to be done within a fixed framework. Improvement, however, is partly a matter of ridding ourselves of this framework and starting to think on entirely new lines (Bergman and Klefsjö, 1994). To a great extent, successful companies are using special programs for quality improvement. The first applications have been started with workers in their own work environment, dealing with problems that arise in their daily work. These programs worked on the basis of improvement teams or quality control circles (QC circles, see section 6.4), small groups of employees who discuss different problems and come up with suggestions for improvement. QC circle activities were not only confined to manufacturing industry, however they were initiated there. There are, for example, QC circles in sales organisations, department stores, hotels, banks and restaurant business.

While all activities in the past focused more or less on the primary production process, Total Productive Maintenance (TPM) is typically resource oriented. In the seventies, the concept of TPM came up in the manufacturing industry. TPM encompassed all production equipment. Equipment may break down, potentially disrupting a firm's entire operation (Melnyk and Denzler, 1996). Typical losses are then downtime, speed losses and product defects. TPM is a method designed to eliminate these losses by identifying and attacking all causes of equipment breakdowns and system downtime. Instead of accepting maintenance as necessary, TPM tried to achieve the ambitious goal of zero breakdowns. Five principles guided TPM programs:

1. maximise equipment effectiveness (reduce downtime to zero),
2. establish a thorough system of preventive maintenance for the entire life span of equipment, from design and acquisition to disposition,
3. implement TPM in all organisational areas,
4. involve every single member of the organisation, from the top manager to the worker on the shop floor,
5. assign responsibility for the program to small autonomous groups of employees rather than managers.

Based on these five principles TPM tried to encompass all kind of activities that influence equipment uptime: keeping a well-organised shop floor, proper operating procedures, improving maintenance plans, improving weaknesses in equipment design and improving operation and maintenance skills. Through these activities TPM attacked variance and waste by preventing problems and eliminating causes of defects.

A rather extreme improvement concept has been introduced in 1993 by Hammer and Champy, which they called *business reengineering*. Reengineering (also known as business process redesign) is a type of improvement with the potential to dramatically improve quality and speed

of work and to reduce its cost by fundamentally changing the processes by which the work is executed (Dean and Evans, 1994).
Reengineering goes beyond the process level by including the entire value chain. It also goes beyond reorganisation by including fundamentally redesign and starting from scratch. It is often used when the improvements needed are so great that incremental changes to operations will not be sufficient. Think about improvements from 50% to 1000%.
The general principles of business process redesign (BPR) include:

1. reduce handoffs: every time a process is handed from one person or group to another, errors can occur;
2. eliminate steps: the best way to save time on a step is to avoid it when possible. If a step does not add customer value, stop doing it;
3. perform steps in parallel rather than in sequence: unless one operation cannot be done until another is finished, do both at once;
4. involve key people early in order to avoid doing things over when key people do not give their input until the process is under way. Doing it is consistent with the total quality management principle: "do it right the first time".

6.2 Quality gurus on improvement

Deming (1986) emphasised that everyone learned a common method of describing and attacking problems. This commonality is an absolute requirement if personnel from different parts of the company must work together on company-wide quality improvement (Edge, 1990). Deming therefore introduced a framework: the Deming cycle, where all parties can discuss problems and suggest improvements (Figure 6.2).

The plan-do-study-act (PDSA) cycle, also referred to as either Stewart cycle or Deming wheel, is the conceptual basis for continuous improvement activities (Stevenson, 1999; Melnyk 1996). Representing the process with a cycle underscores its continuing nature. Going upwards suggests higher levels of quality to be reached. There are four basic steps in the cycle:

1. *Plan*. Begin by studying the current process and document that process. Then collect data to identify problems. Next, survey data and develop a plan for improvement. Specify measures for evaluating the plan.

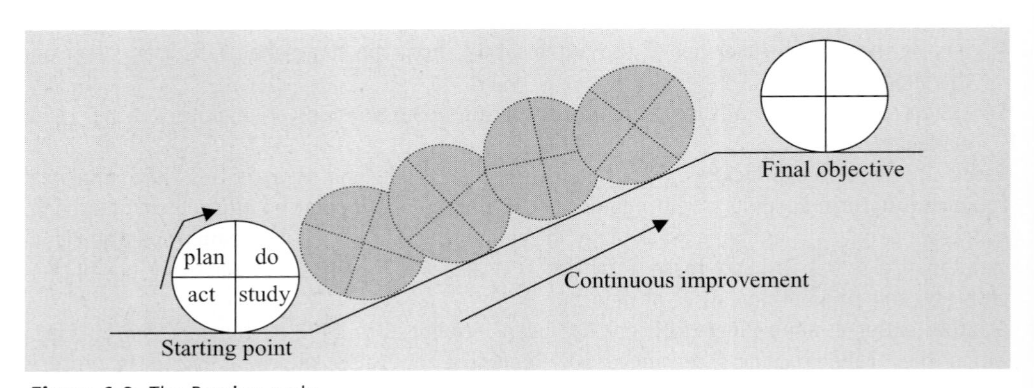

Figure 6.2. The Deming cycle.

2. *Do*. Implement the plan, on a small scale if possible. Document changes made during this phase. Collect data systematically for evaluation.
3. *Study* (check). Evaluate the data collected during the 'do' phase. Check how closely the results match original goals of the plan phase.
4. *Act*. If the results are successful, standardise the new method and communicate the new method to all people associated with the process. Implement training for the new method. If results are unsuccessful, revise the plan and repeat the process or cease the project.

In replicating successful results elsewhere in the organisation, the cycle is repeated. Similarly, if the plan was unsuccessful and you still wish to make further modifications, repeat the cycle. Employing this sequence of steps provides, according to Deming, a systematic approach to continuous improvement.

According to *Juran*(1990) quality planning is necessary to establish processes that are capable of meeting quality standards. Quality control is necessary in order to know when corrective action is needed and quality improvement will help to find better ways of performance (Juran's trilogy). A key element of Juran's philosophy is commitment of management to continual improvement. His approach urged companies to "break the chain of the past" and reach improved quality levels on a continuous basis (Figure 6.3).

Juran described improvement as a *"managerial break through"* being the organised creation of beneficial change (Edge, 1990). It is this organised, step-by-step approach to creating management breakthrough that is the hallmark of the Juran quality improvement process. Juran motivated that quality improvement is not a one-off process. To attack quality problems

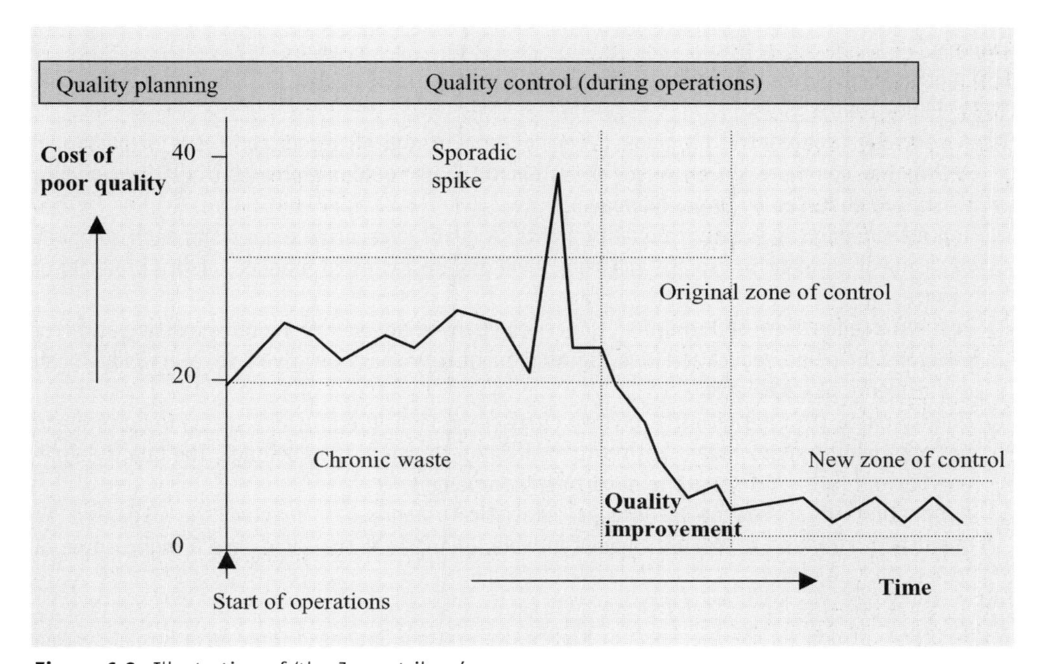

Figure 6.3. Illustration of 'the Juran trilogy'.

successfully, organisations must develop the attitude of creating annual improvements year after year. It is an attitude that refuses to accept today's quality levels as good enough.
The quality improvement effort continues forever and must become formalised into the company's planning and operational practices bringing them into "new zones of control" on a higher quality level. In order to reach this, Juran proposed ten steps for quality improvement (Juran, 1951):

1. Build awareness for the need and opportunity for improvement;
2. Set goals for improvement;
3. Organise people to reach the goals;
4. Provide training throughout the organisation;
5. Carry out projects to solve problems;
6. Report progress;
7. Give recognition;
8. Communicate results;
9. Keep score;
10. Maintain momentum by making annual improvement part of the regular systems and processes of the company.

Ishikawa was strongly influenced by Deming and Juran, although he made significant contribution of his own to quality management (Stevenson, 1999). Among his key contributions were development of the cause-and-effect diagram (also known as a fishbone diagram) for problem solving and implementing *quality circles* that involve workers in quality improvement (see 6.4 for quality circles). He was the first expert who paid attention to the *internal customer*, *i.e.* the next person in the process. He was a strong supporter of the need for companies to have a shared vision in order to join everyone in the organisation in a common goal. In addition, he is widely recognised for his efforts to make quality control "user friendly" for workers.
His contribution can be placed in the context of the features of quality work in Japan (Ishikawa, 1985; Edge, 1990):

1. A company-wide control program. Every department and all levels and types of personnel within the company are engaged in systematic work, guided by written quality policies, which are endorsed by upper management.
2. Top management is subject to quality audits. A quality executive team visits each company department to identify, isolate and help solve any obstacles to produce quality products or services.
3. Industrial education and training. Since company-wide quality requires participation of everyone involved, education and training in quality must be given to everybody in all departments at each level.
4. Quality circle (QC) activities. A quality circle is "a small group which meets voluntarily to perform quality improvement within the workshop to which they belong". The quality circle provides a forum to discuss the department's problems.
5. Application of statistical techniques. Statistical techniques enabled the Japanese to quantify quality and to make it understandable, giving it at the same time a high degree of "scientific" objectivity.

Imai (1986) introduced the elements of Kaizen. Kaizen is a Japanese term for a formal system to promote continuous improvement. It begins with the notion that an organisation can assure

its long-term survival success only when every member of the organisation actively pursues opportunities to identify and implement improvements every day.
Kaizen sets no conditions for the sizes of improvement. In fact, it often favours small, incremental improvements over large innovations. Pursuit of small improvements keeps people thinking about the process and its current operation. They identify potential improvements by understanding the functions of the current system and its weaknesses or relative inefficiencies. Furthermore, small improvements gain returns for the firm without the need for large investments.
Kaizen refines its emphasis on daily incremental improvement to develop three guiding principles:

1. Process view of the system; the basic unit for analysis is the process.
2. Success comes from people; any successful Kaizen program relies heavily on people's knowledge and their insights and intuition.
3. Constant sense of urgency; a successful Kaizen program depends on awareness of the need for change; everyone must feel that they can improve their performance (Melnyk and Denzler, 1996).

Like Deming and Juran, *Crosby* (1979) viewed the march toward quality as a never-ending journey (Melnyk and Denzler, 1996). Before beginning this journey, managers must first assess their firm's current quality culture. The management maturity grid (see Chapter 8) can support this self-assessment by positioning the firm in one of five states of quality awareness and quality culture. Once managers have determined their position, they can begin improving quality by implementing Crosby's 14 steps for quality improvement.
This program emphasises prevention and elimination of process faults rather than inspection. The program seeks to improve product quality by changing the corporate culture. Detailed statistical tools, while important, remain secondary to this general change in culture. The program accepts no standard short of zero defects - designing and building products free of any quality problem. The 14 steps are:

1. create management commitment
2. establish a quality improvement team with clear access to top management
3. establish clear measures of quality stated in meaningful terms that relate specifically to individual activities
4. identify the price of non-conformance and the price of conformance by collecting data on cost components
5. promote awareness of quality
6. take corrective action
7. zero-defects planning: how to move the company from correcting problems to totally eliminating them
8. train managers and employees at all levels, because education is the primary catalyst for change
9. schedule zero-defects day to signal that the company is moving to its goal of zero defects
10. set goals that represent progress towards this target
11. encourage employees to remove causes of error
12. give recognition to motivate and to set examples
13. establish a quality council for information and co-ordination, and
14. do it all over again!

6.3 Quality improvement tools

As a basis for improvement work, data are required together with an analysis of these data (Bergman and Klefsjö, 1994). In Japan it was quickly realised that everyone in a company had to participate to the improvement work. This meant that statistical tools, which have to be used, had to be simple but yet efficient. Seven methods, or "tools", were put together by Ishikawa, amongst others. These methods are called "the seven QC-tools" where QC means Quality Control. Since the beginning of the 1960s these tools have been taught to workers and foremen in Japanese industries who used them systematically for problem solving. The tools aid in data collection and interpretation and provide a basis for decision making. Figures 6.4 and 6.5 give an overview of tools that can be used to support quality improvement processes.

Check sheet
A check sheet is a simple tool frequently used for problem identification. It provides a format that enables users to record and organise data in a way that facilitates collection and analysis. Check sheets are designed on the basis of what users are attempting to learn by collecting data.

Flowchart
A flowchart is a visual representation of a process. As a problem-solving tool, a flowchart helps investigators in identifying possible points in a process where problems occur. A flowchart consists of procedures and decision points.

Scatter diagram
A scatter diagram can be useful in deciding if there is a correlation between the values of two variables, using one variable to make a prediction about another variable or characteristic.

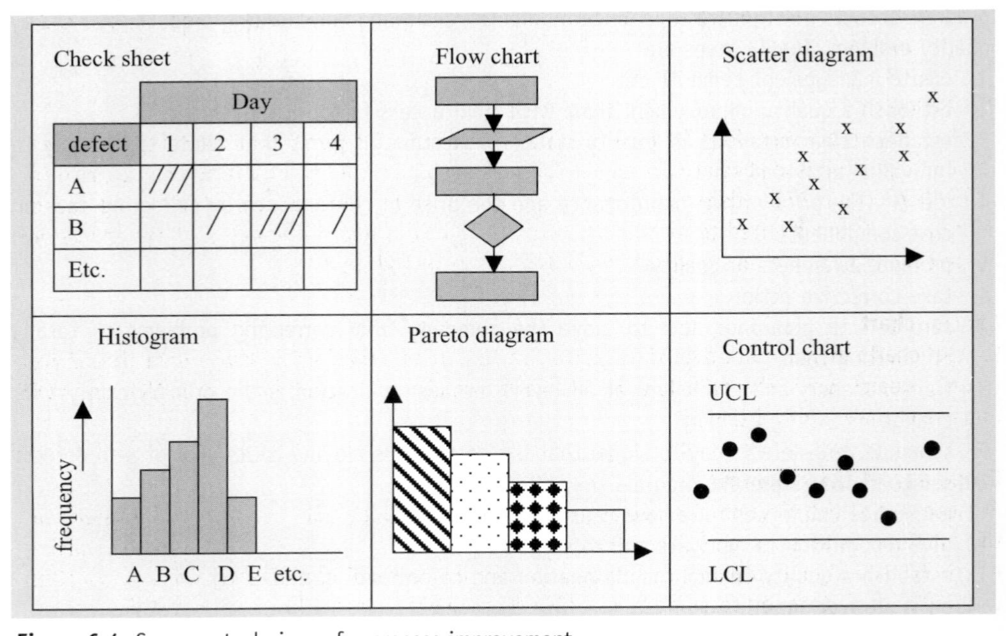

Figure 6.4. Common techniques for process improvement.

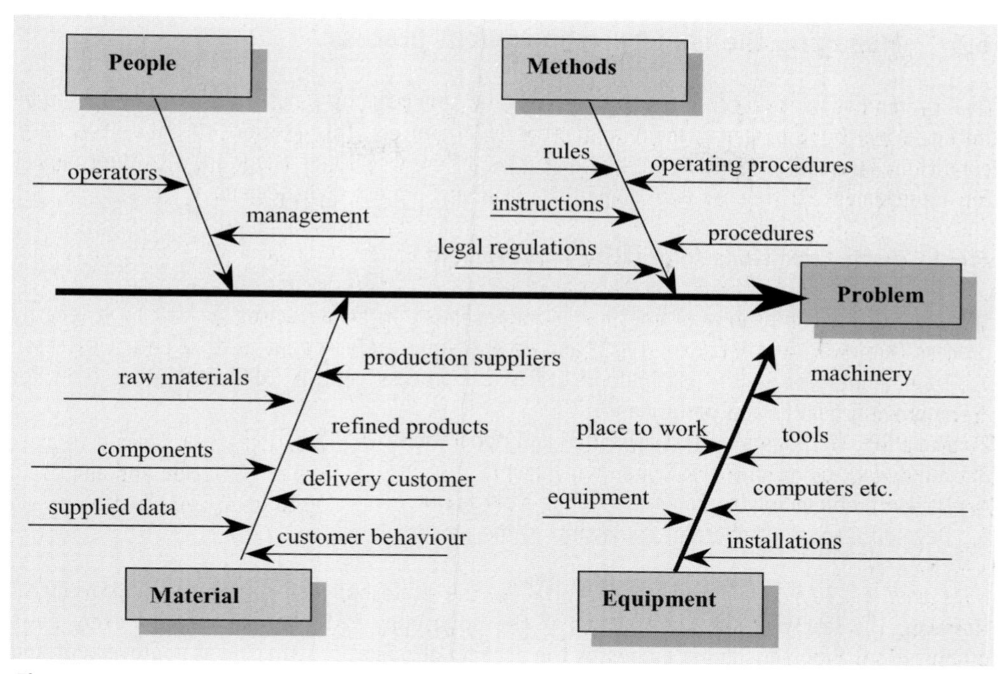

Figure 6.5. Cause-effect diagram of Ishikawa.

Histogram

A histogram can be useful in obtaining a sense of the distribution of observed values. Among other things, one can see if distribution is symmetrical, what the range of values is and if there are any unusual values.

Pareto analysis

Pareto analysis is a technique for focusing one's attention on the most important problem areas. The concept implies that relatively few factors generally account for a large percentage of total cases (*e.g.* complaints, defects). The idea is to classify the cases according to degree of importance. Often referred to as the 80-20 rule, the Pareto concept states that approximately 80 percent of the problems originate from 20 percent of the items.

Control chart

Control charts display data over time as well as computed variations in those data. The charts visually depict when the data falls outside a previously set acceptable range (UCL and LCC: upper and lower control limit).

Cause-and-effect diagram

A cause-and-effect diagram or Ishikawa diagram offers a structured approach to the search for the possible cause(s) of a problem (Figure 6.5). This tool helps to organise problem-solving efforts by identifying categories of factors that might be causing problems. This tool is often used after brainstorming sessions to organise the generated ideas (Stevenson, 1999).

6.4 Managing the quality improvement process

Quality improvement processes sometimes come up spontaneously, but normally they should be initiated by the top management and properly organised. This section describes the basic conditions for quality improvement and discusses different forms of teamwork with their impact for improvement. It finishes with some remarks about project management.

6.4.1 Basic conditions for quality improvement

Implementing a philosophy of continuous improvement in an organisation may be a lengthy process (Krajewski and Ritzman, 1999) and several steps are essential to its eventual success:
1. Train employees in the methods of statistical process control (SPC) and other tools for improving quality and performance;
2. Make SPC methods a normal aspect of daily operations;
3. Build work teams and employee involvement;
4. Utilise problem solving tools within the work teams;
5. Develop a sense of operator ownership in the process.

Note that employee involvement is central to the philosophy of continuous improvement. However, the last two steps are crucial if the philosophy is to become part of every day's organisation. Problem solving addresses the aspects of operations that need improvement and evaluates alternatives for achieving improvements. The sense of operator ownership appears when employees feel as if they own the processes and methods they use and take pride in quality of the product or service they produce. This sense can derive from participating to work teams and to problem solving activities, which give employees a feeling that they have some control over their workplace and tasks.

A potential barrier to improvement is the form of organisation (Edge, 1990). Traditional organisation structures (based on a series of departments each with their own goals) and specialists and command structures, have a natural tendency to focus on their own priorities, such that overall business and company-wide objectives are often neglected.
The formal authority structures present within each department create a complex communication network, which makes it more difficult to keep people informed. Only top management can impose the need for change in such circumstances. Therefore, there is the need to create simpler, flatter structures with added flexibility, which ensure involvement from the total workforce. The result is creation of shorter links to customers and suppliers, such that changing circumstances can quickly perceived and dealt with.
A second potential barrier to improvement is lack of information. People must have ready access to reliable and accurate information and have the organisational scope to act on it. This information must embrace manpower, machines, methods and materials and the company environment. The concept of information as a resource is already described in section 3.5.1 about organisational conditions.

6.4.2 Teamwork

Twenty-five years ago, the decision of companies such as Volvo, Toyota and General Foods to introduce teams into their production processes made news because no one else was doing it

(Robbins, 2000). Today, it is just the opposite. It is the organisation that does not use teams that has become newsworthy. How can the current popularity of teams be explained? Evidence suggested that teams typically do better than individuals when tasks require multiple skills, judgement and experience. As organisations have restructured themselves to compete more effectively and efficiently, they have turned to teams as a way to better utilise employee talents. Management observed that teams are more flexible and responsive to changing events than traditional departments or other forms of permanent groupings. Teams have the capability to quickly assemble, deploy, refocus and disband. Furthermore, teams facilitate employee participation in decision-making, thus stimulating employee motivation.
Teams are especially important in Total Quality Management (see for TQM concept Chapter 8). Teamwork enables various parts of the organisation to work together in meeting customer needs that can seldom be fulfilled by employees limited to one speciality. The TQM philosophy recognises the interdependence of various parts of the organisation and uses teams as a way to co-ordinate work (Dean and Evans, 1994).

Formally defined, a *team* is a small group of people with complementary skills, who work together to achieve a shared purpose and hold themselves mutually accountable for its accomplishment (Schermerhorn, 1999).
The difference with a *workgroup* is that a workgroup is a group that interacts primarily to share information and to make decisions to help each other to perform within each member's area of responsibility (Robbins, 2000). Workgroups have no need or opportunity to engage in collective work that requires joint effort. So their performance is merely the summation of all group member's individual contributions: there is no positive synergy - the creation of a whole that is greater than the sum of its parts. The presence of synergy means that a team is using its membership resources to the fullest and is achieving through collective performance far more than could otherwise be achieved (Schermerhorn, 1999). The usefulness of teams includes the following contributions:

- Increasing resources for problem solving
- Fostering creativity and innovation
- Improving the quality of decision making
- Enhancing members' commitment to tasks
- Raising motivation through collective action
- Helping control and discipline members
- Satisfying individual needs as organisations grow in size.

However teams have their disadvantages as well. Problems that are commonly encountered in teams include:

- Personality conflicts: individual differences in personality and work style can disrupt the team
- Task ambiguity: unclear agendas and/or ill-defined problems can cause teams to work too long on the wrong things
- Poor readiness to work: time can be wasted when meetings lack purpose and structure
- Poor teamwork: failures in communication, conflict and decision making may limit performance and/or hurt morale.

The trend toward greater empowerment is associated with several developments in the utilisation of teams. In Chapter 4, cross-functional teams in product development already have been described. Managers seek to expand opportunities for broad-based participation in workplace affairs. Five forms of teamwork can be distinguished differing in level of empowerment (Schermerhorn, 1999; Robbins, 2000):

1. Committees and task forces

Committees and task forces bring people together outside of their daily job assignments to work in small teams for a specific purpose. They typically operate with task agendas and are led by a designated head or chairperson who, in turn, is held accountable for the results. A committee usually operates with an ongoing purpose and its membership may change over time. An example is the quality steering committee being responsible for developing quality policy and quality systems.

A task force usually operates on a more temporary basis. Its official tasks are very specific and time defined. Once the stated purpose has been accomplished the taskforce may disband.

2. Cross-functional teams

Cross-functional teams are made up of employees at about the same hierarchical level, but from different work areas, who come together to accomplish a task. They are expected to exchange information, to develop new ideas, solve problems, co-ordinate complex projects, and not to be limited in performance by purely functional concerns and demands.

3. Quality circle's

A Quality Circle (QC) is a group of workers that meets regularly to discuss and plan specific ways to improve work quality. Usually it consists of 6 to 12 members from a work area. After receiving special training in problem solving, team processes and quality issues, members of the quality circle try to come up with suggestions that can be implemented to raise productivity through quality improvement. Juran (1990) pointed out the differences between quality (control) circles and project teams (task forces and cross-functional teams) - see Figure 6.6. Problems with Quality Circles raised where basic conditions like training and top management commitment were not fulfilled. This is the main reason for Quality Circles not to have continuous success.

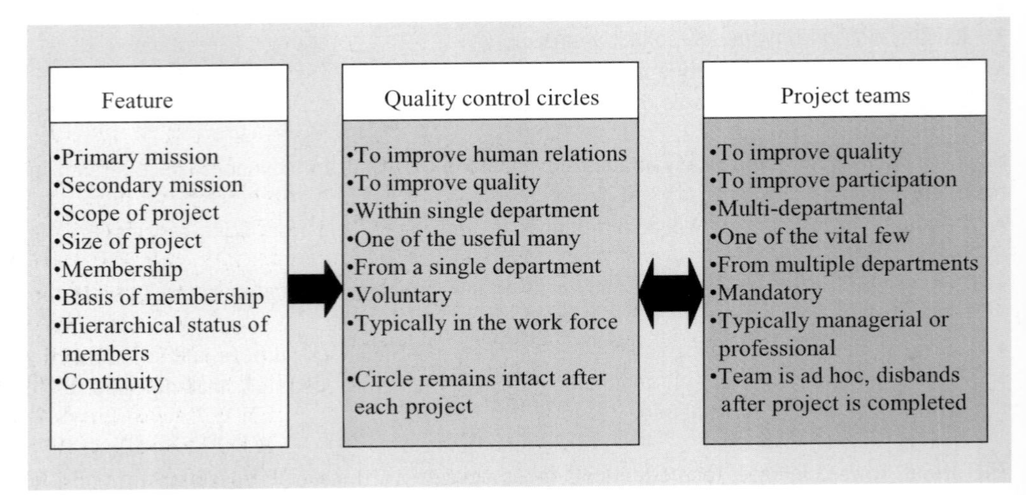

Feature	Quality control circles	Project teams
Primary mission	To improve human relations	To improve quality
Secondary mission	To improve quality	To improve participation
Scope of project	Within single department	Multi-departmental
Size of project	One of the useful many	One of the vital few
Membership	From a single department	From multiple departments
Basis of membership	Voluntary	Mandatory
Hierarchical status of members	Typically in the work force	Typically managerial or professional
Continuity	Circle remains intact after each project	Team is ad hoc, disbands after project is completed

Figure 6.6. Contrast between quality control circles and project teams.

4. Virtual teams

Virtual teams use computer technology to tie together physically dispersed members in order to achieve common goals. Virtual teams can do all things that other teams do - share information, make decisions, and complete tasks. The three primary factors that differentiate virtual teams from face-to-face teams are a) the absence of para-verbal and non-verbal cues, b) limited social context, and c) the ability to overcome time and space constraints. Virtual teams often suffer from less social relationship and less direct interaction among members. Not surprisingly, virtual team members report less satisfaction with group interaction, but on the other hand virtual teams allow people to work together who otherwise might never be able to collaborate.

5. Self-managing teams

Self-managing teams are generally composed of people who take over responsibilities of their former supervisors. Typically, these responsibilities include collective control over the pace of work, determination of work assignments, organisation of breaks and collective choice of inspection procedures. Fully self-managed teams select their own members and the members evaluate each other's performance. As a result, supervisory positions become less important and may even be eliminated. Self-managed teams operate with participation, decision-making and multi-tasking, in which team members each have the skills to perform several different jobs. As shown in Figure 6.7, typical characteristics of self-managing teams are as follows:

- Members are held collectively accountable for performance results
- Members have discretion in distributing tasks and scheduling work within the team

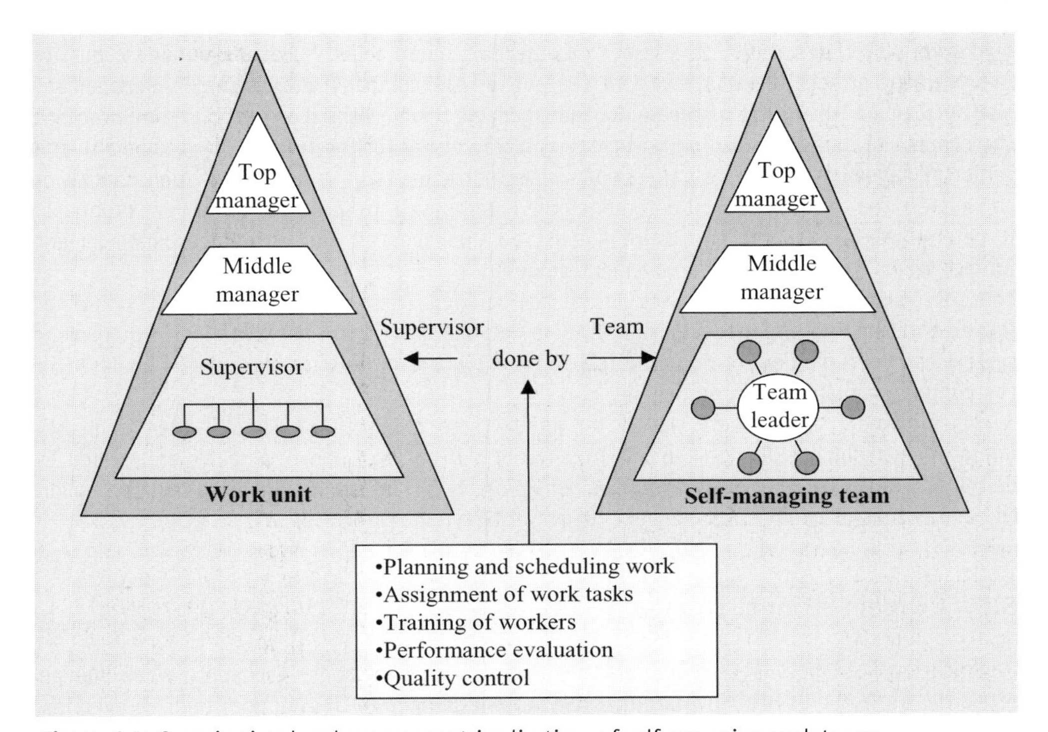

Figure 6.7. Organisational and management implications of self-managing work teams.

- Members are able to perform more than one job
- Members train one another to develop multiple job skills
- Members evaluate one another's performance contributions
- Members are responsible for the total quality of team products.

The overall research on effectiveness of self-managing teams is not uniformly positive. For example, individuals in these teams do tend to report higher levels of job satisfaction. However, employees on self-managing teams seem to have higher absenteeism and turnover rates than employees in traditional work structures (Robbins, 2000).

The five forms of teams just described are examples of formal groups in the organisation. In contrast with them, there are informal groups, which are also important in every organisation. These informal groups are not recognised on organisation charts and are not officially created. They emerge as a part of the informal structure and from natural and spontaneous relationships among people. Informal groups can have a positive impact on work performance. In particular, the relationships and connections made possible by informal groups may actually help speed the workflow or allow people to "get things done" in ways not possible within the formal structure. Furthermore members of informal groups often find that the groups offer social satisfaction, security, support and a sense of belonging (Schermerhorn, 1999).

6.4.3 Project approach

According to Juran, quality improvement effort must go on forever and must become formalised into the company's planning and operational practices, just like corporate budget and annual marketing plan (Edge, 1990). To attain this objective, Juran called for establishment of quality improvement project teams starting at the top of the company with a senior management quality team. All members of the senior management team, without exception, should accept the concept of annual quality improvement as a way of life for the future. Key members of the senior management should be part, among other participants, of a quality steering council or committee. The steering council determines the broad policy for quality improvement programs, it determines how appropriate training must be implemented and encourages management project teams to initiate the improvement process.
These project teams pass a series of phases ranging from start-up to project planning, implementation and project evaluation. A more detailed description of the project management system has been previously discussed in Chapter 4, where project management is analysed in the context of quality design processes.
A typical project team meets weekly for about three months, and should reach a quality improvement solution. Once a team is finished it is dismissed, and new teams are formed to define and tackle another quality improvement project. In large companies it is not uncommon to have dozens of quality improvement teams working simultaneously.

6.5 Organisational change

Organisational change is very important for quality management, especially when human capacities are to be used at full extent. This will be explained in this section, then an overview will be given of organisational change strategies and their advantages for quality management.

6.5.1 Quality management and organisational change

The basis of continuous improvement philosophy is the belief that virtually any aspect of an operation can be improved and that people most closely associated with an operation are in the best position to identify changes that should be made (Krajewski and Ritzman, 1999).
However, in studies on implementation of advanced manufacturing technologies, it was observed that the most difficult aspects in implementation of new technology and systems were changing the organisation and people (Edge, 1990). So organisational change is fundamental to quality management, wherein improvement processes continuously bring changes in technologies and systems. As Dean and Evans (1994) stated "change is a way of life". They also stressed the importance of cultural change, *i.e.* the change in beliefs and values shared by people in the organisation.

Choosing between *top-down* change and *bottom-up* change, the latter is most times preferable in quality management. In top-down change, strategic changes are initiated which have a comprehensive impact on the organisation and its performance capabilities. However, the change that is driven from the top runs the risk of being perceived as insensitive to the needs of lower level personnel. It can easily fail if implementation suffers from excessive resistance and insufficient commitment to change. The success of top-down change is usually determined by the willingness of middle and lower level workers to actively support top management initiatives.
In bottom-up change the initiatives for change come from persons throughout the organisation and are supported by efforts of middle and lower level managers. Bottom-up change is essential to organisational innovation and is very useful in terms of adapting operations and technologies in order to change requirements of work. Empowerment, involvement and participation enable this bottom-up change.
An important condition for change is always a change-oriented leadership. Figure 6.8 shows the contrast between a change leader and a status quo manager. The former is forward looking, proactive and embraces new ideas; the latter is backward looking, reactive and comfortable with habit (Schermerhorn, 1999).

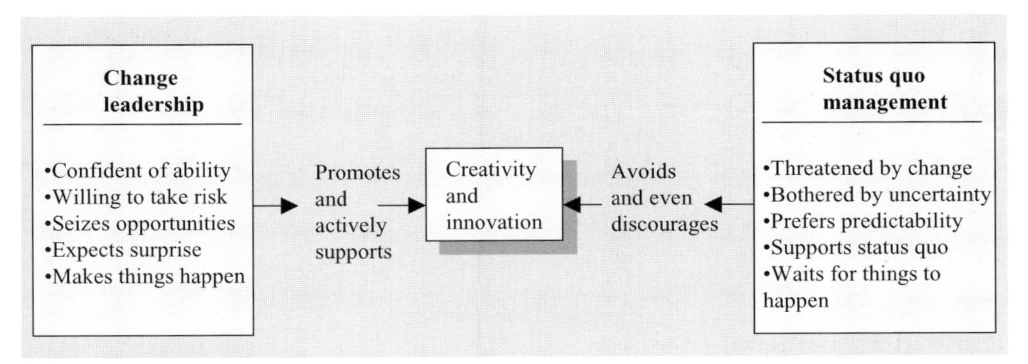

Figure 6.8. Change leadership versus status quo leadership.

6.5.2 Organisational change strategies

When change occurs, commonly there will be resistance against change. Some of the major reasons for resistance to change include uncertainty and insecurity, reaction against the way change is presented, threats to vested interests, cynicism and lack of trust, as well as lack of understanding. Resistance to change may also result of reaction against being controlled or getting less autonomy or power.
Dealing with change objectives and resistance to change requires a change strategy. Elements of change strategies are education, communication, participation, facilitation, negotiation and compulsion. Different strategies prefer different elements depending on the specific situation (Gatewood, 1995). Behavioural scientist Kurt Lewin (1952) provided one of the earliest models of change. Lewin viewed the change process as a modification of forces that keep a system's behaviour stable. His model described change as consisting of three phases wherein driving forces, which are pushers for change, compete with restraining forces, which strive to maintain the status quo. These three phases are (see Figure 6.9):

1. *Unfreezing* involves disrupting forces that maintain the existing state or level of behaviour (state A). This might be done by introducing new information to show discrepancies between state A and one desired by the organisation - that is, a performance gap.
2. *Moving* entails a transition period (state B) during which behaviours of the organisation or department are shifted to a new level (desired state C). It involves developing new behaviours, values and attitudes through changes in structure, technology, strategy and human processes.
3. *Re-freezing* stabilises the organisation at a new state of behavioural equilibrium (state C). This is accomplished through use of supporting mechanisms to reinforce the new organisational state, such as culture, norms, policies and structures.

Lewin's model thus suggests that managers should find ways to unfreeze existing equilibrium before any change will occur (Gatewood, 1995).

Change agents (person or group being responsible for change) use various approaches when trying to get others to adapt change. Figure 6.10 summarises three common change strategies, each of them acting differently on the unfreezing, moving and re-freezing phases (Schermerhorn, 1999).

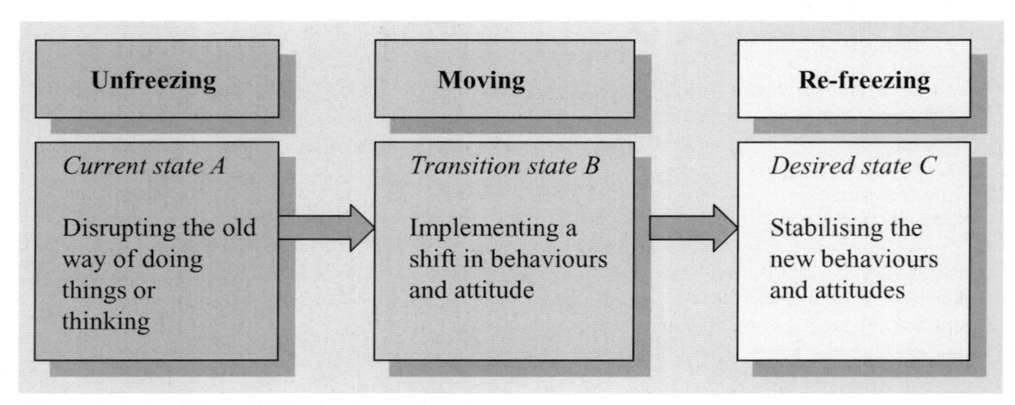

Figure 6.9. Lewin's model for implementing change.

Force-coercion strategies

A force-coercion strategy uses the power bases of legitimacy, rewards and punishments as primary inducements to change. As Figure 6.10 shows, likely outcomes of force-coercion are immediate compliance but little commitment. Force-coercion can be pursued in at least two ways, both of which can be commonly observed in organisations.

In a *direct forcing* strategy, the change agent takes direct and unilateral action to "command" that change takes place. This involves use of formal authority or legitimate power, offering special rewards, and/or threatening punishment.

In *political manoeuvring,* the change agent works indirectly to gain special advantage over other persons and thereby make them change. This involves bargaining, obtaining control of important resources, or granting small favours.

In both versions, the force-coercion strategy produces limited results. Although it can be implemented rather quickly, most people respond to this strategy out of fear of punishment or hope for a reward. This usually results in only temporary compliance with the change agent's desires. The new behaviour continues only as long as the opportunity for rewards and punishments is present. For this reason, force-coercion is most useful as an unfreezing device that helps people break old patterns of behaviour and gain initial motivation to try new ones.

Rational persuasion strategies

Change agents using a rational persuasion strategy attempt to bring about change through persuasion backed by special knowledge, empirical data, and rational arguments. The likely outcome is eventual compliance with reasonable commitment. This is an informational strategy

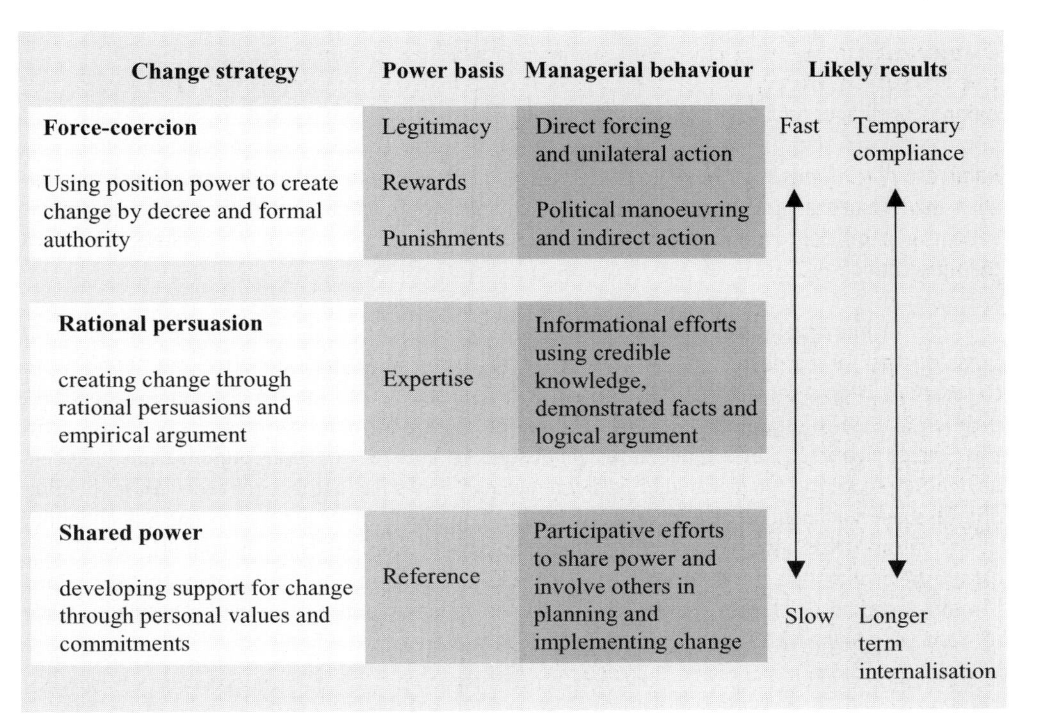

Figure 6.10. Alternative change strategies and their managerial implications.

that assumes that facts, reason and self-interest will guide rational people when deciding whether or not to support a change.
A manager using rational persuasion must convince others that the cost-benefit value of a planned change is high and that it will leave them better off than before. Accomplishing this depends to a large extent on the presence of expert power. This can come directly from the change agent if she or he has personal credibility as an "expert". If not, it can be obtained in the form of consultants and other outside experts or from credible demonstration projects. When successful, a rational persuasion strategy unfreezes and re-freezes a change situation. Although slower than force-coercion, it tends to result in longer lasting and internalised change.

Shared power strategies
A shared power strategy engages people in a collaborative process of identifying values, assumptions and goals, from which support for change will naturally emerge. The process is slow, but it is likely to yield commitment. Sometimes called a *normative-re-education strategy,* this approach is based on empowerment and is highly participative in nature. It relies on involving others, in examining personal needs and values, it relies on group norms, and operating goals. Power is shared by the change agent and other persons as they work together to develop a new consensus to support needed change.
Managers using shared power as an approach to planned change need reference power and skills to work effectively with other people in group situations. They must be comfortable allowing others to participate in making decisions that affect the planned change and the way it is implemented. Because it entails a high level of involvement, a normative-re-education strategy is often quite time-consuming, but it is likely to result in a longer lasting and internalised change.

Among consulting professionals, *organisation development*, or OD for short, is known as a shared power strategy that fits well with the concept of Total Quality Management and employee involvement. Organisation development (Figure 6.11) is supposed to help organisations cope with environmental and other pressures for change, while improving their internal problem-solving capabilities. In this sense, OD translates the need for continuous improvement to a planned change process.

In organisation development two goals are pursued simultaneously. The *outcome goals of OD* focus on task accomplishments, while the *process goals of OD* focus on the way people work together. It is this second goal that strongly differentiates OD from more general attempts at planned change in organisation. You may think of OD as a form of "planned change plus", with the "plus" meaning that change is accomplished in such a way that organisation members develop a capacity for continued self-renewal. OD tries to achieve change while helping organisation members become more active and self-reliant in their ability to continue changing in the future. What also makes OD unique is its commitment to strong humanistic values and established principles of behavioural science. OD is committed to improving organisations through freedom of choice, shared power and self-reliance, and by taking the best advantage of what we know about human behaviour in organisations (Schermerhorn, 1999).

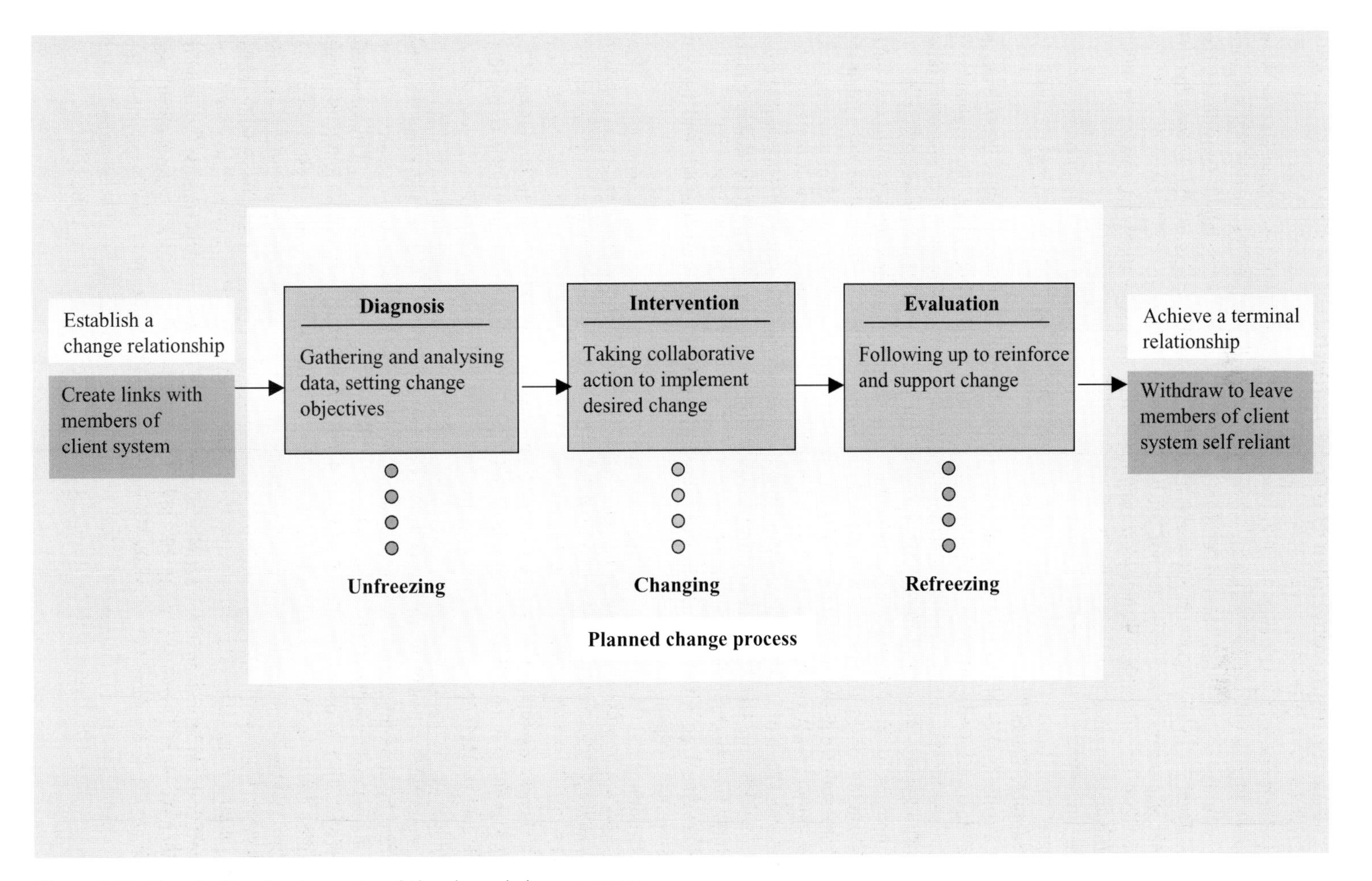

Figure 6.11. Organisation development and the planned change process.

6.6 Quality improvement in the food industry

Quality improvement is still not a success story in the food industry. Sometimes in food companies success stories are told about *e.g.* good functioning of Total Productive Maintenance (TPM), whereas on the other hand experiments with *e.g.* Quality Circles failed. Considering the typical situation in the food industry shows that the following aspects maybe reasons for the fact that improvement activities are not yet very common in food:

1. Low level of education, sometimes combined with language problems, makes it more difficult to involve operators or employees in problem solving activities.
2. The production circumstances are often hindering, like noise, bad smell, high humidity etc. These conditions are dissatisfying and are not motivating if they are not reduced to acceptable levels.
3. As a consequence of the functional organisation structure, most knowledge is centralised in the specialised departments. This implies that the responsibility for improvements is assessed by specialists coming from those departments. On the other hand, operators have much experience and knowledge, but are often not involved in improvement activities.
4. Equipment is often not very accessible for improvement. The knowledge is allocated with the supplier of the equipment, who also carries out modifications. The user of the equipment often cannot put any requests or needs and is restricted to the assortment of equipment as offered by (few) suppliers.
5. Poor feed back of information about quality performance results. Especially the lowest level of operators in the organisation are not well informed or only indirectly about quality results, which is not very stimulating.
6. Improvement activities are usually not rewarded. Especially acknowledgement and support of top management are missing.
7. Soft organisation methods are not very common in the food sector. In general, the culture is more solution-oriented and pro-active, whereas for discussion and analysis less time is spent.

From a techno-managerial perspective, quality improvement processes can be encouraged by focusing on those issues that are strategically relevant for the organisation from top to bottom, *i.e.* the core competences of the company. In the food industry, core competences mostly have a technical or technological nature. Therefore, the techno-managerial approach suggests to focus improvement processes on those core competences, which must at the same time be supported by appropriate management activities. Below some illustrations are given:

- A managerial condition for improvement is appropriate measurements and information systems. This information should include the critical quality and safety points that are related to the technological core competences. Furthermore, information must be easily accessible and applicable for statistical tools of analysis (like Pareto analysis, standard deviation etc), which enables not only the involved people but also the management to take appropriate decisions.
- Another essential management activity with respect to improvement is training. Training of operators and employees must be aimed at developing relevant technological and managerial knowledge for the specific quality control points they are responsible for. Moreover, training on simple statistical techniques is required to be able to organise and process data effectively, in order to use it as a basis for analysis and argumentation.

- Team building is another important management task but cannot be forced and requires good leadership capacities. Managers should not just install teams, but should facilitate team building in such a way that teams can grow spontaneously.

In the food industry, a restricted number of companies currently work with self-managing teams and use statistical tools to support improvement activities. However, to reach this situation, these companies consciously worked for years on organisational change processes. In fact, development of a culture where improvement processes are common requires much time.

More recently, initiatives came up to broaden improvement within the agri-food production chain. Following the notion that quality at the end of the chain strongly depends on quality performance in every part of the chain reveals that joint improvement activities should be organised. Figure 6.12 illustrates how external relationships include the possibility of joint improvement, based on customer requirements (Kehoe, 1995).
From the techno-managerial approach, it could be argued that a profound analysis of technological interdependencies on core quality attributes provides a relevant basis for discussing managerial collaboration and organisational relationships.

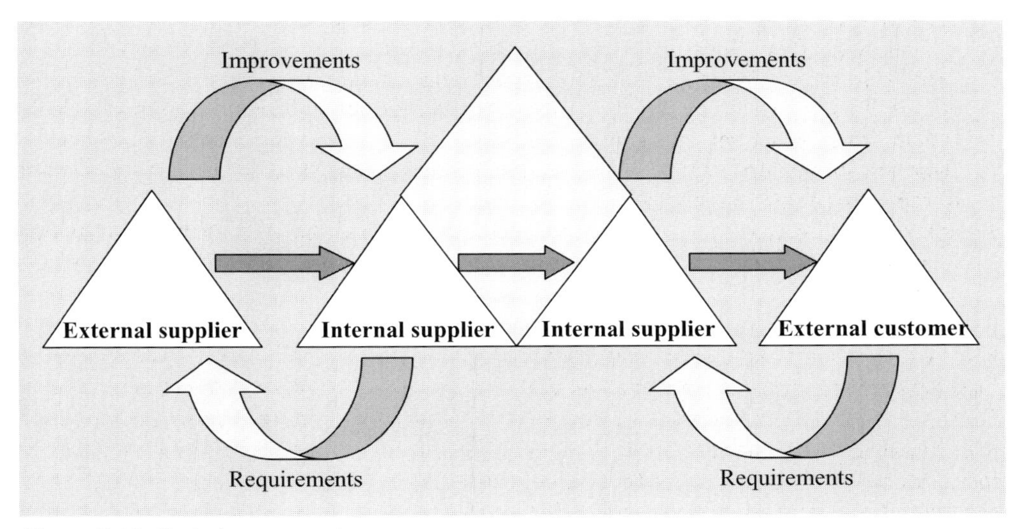

Figure 6.12. Chain improvement.

7. Quality assurance

The objective of quality assurance is to guarantee that quality requirements, such as product safety, reliability, service etc., are realised by the quality system. On the other hand quality assurance should provide confidence to customers and consumers that quality requirements will be met (ISO 1998). A quality system is defined as the organisational structure, responsibilities, processes, procedures and resources that facilitates the achievement of quality management (NNI, 1999). In the food industry several quality assurance systems (QA-system) and norms have been developed, which can be used as guidelines for development of a quality system to assure the established quality requirements. Common QA-systems in the food are the Good Practices (*e.g.* Good Manufacturing Practice (GMP), Good Hygienic Practice (GHP) etc), HACCP (Hazard Analysis Critical Control Points), ISO (International Standard Organisation) and combined systems such as BRC (British Retail Consortium). The QA-systems differ in their quality focus (*e.g.* food safety, supply guarantee, total quality) and their approach (Hoogland *et al.*, 1998; Waszink *et al.*, 1995). With respect to approach, GMP and HACCP mainly focus on assurance by technological requirements, whereas ISO is more focused on management. Figure 7.1 illustrates how the common QA-systems are mapped on their technological and management focus.

This chapter describes general principles of the common QA-systems used in the agri business and food industry. Moreover, managerial aspects involved in applying and managing quality systems have been described. The sections on common QA-systems include short history, legislative aspects and principles of respectively Good Manufacturing Practice codes (GMP), Hazard Analysis and Critical Control Points (HACCP), and the ISO series. In addition, specific QA-systems in the agri-food chain are described. The concept and application of total quality management (TQM) is described in Chapter 8, because it is more a policy than a specific system with concrete guidelines.
Quality assurance activities often lead to structural interference in the organisation and/or technology. The type of interference depends on the objective of quality assurance (*e.g.* to guarantee product safety, supply, service quality etc.). As a matter of fact, introduction and implementation of a quality system is not a simple procedure. It requires insight in relevant

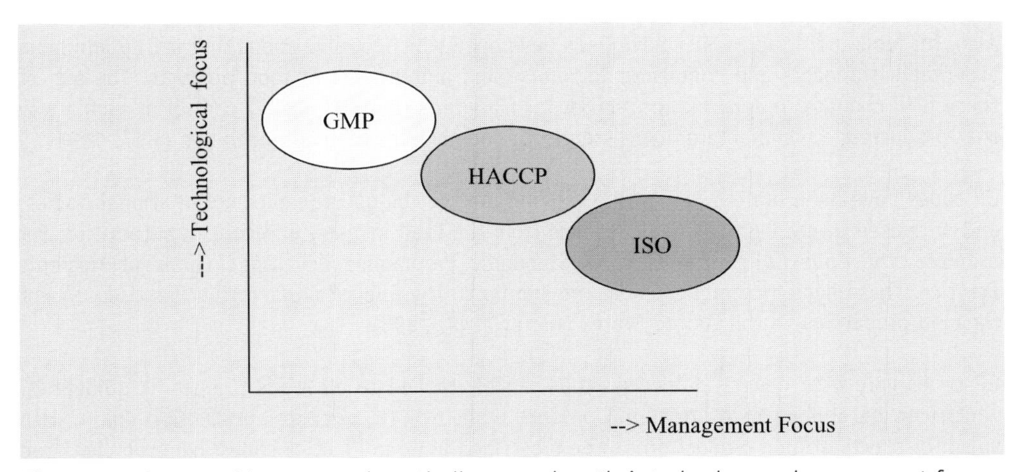

Figure 7.1. Common QA-systems schematically mapped on their technology and management focus.

managerial processes. Most of these processes have been described in detail in previous chapters, i.e. in Chapter 3 (in specific leading, organising and quality behaviour) and in Chapter 6 (improvement processes and organisation changes). In this chapter some attention is paid to the managerial aspects specific for the quality assurance process.
A topic closely related to QA-systems is auditing and certification. Although not yet a diagnostic instrument exists to measure actual performance of QA-systems, auditing is applied to assess the functioning of the QA-system. The principles of auditing and certification are shortly described. In the last section the typical situation in the food industry with respect to quality assurance is shortly described and considered from the techno-managerial perspective.

7.1 Good practices (GP)

Good practices (GP) involve guidelines that are aimed at assuring minimum acceptable standards and conditions for processing and storage of products. The GP's have no legal status but are often advised as basic condition for other systems. There are several types of GP's, that are globally described in the next section.

7.1.1 Short history and legislation

Development of good practice (GP) codes was initiated in the 60's in anticipation to a world-wide need for clarity about safe and hygienic manufacturing procedures. The most common GP code is Good Manufacturing Practice (GMP). For pharmaceuticals, GMP guidelines have been developed in anticipation to the resolution adopted by the World Health Organisation (WHO), concerning correct preparation and quality control of pharmaceuticals. These GMP codes include the following four topics:

- *Means*, which includes prescriptions on buildings, equipment and utilities
- *Materials*, which refers to requirements for raw materials and supporting materials
- *Methods*, which includes standard operating procedures (SOP) and work instructions
- *People*, which refers to knowledge and experience of employees (*e.g.* job description, training programmes)

Also, for food and drinks, GMP's have been developed which involve codes that set up minimum acceptable standards and conditions for processing and storage of food products. The aim of these GMP codes is to combine procedures for manufacturing and quality control in such a way that products are manufactured consistently to comply with their intended use (IFST, 1991).

GP codes have been drafted, and still are being drafted, by governments, international bodies, and organised interest groups like consortia from the food industry and control authorities. For example, the Codex Alimentarius Food Standards Programme developed Codes of Hygienic Practice. These hygienic practice codes are similar to the Good Manufacturing Practice concept for food production in the United States (Garret *et al*, 1998).

GP codes vary from general guidelines that can be applied to all foods to specific guidelines, which can be applied to a horizontal product branch or to a vertical production chain. Also specific aspects of food production can be prescribed in Good Practice codes. For example, Good Hygienic Practice (GHP) codes specifically focus on all hygienic aspects of food production. GP codes can also refer to other topics than food production, like Good Laboratory Practices (GLP), Good Clinical Practices (GCL) or Good Agricultural Practices (GAP). Moreover, GP codes can have a national or international recognition. Some examples are mentioned in Table 7.1.

Legislation

GP codes have no legal status but are rules of self-discipline. However, often elements of GP codes have been enacted in national regulations on Food hygiene. Moreover, when the GP code is a branch-code, it can be enforced (formal or informal) by the branch to implement the code. For example, the Dutch feed producers are not obligated to implement GMP codes for animal feed production. However, various quality programs and codes for production of meat, milk and eggs, stipulate that suppliers of animal feed must produce according to GMP. In this way, Dutch feed manufacturers are forced to implement the code (Hoogland *et al*, 1998).
Furthermore, GP codes must be applied in view of current European (and if necessary) national regulations, which means that legislative norms and requirements must be taken into account. Finally, Good Practices is not a static concept but an evolutionary mechanism by which overall improvements can be made and maintained (IFST, 1991; FDA, 1994).

7.1.2 Good Manufacturing Practice codes for food production

To our knowledge there is no acknowledged univocal guideline for the structure of GMP codes. However, some topics, commonly included in general GMP codes for food, are personnel, buildings, equipment and utensils, manufacturing process, storage and distribution. Depending on the comprehensiveness of the GMP's, additional topics are included like, recovery of

Table 7.1. Examples of good practices in the agri-food branch.

Title	Focus	Impact	Publishing organisation
Food and Drink GMP	General	UK	Institute of Food Science and Technology, 1991
GMP in manufacturing, packaging or holding human food (21 CFR 110)	General	USA	Food and Drug Administration, 1994
GMP for quality control procedures assuring nutrient formulas of infant formulas (21 CFR 106)	Product specific	USA	Food and Drug Administration, 1994
GMP for thermally processed low-acid foods in air-tight containers (21 CFR 113)	Product specific	USA	Food and Drug Administration, 1994
Draft code of Hygienic Practice for milk and milk products	Branch specific	International	Codex Alimentarius, 1998
Code of hygienic practices for dried milk	Branch specific	International	Codex Alimentarius, 1983
Code for manufacturing of vegetables and fruits	Branch specific	The Netherlands	Dutch Commodity Board for fruits and vegetables, 1988
GMP for the animal feeding branch	Branch specific	The Netherlands	Dutch Commodity Board for animal feed, 1998

materials, documentation, complaint and recall procedures, labelling, and/or infestation control (IFST, 1991; FDA, 1994). Within these topics the following aspects are usually included in the general GMP codes:

- Which *facilities and resources* should be provided, such as
  - appropriate qualified personnel
  - adequate premises and space
  - correct and adequately maintained equipment
  - specified raw materials, packaging materials and (operational) procedures
  - cleaning schedules
  - appropriate management and supervision
  - adequate technical, administrative and maintenance support
  - suitable storage and transport facilities
- *Hygienic aspects* of food production, which includes personnel hygiene, cleanability of equipment, utensils, and buildings, use of sanitising agents, equipment design, pest control, and all kinds of protection measures against microbial, physical and/or chemical contamination.
  For example, with respect to hygiene of personnel, measures must be taken for disease control and cleanliness (*e.g.* wearing special clothing, gloves and caps, washing hands, removing objects like rings, watches, no drinking and eating on the production floor etc). Moreover, personnel shall be educated and trained on all hygiene aspects.
  The hygienic aspects can be explicitly described in a separate chapter, or can be integrated in all parts of food production (IFST, 1991; FDA, 1994)
- *Requirements* for raw materials, ingredients, intermediate products and finished products like compliance to specifications, process and storage conditions (*e.g.* temperature, separate (storage) locations) and identification (by *e.g.* labelling).
- *Control and inspections* with established procedures for sampling and analyses, use of accurate and checked measuring equipment with experienced staff.
- *Keeping of records and documentation* for control and traceability aims.
- *Management topics* are often mentioned, like assessment of authorities and responsibilities for specific tasks.

GMP's for specific food groups contain additional aspects, typical for that product group. For example, for preparation of chilled foods additional requirements are set for preparation rooms. The rooms should be completely isolated, there must be no interchange of staff or equipment between areas, the air must be filtered, and the temperature must be kept at 8°C at normal pressure (IFST, 1991).
In general it can be said that GMP codes in the food production can function as a proper basis for application of other specific quality systems.

7.2 Hazard Analysis Critical Control Points (HACCP)

Hazard Analysis Critical Control Points (HACCP) is a systematic approach to the identification, evaluation and control of those steps in food manufacturing that are critical to product safety. It is an analytical tool that enables management to introduce and maintain a cost-effective, ongoing food safety programme. The basic objective of the HACCP concept is assuring production of safe food products by prevention instead of by quality inspection (Leaper, 1997; NACMCF, 1998).

In this chapter, different aspects of HACCP will be discussed including short history and legislation, the procedure and principles for development of an HACCP plan, and illustrations of a HACCP plan. Moreover, quantitative risk analysis as supporting tool for HACCP is explained in more detail.

7.2.1 Short history and legislation

The basis for the HACCP system originated from the need for safe food supply for manned space flights by the NASA (National Aeronautics and Space Administration) in 1959. They required food with a quality level as close to 100 % assurance as possible, because any illness or injury in space might end in an aborted or catastrophic mission. The quality control at that time could not give that high level of assurance. Therefore, a system was developed in collaboration with Pillsbury Company that participated in the Space program. A system that controlled raw materials, the process, the environment, personnel, storage and distribution, beginning as early in the system as possible. This way of control combined with record keeping had to ensure that the final packaged product did not require any other testing than for monitoring purposes.

Although started in 1959, the HACCP system was first formally presented to the general public in 1971. It was not until 1985 that the HACCP system was seriously considered for broad application in the food industry (Bouman, 1996). In 1993, the European Union (EU) officially recognised the HACCP methodology as a standard production method for food manufacturers to implement and maintain a production control system. The recognition was established in 'The hygiene of Foodstuffs Directive 93/43/EEC'. Since that time, Member States had 30 months to implement this Directive in their laws. As a consequence, all food-manufacturing facilities within the EU were forced to work in accordance to the HACCP methodology from the 1st of January 1996. In the Netherlands, the Directive is implemented in the Commodities Regulation Food Hygiene articles 30 and 31.

HACCP is basically designed for application in all parts of the agri-food production, ranging from growing, harvesting, processing, manufacturing, distribution and merchandising to preparing food for consumption (NACMCF, 1998). However, application of HACCP is yet only regulatory obliged for each company that prepares, and/or processes, and/or handles, and/or packages, and/or transports and/or trades food. So, besides the food industry, also small and medium enterprises, traditional manufacturing enterprises and hotel and catering industry are included. Nevertheless producers of fresh, unprepared animal and vegetable products (*e.g.* breeding, farming) are not yet compelled to implement a safety system according to the HACCP principles.

There are different ways for companies to comply with the legislation on food safety, i.e. application of branch hygiene codes or development of one's own HACCP plan (Inspectorate for Health protection, Food Commodities and Veterinary Affairs, 1995). Branch organisations are allowed to develop hygiene-codes based on the HACCP principles. Companies, which have processes similar to the ones described in the branch hygiene-code, can confine to application of the rules described in that code. If companies' processes deviate from branch hygiene-code, then the additional processes must be evaluated according to the HACCP principles. Branch organisations can also develop framework-codes, which are more global than the hygiene-codes. The framework-codes can be helpful in the development of a one's own HACCP plan.

7.2.2 Developing an HACCP plan

Several articles have been published, which describe the HACCP principles and procedures for development and implementation of an HACCP plan (ICMSF, 1989; MFSCNFPA, 1993a,b; European Commission, 1996; Codex Alimentarius, 1997; Early, 1997; Leaper, 1997; NACMCF, 1992, 1998).
An *HACCP plan* is the written document that is based on the principles of HACCP and that delineates procedures to be followed. The *HACCP system* is defined as result of implementation of the HACCP plan.

Officially the 'Hygiene of Foodstuffs Directive 93/43/EEC' describes five principles for ensuring food safety but most literature sources refer to seven principles containing two additional principles (Figure 7.2). These seven principles include:
1. *Hazard analysis*, i.e. potential hazards along the production chain must be identified and analysed.
2. *Critical control point (CCP) identification*, i.e. CCP's must be identified, which must be monitored to avoid or minimise occurrence of hazards.
3. *Establishment of critical limits* in order to control hazards at each CCP
4. *Monitoring procedures*, i.e. surveillance systems for regular monitoring or observation of critical control points
5. *Corrective action* must be established including measures which should be taken whenever an inadmissible deviation is recorded at critical control points
6. *Verification procedures* must established for verification of correct functioning of the HACCP system
7. *Record-keeping and documentation* relating to the HACCP plan must be developed for effective management

Several procedures, which facilitate the development and introduction of a HACCP plan, have been described. The Codex Alimentarius (1997) described a 12-stages procedure (Figure 7.2). Similar procedures have been proposed by Leaper (1997), Early (1997), and the National Advisory Committee on Microbiological Criteria for Foods (NACMCF, 1998). The latter proposed some additional requirements, before applying HACCP. These requirements and the 12-stages procedure are explained below.

A. Before applying HACCP

1. Prerequisite programmes

According to the revised document of NACMCF (1998), an HACCP system should be built on a solid foundation of prerequisite programmes. These programmes are often accomplished through application of *e.g.* Good Manufacturing Practice codes or Food Hygiene codes. These programmes should provide basic environmental and operating conditions that are necessary for production of safe, wholesome food. Prerequisite programmes include for example:
- *Facilities* including location, construction and maintenance according to sanitary principles. Product flow must be linear and traffic must be controlled to minimise cross contamination.

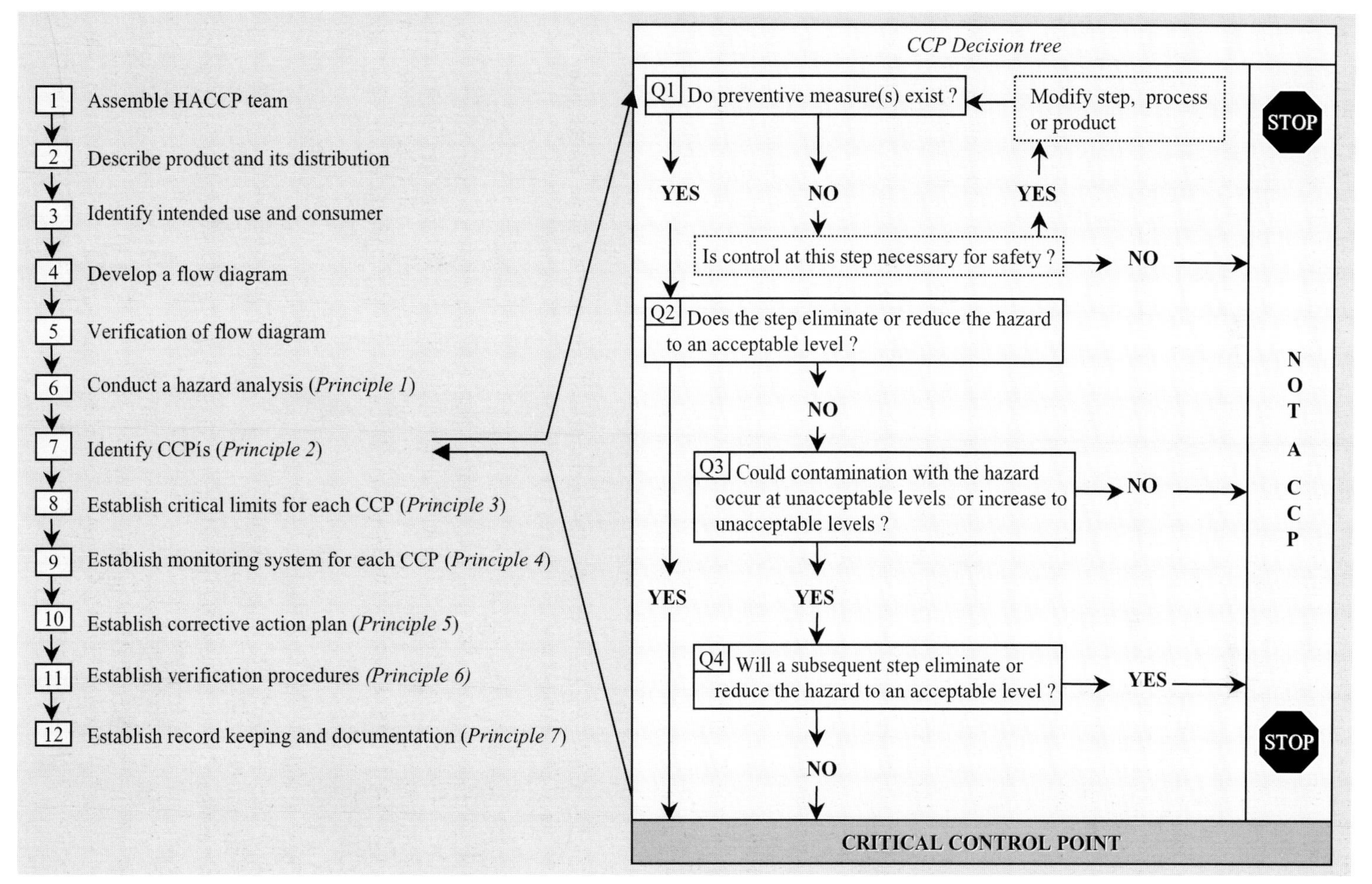

Figure 7.2. Procedure for the development of an HACCP plan.

- *Cleaning and sanitation* programme, which includes all written procedures and a master scheme for cleaning and sanitation of equipment and facilities; the programme must be operational i.e. followed by the personnel
- *Training,* all employees should receive documented training in personal hygiene, GMP, cleaning and sanitation procedures, personal safety, and their role in the HACCP programme.
- *Traceability* and *recall,* all raw materials and products must be lot-coded and a recall system must be present in order to perform rapid and complete traces and/or recall if required.
- *Pest control,* effective pest control programmes must be present.

Many of these regulations are already stated in (inter) national and/or branch regulations and guidelines. All these prerequisite programmes must be documented and audited on regular basis.

2. Training and education

The success of an HACCP system depends greatly on proper understanding of the HACCP principles by both management and employees. Therefore, education and training on the importance of HACCP, the role of people in producing safe foods, and how to control food borne hazards in all production stages is required (Barendsz, 1994; Codex Alimentarius, 1997; NACMCF, 1998).

3. Management commitment

Last but not least, for a successful implementation of an HACCP system, management must be committed to the HACCP approach and indicate awareness of the benefits and costs of HACCP (NACMCF, 1998). Furthermore, in order to perform the HACCP study, management must provide necessary team members for a number of periods and financial support. In addition, Barendsz (1994) mentioned that prior to implementing HACCP, quality policy and objectives, as part of the companies' policy, should be clearly established.

B. The 12-stages procedure for the development and introduction of an HACCP plan

The 12-stage procedure of HACCP includes the following steps (Figure 7.2):

1. Assembling of an HACCP team

An HACCP study requires multidisciplinary skills and all relevant departments involved in food production should be represented. This means individuals with specific knowledge and expertise appropriate to the product and process, but also people directly involved in daily activities, as they are familiar with variability and limitations of the process. The team should, at least, have the following constitution:

- A *quality assurance/quality control specialist,* who has knowledge on microbiological and/or chemical hazards and associated risks for the particular product group.
- A *production specialist,* who is responsible, or is closely involved with, the production process under study.
- An *engineer,* who has knowledge on hygienic design and engineering operation of process equipment under study
- *Other specialists* can be added to the team like buyers, operators, packaging experts, distribution experts, and a hygiene manager.
- *A member of the management* to ensure management commitment

In small and medium enterprises, often not all expertise is available and it is recommended to involve external support or information to ensure that the team meets all required skills.

2. Description of the product and its distribution
The team should make a full description of the product and its distribution. The description should include:
- Composition and physical features of final product (a_w, pH, gas tension)
- Process information (*e.g.* production methods used)
- Method of packaging
- Required shelf life
- Storage and distribution conditions along the chain (*e.g.* frozen, refrigerated, shelf-stable)
- Legislative product requirements
- Instructions for use and storage by consumers

3. Identification of intended use and consumers
To encompass any special considerations, the intended use of the product by consumers should be defined, *e.g.* is the product intended for special groups, which put higher demands on food safety like babies or people with reduced resistance? Likewise, intended use should be established when applying HACCP in animal feed production, *e.g.* young animals, sort of animal. In this stage also the potential for abuse by consumers should be considered. For example, which pathogens might multiply under excessively warm storage conditions inside the consumers' home or business?

4. Development of process flow diagrams
Prior to the actual hazard analysis it is necessary to examine the production process thoroughly. Therefore, a process flow diagram must be drafted, which provides a unambiguous, simple outline of all steps involved in the process, i.e. in words and without engineering drawings. There are no rules for presentation of flow diagrams, except that each process step and the sequence of steps should be clearly outlined. In the process diagram, sufficient technical data for the study must be provided. Typical data that can be included are:
- All raw materials/ingredients and packaging used (including relevant microbiological, physical and/or chemical data)
- Time/temperature history of all raw materials, intermediate and final products
- Process conditions like, flow rate, temperature, time, pH, gas tension etc.
- Storage and distribution conditions
- Product loops for recycling or rework
- Routes of potential cross-contamination
- High/low risk area segregation
- Overview of floors and layout of equipment
- Features of equipment design
- Efficacy of cleaning and disinfection procedures
- Personal hygiene practices
- Consumer-use instructions

The flow diagram should at least cover all steps in the process, which are directly under control of the company (Leaper, 1997).
Furthermore, to simplify communication with third parties (*e.g.* for inspections or audits) it is recommended to use international standards in a flow diagram. The international standards for abbreviations and symbols to draft flow diagrams are, in the Netherlands, described in NEN 1422, NEN 3283 and NPR 3592.

5. *On-site verification of flow diagram*
In case of an existing line, the HACCP team should inspect the operation process to verify that each step in the flow diagram is an accurate representation of the actual situation. Inspections of night- and weekend shifts should also be carried out. In case the analysis is being applied to a proposed line and verification will not be possible, the team must ensure that the flow diagram represents the most likely processing options.

6. *Conducting of a hazard analysis (Principle 1)*
In this stage the HACCP team must perform the hazard analysis. In practice, hazard analysis appeared to be one of the most difficult steps in the HACCP procedure because appropriate identification of potential hazards and assessment of their risk is rather complex and requires much technological knowledge and information. The result of this step is a list of significant hazards, which must be controlled in the process. By applying the CCP-decision tree, as discussed in the second principle, it can be assessed where these hazards must be controlled

Hazard analysis consists of hazard identification, hazard analysis (evaluation), and listing of relevant preventive measures.
In *hazard identification* a list must be composed, which contains all potential hazards that are reasonably likely to cause injury or illness if not properly controlled.
Within this scope, the word hazard is limited to safety. Hazards can be distinguished in biological (*i.e.* microorganisms and their metabolites), chemical (*e.g.* residues of agricultural chemicals, packaging materials, additive or environmental residue), and physical agents (*e.g.* pieces of glass, metal, wood etc.).
For conductance of the hazard identification, the HACCP team must review all potential hazards due, among other things to::

- *Raw materials*, *ingredients* and *semi-finished products* (e.g. residues of pesticides, initial content of micro-organisms)
- Where can *contamination* occur by equipment, personnel or environment?
- *Facility design*, e.g. is an adequate separation of raw and processed materials guaranteed?
- *Equipment design*, e.g. are appropriate time-temperature conditions obtained, can it be properly cleaned and sufficiently controlled?
- *Surviving* of process steps, e.g. heat-resistant toxins
- *Packaging*, e.g. is packaging resistant to damage, temper-evident, does it contain proper labelling and instructions?
- *Conditions* of the food which favour microbial growth, e.g. composition, pH, a_w
- *Deviations* in managing the process, e.g. process delays, technical trouble (NACMCF, 1998)

In *hazard analysis* (evaluation), the HACCP team must evaluate potential hazards, wherever possible, on:

- Severity of potential hazard, i.e. magnitude and duration of illness or injury
- Likely occurrence of hazard, which is usually based on a combination of experience, epidemiological data, and information in the technical literature
- Qualitative and/or quantitative evaluation of the presence of hazards
- Number of people potentially exposed to the hazard
- Age/vulnerability of those exposed
- Survival or multiplication of micro-organisms of concern
- Production or persistence in foods of toxins, chemical or physical agents

- Conditions leading to the above (SCF, 1997; NACMCF, 1998).

In other words, which hazards are of such a nature that their elimination or reduction is essential for production of a safe food?

This procedure provides a *qualitative* analysis of risks of identified hazards. However, in practice, there are often differences of opinions even among experts. Nevertheless, different assessors of risks should perform the process in a uniform way. Therefore, the Scientific Committee for Foods (SCF, 1996) and Codex Alimentarius (1998) proposed guidelines, with the essential elements of a (semi) *quantitative* risk assessment. This will be discussed in more detail in the next section.

Finally, the team must consider which *preventative* or *control measures* should be applied for each hazard. Control measures have been described as those actions and/or activities that are required to prevent or eliminate hazards or to reduce their occurrence to an acceptable level. It is possible that one or more control measures may be required to control a specific hazard that occurs at different stages of the production process (Leaper, 1997). For example, *Listeria monocytogenes* can be present in raw materials and ingredients but it can also arise from environmental contamination during ingredient assembly at the end of the production chain. In the first situation, the hazard can be often controlled by appropriate application of the heat treatment, whereas in the other situation no more heat (or other inactivation) treatment is applied to control the pathogen. In this situation other (additional) control measures are required like e.g. barrier hygiene or appropriate temperature control in the storage and distribution chain.

Control measures need to be underpinned by detailed specifications and procedures to ensure their effective implementation and performance (e.g. detailed cleaning schedules, ingredient specifications, personnel hygiene policy, barrier hygiene (Leaper, 1997).

Control measures not necessarily have a technological nature (e.g. measuring of physical parameters like temperature, pH etc) but also education and training to improve knowledge and experience can be a form of control measures (Janssen, 1991).

7. Determination of Critical Control Points (Principle 2)

A critical control point (CCP) is a step (i.e. point, procedure, operation, or stage in the food production system) at which control can be applied, and where control is essential to prevent or eliminate a food safety hazard or to reduce it to an acceptable level.

CCP's are unique for each process at each facility. There is no such thing as a standard CCP returning in every food production process. There is no limit on the number of CCP's that may be identified in the flow diagram.

CCP's are determined by applying the CCP decision tree, at each step of the process, for each hazard established in hazard analysis (Figure 7.2). The CCP decision tree is a sequence of questions to determine whether a step is a CCP or not. The decision tree exists of four questions (Q1-Q4) which all can be answered with yes or no. The answer leads to a subsequent question or marks the process step as CCP or No CCP (Stop).

8. Establishment of critical limits for each CCP (Principle 3)

Each CCP will have one or more preventive measures that must be controlled in order to assure prevention, elimination or reduction of hazards to an acceptable level. For each preventive measure, critical limits (target plus tolerances) must be established. Critical limits can be set by legal and/or other requirements, or can be based on information from hazard analysis or quantitative risk analysis.

9. Establishment of a monitoring system for each CCP (Principle 4)
Monitoring is the scheduled measurement or observation of a CCP relative to its critical limits. It is required to assess if the CCP is under control and to provide written documentation for verification. The monitoring system and procedure for each CCP:
- Must be able to detect loss of control at the CCP concerned
- Must contain information about how to adjust the process before a deviation occurs
- Includes the assignment of a person with relevant knowledge, who evaluates and signs the monitoring data
- If monitoring is not continuous then also the frequency should be mentioned.

For monitoring purposes usually physical and chemical tests and visual inspection are applied. Microbiological testing are seldom an effective means of monitoring because of the time required to obtain results (NACMCF, 1998).

10. Establishment of a corrective action plan (Principle 5)
If monitoring data reveals that the process has deviated from the critical limit, then a corrective action must be taken. Actions must ensure that the CCP has been brought under control. Corrective action includes:
- Determination and correction of the cause of non-compliance (non-conformance)
- Characterisation of the non-compliant product (non-conformance product)
- Recording of taken corrective actions

The corrective action plan must provide information about *which* actions should be taken when the process exceeds critical limits, and *who* is responsible for implementation and recording of corrective actions.

11. Verification of the HACCP plan (Principle 6)
Verification is defined as those activities, other than monitoring, that determine validity of the HACCP plan and that the system is working according to plan (NACMCF, 1998).
Verification can include the following topics:
- Validation of initial HACCP plan to determine if the plan has been properly developed (e.g. are all potential hazards identified?). Validation can be performed by (external) experts and/or scientific studies.
- Verification of actual execution of the HACCP plan in practice; is the plan correctly followed, are CCP's monitored, are they under control and are corrective actions recorded? The actual HACCP system can be verified by regular inspections, internal and external audits.
- Validation of process steps by sampling and testing of CCP's. For example, deliver applied process conditions required conditions, e.g. time and temperature conditions in the product? Microbial analyses can be conducted to validate food safety, or storage experiments can be applied to confirm product shelf life.
- Also, calibration of equipment to check functioning of measuring equipment is part of verification
- Checking of training and knowledge of personnel responsible for monitoring CCP's

In fact, all relevant records and documentation form basic input for verification of the HACCP system.

12. Establish record keeping and documentation (Principle 7)
Documentation and record keeping are essential for the HACCP system. The approved HACCP plan and HACCP procedures must be documented, whereas relevant data obtained during operation must be recorded. Examples of *documentation* are process flow diagrams, conductance of hazard and CCP analysis. *Record* examples include, information about used ingredients, processing data, specifications of packaging materials, temperature records of storage and distribution, deviation and proceeded corrective action records and employee training records.

7.2.3 Quantitative risk analysis

Currently, the internationally acknowledged HACCP system is used as a qualitative and subjective system, whereby the majority of decisions are based on qualitative instead of quantitative data. Several studies emphasised that a more quantitative approach to HACCP would improve the scientific basis for hazard analysis. It would result in a better selection of critical control points and their corresponding control measures (Buchanan, 1995; Notermans and Louve, 1995; Notermans and Mead, 1996; Gerwen et al, 1997; ICMSF, 1998; Northolt and Hoornstra, 1998; Mayes, 1998; Notermans et al, 1998; Van den Braak and Bloemen, 1999).
Moreover, international trade of foods is increasingly considered under the auspices of the SPS agreement (see also chapter 2.4). Continual strive for harmonisation of international guidelines and standards will stimulate assessment of equivalence of food safety protection systems in different countries. Currently, this is performed by qualitative evaluation but there will be an increased reliance on quantitative analysis. Guidelines for risk assessment and management have been developed by the Codex Alimentarius Commission (ICMSF, 1998).
Risk assessment is often considered as a separate process from HACCP. Nevertheless, risk assessment can assist hazard analysis to determine which hazards are of such nature that their elimination or reduction to acceptable levels is essential to the production of safe foods (Codex Alimentarius, 1996; ICMSF, 1998).

Quantitative risk assessment has been extensively used for determination of chemical risks in food, but for microbial risks there is yet restricted experience. As a matter of fact some characteristics, which determine microbial safety of food, are distinctively different from chemical properties, such as:
- Microbial risks are mostly due to a single *exposure* (except for microbial toxins), whereas chemical risks are affected by chronic duration of exposure.
- *Response* to infective pathogens is probably more variable, for different sub-populations depending on immune status, as compared to chemical agents.
- In contrast to chemical agents, micro-organisms are dynamic and can adapt to situations, to e.g. antibiotics or heat treatments. So, risks of specific micro-organisms may change over time (ICMSF, 1998).

However, chemical contaminants are usually not reduced or removed by any of the processing steps. Therefore, chemical risks must be controlled as early as possible in the agri-food chain (Northolt *et al*, 1998).

In order to obtain a uniform approach, the European Scientific Committee for Food (1996) proposed a framework for risk assessment of food borne microbiological hazards. The United States drafted similar principles and guidelines for the Codex Committee on Food Hygiene (Codex Alimentarius, 1996). The guidelines include the following information:

Risk assessment is part of risk analysis (Figure 7.3). Risk analysis is defined as a structured and multidisciplinary approach to identifying risks. It consists of the following elements

- *Risk assessment*, is the scientific evaluation of, known or potential, adverse health effects which result from human exposure to food borne hazards. The purpose of risk assessment is to provide a scientifically based *estimation of risk* of that hazard to any given population.
- *Risk management*, is the process whereby policy alternatives, to accept, minimise or reduce the assessed risk, are balanced and appropriate options are implemented. The aim of risk management is to identify acceptable risk levels and implement control options to comply with public health policy.
- *Risk communication* is an interactive process of information exchange and opinion on risk, among risk assessors, risk managers and other interested parties (SCF, 1996).

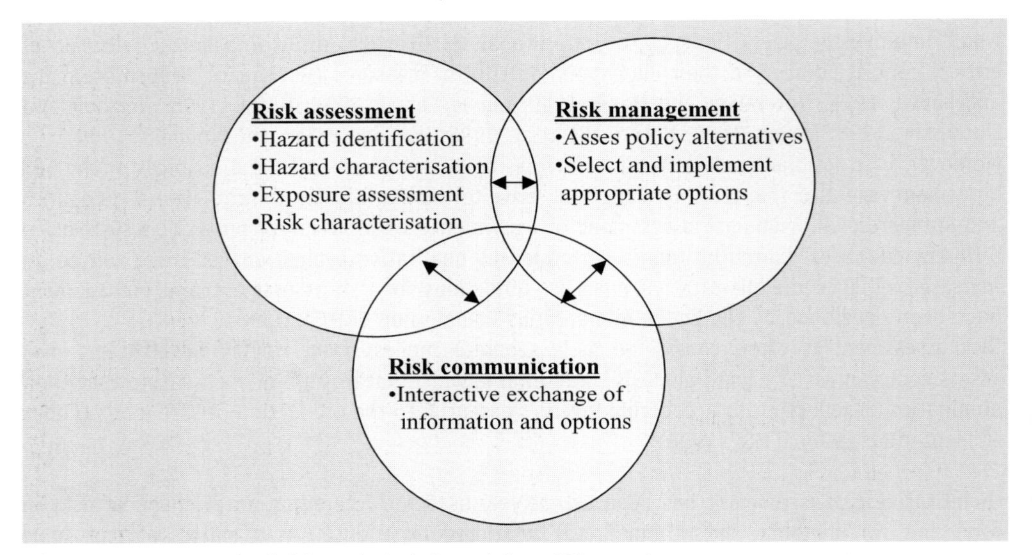

Figure 7.3. Framework of risk analysis (adapted from SCF, 1996).

Risk assessment as proposed by SCF (1996) and Codex Alimentarius (1998) consists of four parts, (1) hazard identification, (2) exposure assessment, (3) hazard characterisation and (4) risk characterisation (Figure 7.4), each will be explained in detail.

1. Hazard identification

Hazard identification is the first step of risk assessment and includes identification of, known or potential, health effects associated with a particular biological, chemical or physical agent. Hazards can be identified by a systematic evaluation of each step in food manufacturing (up to consumption), whereby information and data must be collected from relevant sources, such as scientific literature, databases from food industry and government. Typical data can originate from surveillance, clinical studies, laboratory animal, epidemiological and microbiological studies agencies (Notermans and Mead, 1996; Van den Braak and Bloemen, 1999).

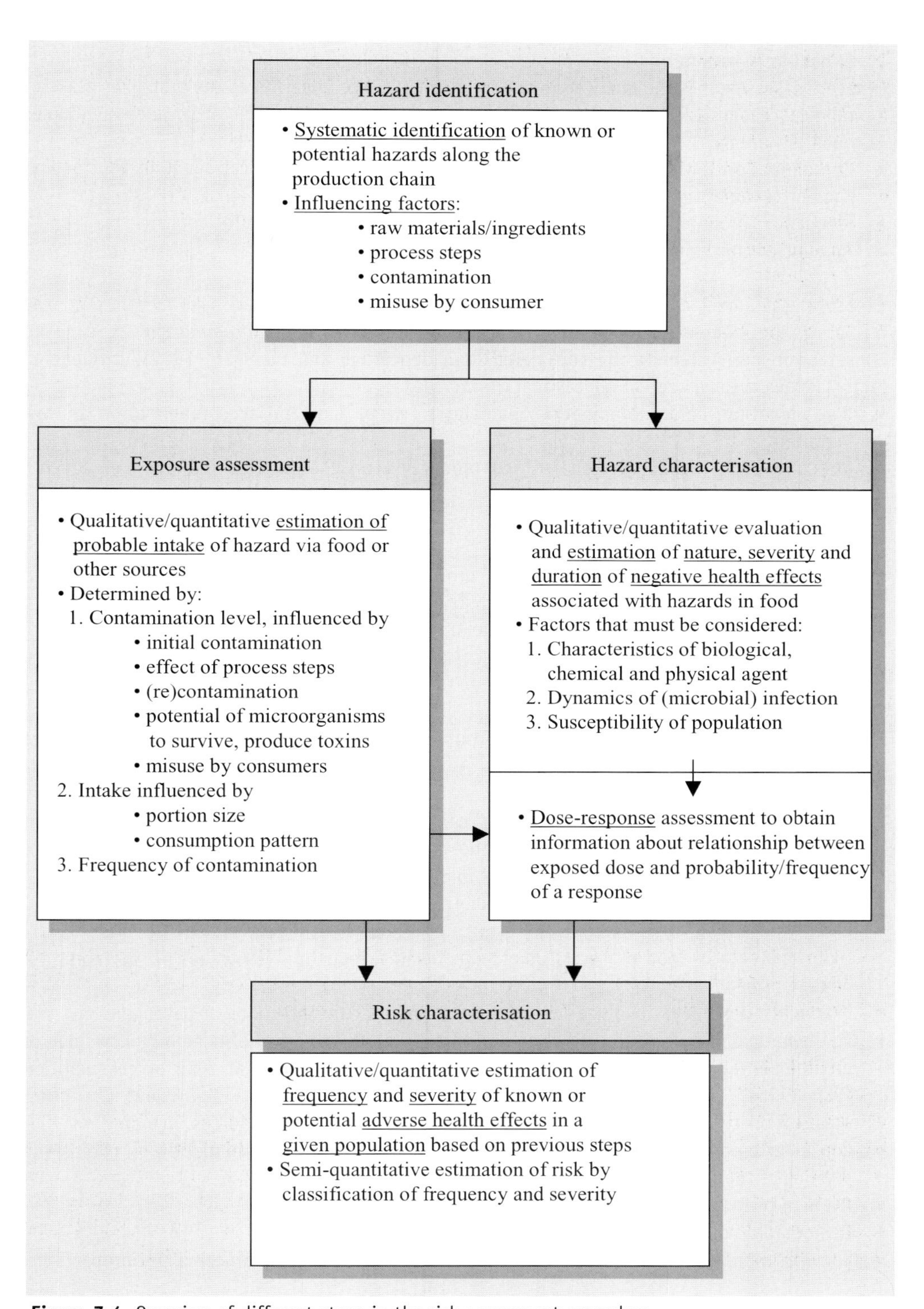

Figure 7.4. Overview of different steps in the risk assessment procedure.

2. Exposure assessment

Exposure assessment is the qualitative and/or quantitative estimation of the probable intake of a biological, chemical or physical agent via food as well as exposure from other relevant sources (e.g. via air, animals). Exposure to a hazard is determined by:

- Contamination level, i.e. amount of biological, chemical or physical agent in the food at time of consumption.
- Frequency (probability) that a food is contaminated by a pathogen of concern
- Amount of contaminated product that is consumed per time unit.

Exposure assessment can be performed by a step-wise procedure in which all factors that may influence the degree of microbial (or chemical or physical) *contamination* at time of consumption are systematically considered (Van den Braak and Bloemen, 1999). Influencing factors relevant for microbial contamination include:

- Contamination before elimination/inactivation step, e.g. initial contamination of raw materials
- Effect of each individual step in food production on the level of pathogen of concern (including final preparation by consumers)
- Possibilities for (re) contamination or cross contamination after elimination/inactivation treatment
- Potential of micro-organisms to survive, growth, and/or produce toxins, which is affected by product characteristics, storage and distribution conditions and packaging methods.
- Contamination due to misuse by consumers.

Also, the *frequency* (probability) of food contamination can be estimated by a step-wise procedure, considering all relevant aspects of food production (Van den Braak and Bloemen, 1999).

Amount of contaminated product that is consumed per time unit is depended on portion size and consumption patterns (SCF, 1996; Codex Alimentarius, 1996).

Several techniques and information sources can be used to obtain relevant information for exposure assessment. In order to assess the contamination level and likely occurrence of contamination the following techniques and information sources can be used:

- Data collection, e.g. on prevalence and distribution of microorganisms in food
- Challenge testing, to estimate likely growth of relevant pathogens in specific foods
- Historical data of food process studies, e.g. Decimal reduction (D) values of survivors to a heat treatment
- Examination of foods involved in outbreaks to assess concentrations of pathogens
- Mathematical modelling to predict growth, inactivation or survival of microorganisms due to environmental conditions, and likely number of microorganisms at time of consumption (SCF, 1996).

Information on consumption patterns may be obtained from several sources including:

- Consumer preferences and behaviour studies, which may provide information on choice and amount of food intake
- Average portion size as consumed per time
- Demographic data like age distribution, susceptible groups, etc.
- Socio-economic and cultural data, which may provide information on consumption patterns.

3. Hazard characterisation

Hazard characterisation is the qualitative and/or quantitative evaluation and estimation of the nature, severity and duration of adverse health effects associated with biological, chemical and physical agent that may be present in food.

In order to make an estimate, factors that can influence severity and duration of adverse health effects must be evaluated. A step-wise procedure to evaluate relevant factors has been proposed by Van den Braak and Bloemen (1999). According to the SCF proposal (1996) factors that should be considered include characteristics of hazard, dynamics of infection, and susceptibility of the population.

a. *Characteristics* of microbial, chemical and/or physical hazard:
   Typical microbial aspects include, amongst others,
   - Tolerance to environmental conditions can affect the survival chances of micro-organisms
   - Microbial replication characteristics, e.g. self replication, generation time
   - Virulence factors like, ability to circumvent host's immune response, the ability to attach to the surface, ability to synthesise various toxins.

   Typical chemical aspects include,
   - Mutagenicity of the agent
   - Carcinogenicity of the agent
   - Toxicity of the agent
   - Allergy
   - Intolerance
   - Decomposition rate of the chemical agent

   Typical physical aspects include size, sharpness, hardness, and decomposition rate of the physical agent.

b. *Dynamics of infection* is only relevant for biological hazards and includes
   Latency, i.e. delayed onset of the clinical infection after exposure by the hazard
   Disease pattern, e.g.
   - Acute, chronic, persistent or latent disease
   - Severity and duration of illness
   - Incubation period
   - Possible disease outcome, such as recovery, mortality, chronic sequelae
   - Persistence of the micro organisms in certain individuals resulting in continued excretion and risk of spread
   - Rate of infection

c. *Susceptibility of the population*

Extent of infection can differ markedly between different (sub)-populations. Infection can be more severe for groups with reduced resistance. Factors that can influence sensitivity of the (sub)-population involve:

Genetic factors that influence immune response

Immune status of (sub)-population e.g. pregnant women, children, and elderly can have a reduced immune status as compared to healthy adults.

Breakdown of physiological barriers can lead to increased susceptibility (e.g. use of antibiotics, concomitant infections).

Combining exposure assessment and hazard characterisation

The evaluation described above provides information about *severity* of response (i.e. the adverse health effect) but gives no information about the *frequency* (probability) that a (sub)-population is negatively affected. The probability that a (sub)-population is actually affected by the hazard can be determined by a dose-response assessment for the appropriate hazard.

Dose-response assessment is the process of obtaining quantitative information on probability of human illness following exposure to a hazard; it is the translation of exposure into harm (SCF, 1996).

For some hazards, *dose-response curves* have been determined. Dose-response curves show the relationship between exposed dose and probability of a response to a specific hazard (Figure 7.5).

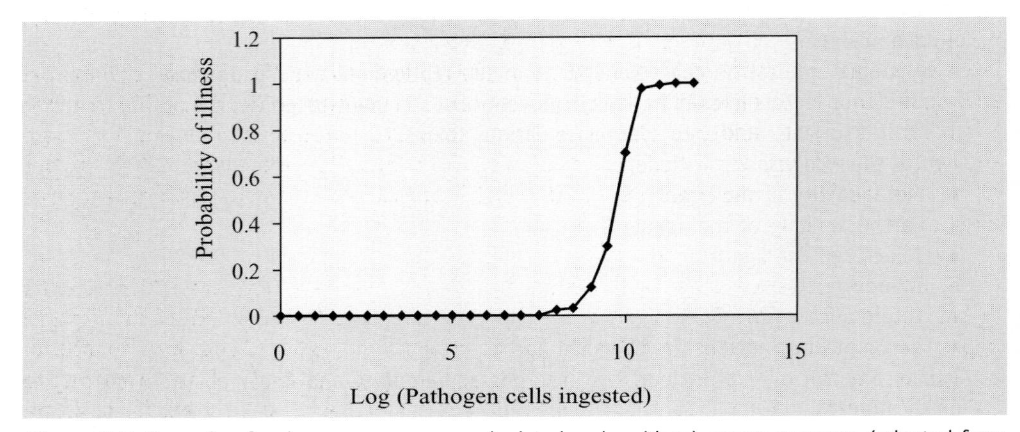

Figure 7.5. Example of a dose-response curve depicted as log (dose) versus response (adapted from Buchanan *et al*, 1997).

Ingested dose strongly influences frequency (probability) and severity of adverse effects. For example, increasing levels of a pathogen in food will generally result in a greater percentage of population becoming ill (increased probability). It will hasten the onset of symptoms and may increase severity of the disease of individuals (ICMSF, 1998).

Information about probability on a certain dose must be derived from exposure assessment (step 2 of risk assessment).

However, it should be emphasised that there is little information on dose-response relationships, especially for microbial hazards. Moreover, data is often obtained from studies with volunteers, which may not reflect the normal exposure to pathogenic organisms among the general population (Notermans and Mead, 1996).

Exposure assessment and hazard characterisation thus provide an estimate of severity of the negative health effect(s) and the probability that these adverse effects occur after exposure to the specific hazard

4. Risk characterisation

Risk characterisation is the last step in the risk assessment procedure. It is the quantitative and/or qualitative estimation, including uncertainties, of frequency and severity of known or potential adverse health effects in a given population based on the previous steps.
The degree of confidence in the final estimation of risk depends on variability, uncertainty, and assumptions identified in all previous steps. The influence of estimates and assumptions can be evaluated by using sensitivity and uncertainty analyses (SCF, 1996; Codex Alimentarius, 1996).

In practice, it is often not possible to make a quantitative estimation of the risk, and therefore a semi-quantitative estimation is can be applied based on the Failure Mode Effect Analysis principles (see also Chapter 4). In the FMEA procedure, severity (S), occurrence/frequency (O), and ability to detect (D) a potential failure (or hazard) are classified using a scale from 1-10. Subsequently, a risk priority number (RPN) is calculated, which is the product of severity (S), occurrence (O), and ability to detect (D), thus RPN=S x O x D.
Likewise, scales, to classify the *frequency/probability (P)* and *severity (S)* of known or potential adverse health effects, have been developed to calculate a risk number (RN=PxS) (ICMSF, 1989; Barendsz, 1997; Maas and Barendsz, 1997; Van den Braak and Bloemen, 1999). Also more simple classifications are used in practice as shown in Table 7.2. (Barendsz, 1997). In this classification table it is not clearly described what is considered as low or high frequency and great or small severity.

It should be emphasised that classifications into great, moderate, etc. are arbitrary. What includes a great severity? Does it mean that lethality is ≥ or ≤ to 0.1 or 1% of the exposed population? Likewise, what is considered as high frequency, is it <1 to 1000, or<1 to 100, or others? As a matter of fact, this is an ethical question about acceptability of risk. According to the ICMSF working group (1998), risk assessment data will serve as a scientific basis for discussions between governments and stakeholders to establish *food safety objectives*. Food safety objectives should specify tolerable levels of risk, which are quantifiable and measurable directly or indirectly. These food safety objectives have not yet been established.
For the current situation in practice, it is important that *within* the company (and preferably also for relevant suppliers and purchasers) there is consensus about the interpretation of categories.

The value of risk assessment for microbial risks has not yet been proven, because there is yet little practical experience (Mayes, 1998). Moreover, the general application of quantitative risk assessment is still restricted. Firs, because the methodology is not yet fully developed and,

Table 7.2. Classification of estimated severity and frequency of known or potential adverse health effects in a given population, using a scale of 1- 4 (low to high).

Severity of hazard	Frequency		
	Low	Moderate	High
Great	3	4	4
Moderate	2	3	4
Low/small	1	2	3

therefore, it is not yet internationally acknowledged. Secondly, reliable data on microbial dose-response relationships are scarce. In addition, there are still scientific questions unanswered like, what is the effect of pathogen distribution in food and type of food on the response? What are differences in response between different groups in the population (Notermans and Mead, 1996; Van den Braak and Bloemen, 1999)?

7.2.4 Illustration of an HACCP plan

The actual HACCP plan, is the written document that is based on principles of HACCP and that delineates the procedures to be followed (NACMCF, 1998). There are no strict guidelines for layout and composition of a quality handbook. However, the handbook must be clear, univocal and user-friendly. The following aspects are commonly described in the HACCP plan or quality handbook:

- *General aspects* include for example
  - Companies information
  - Quality policy
  - Organisational aspects, like organisation structure, description of tasks and responsibilities.
- *Diagrams of process flows* include the diagram of the overall process flow and detailed diagrams/charts of all processes. Some major standard symbols (according to NEN 3283 standard) are presented in Figure 7.6 and an example of a diagram of a process flow is presented in Figure 7.7.

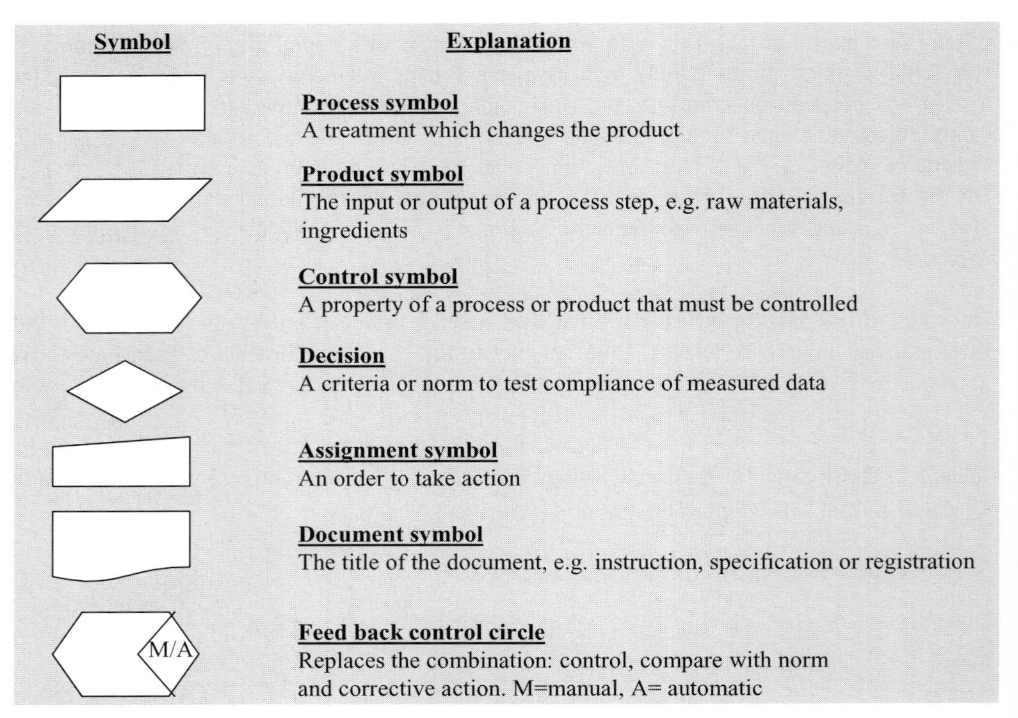

Figure 7.6. Symbols for process charts (according to NEN-3283 standards).

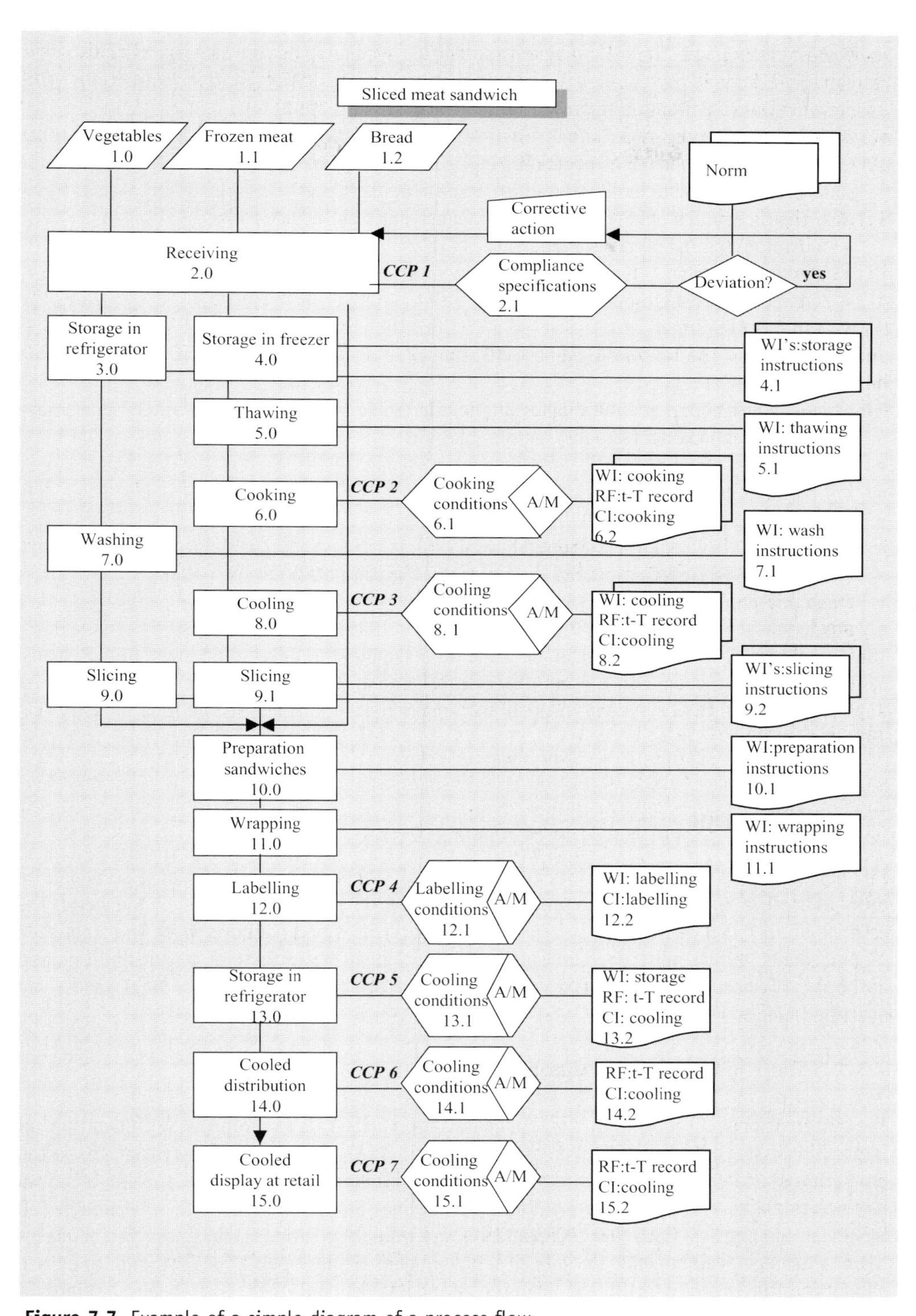

Figure 7.7. Example of a simple diagram of a process flow.

- *Process description*. The flow diagrams should be supported by process descriptions which can be described in detailed process/work instructions (WI), control instructions (CI), specifications (SP) and registration forms (RF) (Janssen, 1991).
- *HACCP plan summary table*. The results of the hazard analysis (and/or risk assessment), the CCP analysis, establishment of monitoring, corrective action plans and verifications can be summarised in a table as illustrated in Figure 7.8.
- Also the procedures of prerequisite programmes must be included like:
  - Cleaning and disinfecting plan
  - Pest control plan
  - Complaint service procedure
  - Product recall plan

Application of HACCP

For examples of elaborated HACCP plans one is referred to literature. Generic HACCP plans have been elaborated for product groups/sectors, like for:

- MAP and sous vide vacuum sealed??products (Snyder, 1993),
- Dry products (MFSCNFPA, 1993a),
- Chilled foods (MFSCNFPA, 1993b),
- Meat and poultry slaughters (Goodfellow, 1995),
- Meat and poultry production and distribution (Tompkin, 1995),
- Fresh and processed seafood (Garret et al, 1995)
- Ready-to-eat meals (Van den Braak and Bloemen, 1999)

These elaborated HACCP plans can be used as example for developing own HACCP plans. Moreover, the European Commission Directorate-general III provided a practical guide for implementation of HACCP system in small and medium-sized business in the food industry (EC, 1996).

7.3 ISO-series

ISO (International Organisation for Standardisation) standards are international standards in order to achieve uniformity and to prevent technical barriers to trade throughout the world. The essence of an ISO-based quality system is that all activities and handling must be established in procedures, which must be followed by ensuring clear assignment of responsibilities and authority. A short history, legislative aspects, and principles of the ISO 9000 family are described below.

7.3.1 Short history and legislation

The International Organisation for Standardisation (ISO) is a world-wide federation of international standard bodies, which was founded in 1947. The major objective of ISO is the promotion of standardisation and related activities throughout the world to prevent technical barriers to trade. The preparation of the International Standards is carried out through ISO technical committees (ISO/TC). ISO standards are voluntary, unless a business sector makes them a market requirement, or unless a government issues regulations making their use obligatory. The main part of the adopted standards is concerned with health, safety or environmental aspects.

Product description

(Refer to) verified diagram of process flow

Step	Hazard(s)	Control measure(s)	CCPs	Critical limit(s)	Monitoring procedure(s)	Corrective action(s)	Record(s)
No. in flow Chart	Specific description of hazard(s)	Description of control measure(s)	Number and description of CCP	Norm plus tolerances	How, frequency,	How	Refer to schedule for record keeping
Three examples referring to Figure 7.7							
2.0	*Bacterial growth of pathogens and spore formers specified for meat bread, vegetables* *Physical hazards : foreign objects in vegetables* *Chemical hazards : pesticide residues in vegetables*	*Analyses of product specifications, check compliance to supplier specifications* *Temperature control for chilled foods* *Check integrity packaging*	*CCP1 receiving control*	*Compliance to specified tolerances in specifications For example:* *Norm: $0 \leq T \leq 7°C$ for chilled foods No foreign bodies No defect packaging*	*Control and registration product specifications and temperature of chilled foods for each batch* *Visual inspection packaging, foreign bodies for each batch*	*Reject lot* *Reject supplier when it occurs more frequently*	*RF-code: receiving control SP-code: product specifications*
10.0	*Microbial contamination Foreign bodies (hair, dirt)*	*Hygiene instructions for personnel*					*GMP-codes: requirements: personal hygiene, cleaning and sanitation procedures, personal safety*
6.0	*Under cooking: Surviving and growth of spore formers in meat e.g. Clostridium botulinum and Bacillus cereus* *Toxin formation by Cl. Botulinum, E. coli 157:H7*	*Control t-T range during cooking*	*CC P cooking*	*E.g. norm: time ≥ 15 minutes and T ≥ 85°C*	*Time and temperature measurement and registration* *Continuous measurement for each batch*	*Extension of cooking time or product removal depending on t-T cooking conditions*	*RF-code: t-T records SP-code: cooking specifications*

Verification: description of verification procedures. For example, bacterial analysis of cooked food, control of HACCP plan is properly followed etc.

Figure 7.8. Summary table of a HACCP-plan.

Most used, and probably best known of all ISO standards is the ISO 9000 series for quality. In 1987, the Technical Committee TC/176 developed the 9000-series, which provided a framework for quality management and quality assurance. These standards were generic and independent on any specific industry or economic sector (Stanley, 1998). The 9000 series consisted of two major groups i.e. standards for *internal* and for *external* quality assurance.
Internal quality assurance is focused on improving efficiency as well as the quality of products and/or services within the organisation. For this purpose, the standard of the ISO 9004:1994 type was recommended. This standard provided guidelines for the development and implementation of quality management. It included typical elements of quality systems but also paid attention to economic aspects (i.e. benefits, costs and risks) of quality assurances (Hoogland *et al*, 1998).
External quality assurance is focused on assuring customers that the products or services meet the required specifications. For this purpose the three 'certification' standards ISO 9001, ISO 9002 and ISO 9003 could be applied. In these standards specific quality system requirements were described, which covered topics ranging from management responsibility to inspection.

In 1990, ISO/TC 176 adopted a two-phase revision process. The first phase allowed limited change to the standards and was completed in 1994. The second, more thorough revision has been completed in 2000. This more thorough revision was developed in anticipation to the needs of users and customers of ISO series. A large survey was carried out in 1997, which indicated that the revised standards:

- should have a common structure based on a process model
- should be simply to use, easy to understand, and use clear language and terminology
- should facilitate self-evaluation
- must be suitable for all sizes of organisations, operating in any economic or industrial sector
- for ISO 9001 should include demonstration of continuous improvement
- for ISO 9001 should address effectiveness while ISO 9004 should address both efficiency and effectiveness
- for ISO 9004 should help in achieving benefits for all interested parties, such as customers, owners, employees, suppliers and society
- should increase compatibility with the ISO 14000 series of Environmental Management System Standards (ISO, 1998).

The introduction of the new ISO 2000 series in food (and other) companies will take time, because it also aims at providing tools for development of an excellent organisation (like Total Quality Management). The development of an excellent organisation will take years. Since the ISO 9001 to ISO 9003 series are still in use and are valuable for some more years, their principles will be described in the next section.

7.3.2 Principles of original ISO 9000 series

The International Standard specified quality system requirements which could be used when a contract between two parties required the demonstration that the suppliers was capable to prevent non-conformance at all relevant stages of that supplier (e.g. ranging from design to service). The systematic approach enabled companies to operate in a consistently, measurable and efficient manner (Stanley, 1998). The original ISO 9000 family was based on the principle 'write down what you are doing', 'do what you described' and 'prove that you did what you described'. In practice, it included that for all relevant topics procedures had to be established, then carried out, and controlled.

The main difference between the three standards concerned the scope of the standards.
- ISO 9001 was the standard for quality assurance for companies involved in the processes design/development, production, installation and servicing;
- ISO 9002 was the standard for quality assurance in the processes production, installation and servicing;
- ISO 9003 was the standard for quality assurance in the processes final inspection and tests.

The original 9001 series included twenty topics, for which procedures had to be established. These topics were expected to be crucial for realising product conformance by the supplier. The topics are shortly described below, but for details is referred to the original standards (ISO, 1987, 1994).
1. *Management responsibility*, herein the supplier is prescribed that
   - a quality policy must be defined
   - the organisation including responsibilities, authority and interrelationships affecting quality must be defined
   - a management representative must be appointed which has authority and responsibility for ensuring that requirements of ISO are implemented and maintained.
2. *Quality system*, the supplier is required to establish and maintain a documented quality system to ensure, that products conform to specified requirements.
3. *Contract review*, the supplier must establish and maintain procedures for contract review and co-ordination of these activities. This is, amongst others, to ensure that requirements are adequately defined.
4. *Design control*, the supplier must establish and maintain procedures to control and verify the product design process in order to ensure that specified requirements are met. It included aspects like design and development planning, documentation of design input and output, and design verification to ensure that the design output meets input requirements.
5. *Document control*, herein suppliers are prescribed to establish and maintain procedures to control all documents and data related to requirements of the ISO standard.
6. *Purchasing*, the supplier must ensure that purchased products and materials conform to specified requirements. Sub-contractors must be selected based on their ability to meet requirements and purchasing data must be clearly documented.
7. *Purchaser supplied product*, the supplier must establish and maintain procedures for verification, storage and maintenance of purchaser supplied product, damaged products or other unsuitable products must be recorded and reported to the purchaser.
8. *Product identification and traceability*, the supplier must establish and maintain procedures for identification of products during all stages of production, delivery and installation.
9. *Process control*, the supplier must identify and plan the production (and installation process) which directly affect quality to ensure that these processes are carried out under controlled conditions.
10. *Inspection and testing*, includes receiving, in-process and final inspection and testing. Also here suppliers must establish and document procedures to ensure conformance to specifications.
11. *Inspection, measuring and test equipment*, the supplier must control, calibrate and maintain inspection, measuring and test equipment.

12. *Inspection and test status*, the supplier must ensure that the inspection and test status of product is identified by using markings, authorised stamps, tags, labels, routing cards, inspection records, test software, physical locations or other suitable means, to indicate the conformance or non-conformance of a product.
13. *Non-conformity review and disposition*, the supplier must define responsibilities for review and authority for disposition of non-conforming products. Non-conformance products may be reworked, accepted by concession, re-graded or rejected.
14. *Corrective action,* the supplier must establish and maintain procedures with respect to corrective actions. Including for example, procedures for investigating cause of non-conformance and corrective actions needed to prevent recurrence.
15. *Handling, storage packaging and delivery*, herein suppliers are prescribed to establish, document and maintain procedures for handling, storage packaging and delivery of product.
16. *Quality records*, the supplier must establish and maintain procedures for identification, collection, indexing, filing, storage, maintenance and disposition of quality records.
17. *Internal quality audits*, the supplier must carry out a comprehensive system of planned and documented internal quality audits to verify if quality activities comply with planned arrangements and determine effectiveness of the quality system.
18. *Training*, the supplier must establish and maintain procedures for identification of training needs and provide training for all those employees that are involved in quality related activities
19. *Servicing,* where it is specified in the contract the supplier must establish and maintain procedures for performing and verifying that service meets specified requirements
20. *Statistical techniques*, the supplier must establish and maintain procedures for identification of appropriate statistical techniques needed for verifying the acceptability of process capability and product characteristics.

Developing and maintaining procedures on all the relevant topics provided confidence to external customers that requirements could be met. Companies could be certified for one of the three ISO (9001-9003) series. Where these series were intended to provide external assurance, the ISO 9004 was aimed at providing internal assurance.

ISO 9004 provided guidelines for development of a quality management system. According to the ISO 9004 series, the primary objective of a company should be quality of its products and/or services. In order to meet these objectives, the company must organise itself in such a way that all technical, administrative and human factors that can influence quality are controlled. The two major aspects of a quality management system included the company's interest and needs and customers' needs and expectations. In addition, risk, cost and benefits of products and/or services for both company and customers must be considered. The ISO 9004 series described a basic set of elements by which quality management systems can be developed and implemented. For details is refered to the original guidelines (ISO 1987, 9004 series, 1994).
The twenty elements of the original ISO 9000 family are incorporated in the revised ISO 2000 version, but are enclosed in a clear process model.

7.3.3 Principles of revised ISO 9001:2000 and ISO 9004:2000

In 1999, the Technical Committee 176 of ISO has proposed a revision for the ISO 9000 family. Major changes and proposed structure of ISO 9001 and ISO 9004 are described below (ISO, 1998, 1999).

Major structural changes

The ISO 9000 family has been re-structured to four primary standards, i.e.

- *ISO 9000*, which includes an introduction to quality concepts and a revised vocabulary. The current vocabulary ISO 8402 will be replaced by the new ISO 9000.
- *ISO 9001*, which includes requirements for quality management systems. The current ISO 9001, 9002, and 9003 are consolidated into one single revised ISO 9000:2000 standard.
- *ISO 9004*, which provide guidelines for development of a comprehensive quality management system going in the direction of an excellent organisation.
- *ISO 10011*, which contains guidelines for auditing of quality systems.

Structure of revised ISO 9001:2000

The primary objectives of the quality management system requirements of ISO 9001:2000 are to achieve *customer satisfaction* by meeting customer requirements through application of the system, *continuous improvement* of the system and *prevention of nonconformity*.

The revised ISO 9001 is structured and organised according to the process model. A process is considered as any activity or operation, which receives inputs and converts it to outputs. Based on this basic concept the 'Quality Management Process' model has been developed by ISO/TC 176 (ISO, 1999), as shown in Figure 7.9. Characteristics of the model are:

- The four major topics of the quality management system are represented, i.e.:
    1. management responsibility
    2. resource management
    3. product and/or service realisation
    4. measurement, analysis and improvement
- The continuous improvement aspect and measuring of customer satisfaction are clearly expressed
- Both products and services are explicitly incorporated
- The model shows interaction between processes, and processes are closed loops. For example, the input for product and/or service realisation is customer-driven. Subsequently, the output is measured by consumer satisfaction measurements. This information is used as feedback to evaluate and validate whether customer requirements are achieved. The management, on its turn, shall ensure that customer requirements are fully understand and met.

The twenty elements of ISO 9001:1994 have been incorporated, with some additional elements, in the new process model. These elements are now clustered into four heading topics (Figure 7.10).

ISO 9001:2000 starts with *requirements* with respect to *quality management system*. The organisation must define and manage processes that are necessary to ensure that products and/or service conform to customer requirements. For implementation and demonstration of such processes, the organisation must develop, document and maintain a quality management system according to the International Standard

In the topic *management responsibility (1)*, requirements are set on commitment and responsibility of (top) management. Major aspects are that (top) management must:

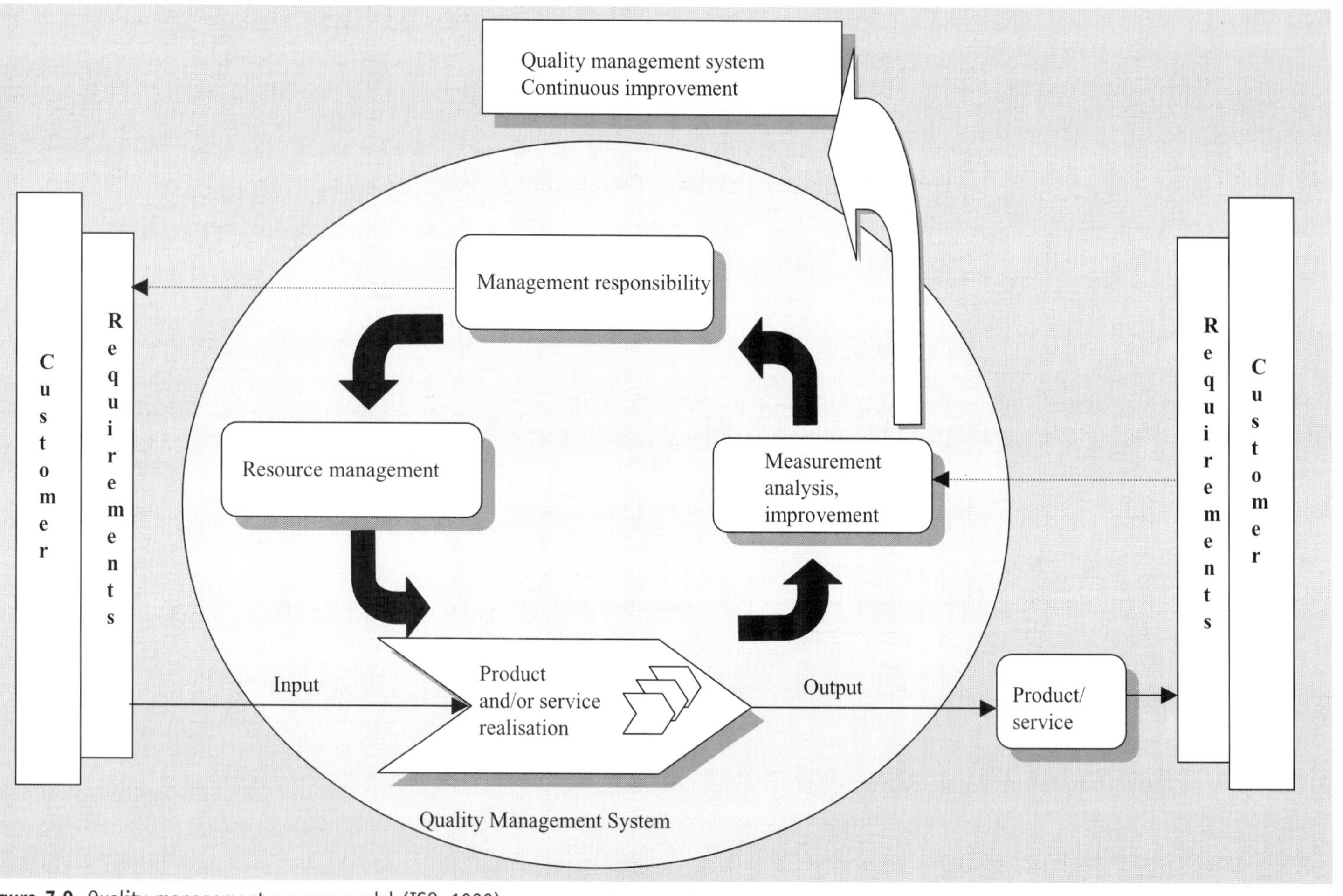

Figure 7.9. Quality management process model (ISO, 1999).

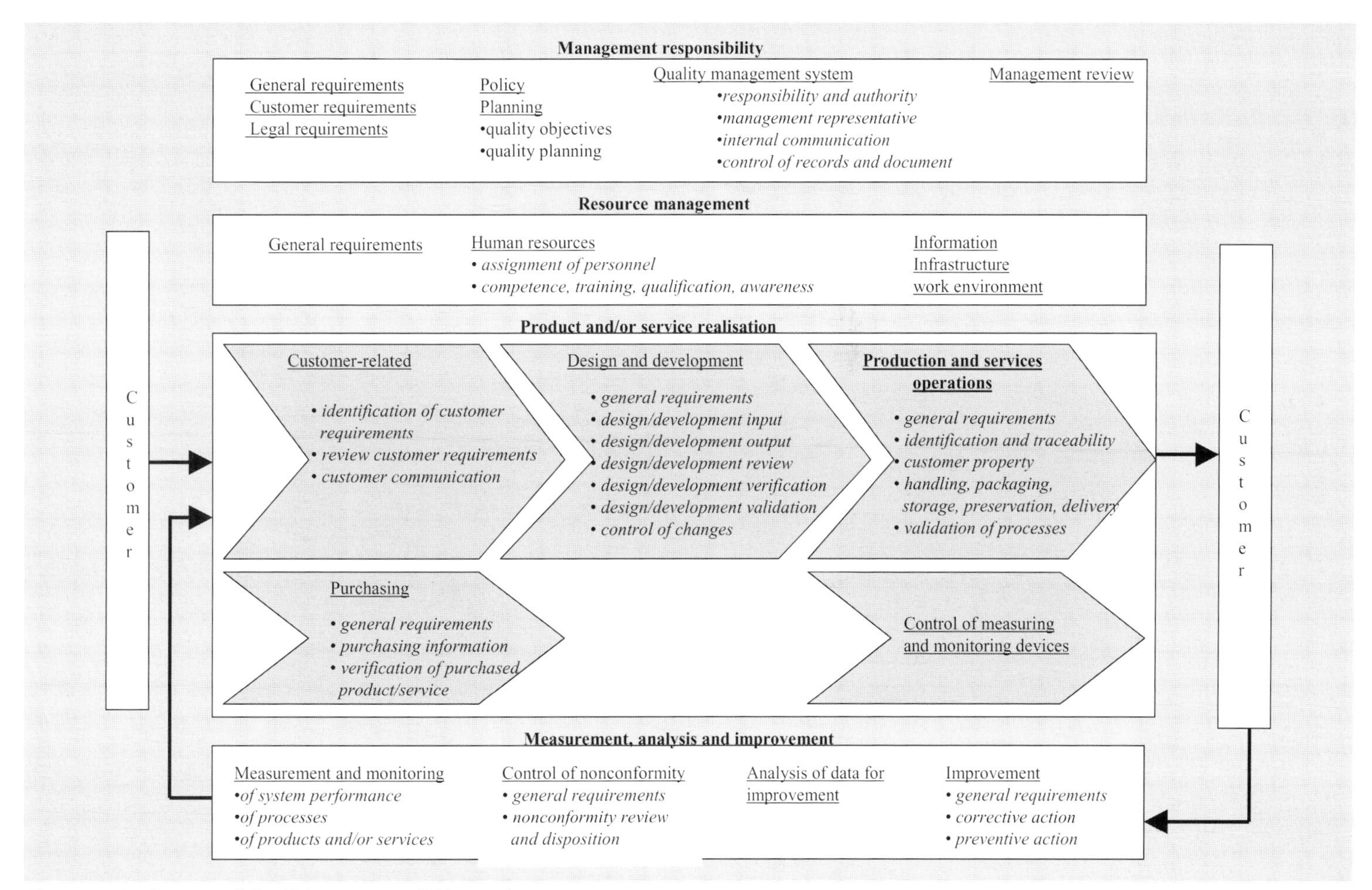

Figure 7.10. Elements of the ISO 9001:2000 (ISO, 1999).

    - *demonstrate commitment* with respect to realisation of customer demands by, amongst others, creating awareness about the importance to fulfil customer demands, by ensuring availability of resources, and by establishing quality policy, objectives and planning
- ensure that *customer needs are determined* and translated to requirements for the organisation
- establish an *appropriate quality policy* to ensure, amongst others, fulfilment of customer needs, commitment to continual improvement, and communication throughout the organisation
- establish *quality objectives* for each function and level in the organisation
- identify and plan activities and resources which are necessary to achieve quality objectives, i.e. *quality planning*
- develop *quality management system* including,
    - assignment of responsibilities and authority
    - appointment of management representatives to ensure implementation and maintenance of the quality management system and awareness of customer requirements
    - procedures for communication
    - preparation of the quality manual
    - development of procedures for controlling documents and records, on system level
- establish a procedure for *management review*, i.e. management must provide a status with modifications related to the quality management system, policy and objectives, which is based on audit reports, customer complaints analysis, status of quality policy and objectives

In the topic *resource management (2)*, requirements are set with respect to
- *human resources*, such as assignment of appropriate personnel, development of procedures to determine training needs and evaluate its effectiveness
- *information*, like procedures for managing information on e.g. process, product, suppliers etc
- *infrastructure*, like workspace, associated facilities, equipment, but also supporting services
- *work environment*, such as health and safety conditions, work methods and ethics

The topic *product and/or service realisation (3)* consists of general requirements and five subtopics.
The general requirements refer to determination, planning and implementation of realisation processes (e.g. criteria and methods to control processes, arrangements of measuring, monitoring and corrective actions, and maintenance of quality records).
The five subtopics are:
1. *Customer-related processes*, which set requirements on identification and review of customer requirements, and customer communication (such as customer complaints, product and/or service information).
2. *Design and development* sets requirements on planning and control of design and/or development of products and/or services. The plans must include stages of design/development processes, review, verification and validation activities and the responsibilities and authorities of personal involved.
3. *Purchasing* puts demands on purchasing-process to ensure that products and/or services comply with organisational requirements. The purchasing documents must contain appropriate information and purchased products and/or services must be verified. The organisation must ensure adequacy of these documents and verification activities.

4. *Product and service operations* refer to planning and control of actual operations. It sets general requirements, for example on, use of suitable production equipment, implementation of suitable monitoring and verification activities, and availability of clear work instructions. Moreover, it put demands on
   - identification and traceability (where and how to record)
   - handling, packaging, storage and preservation and delivery
   - validation of processes, i.e. to detect deficiencies, which may become apparent only after the product is in use or the service is delivered
5. *Control of measuring and monitoring devices* sets requirements on control, calibration and maintenance of measuring and monitoring devices.

In the topic *measurement, analysis and improvement (4)* general requirements refer to planning and implementation of these measuring, analysis and improvement processes to ensure that the quality management system, processes and products and/or services comply with requirements.
- *Measuring and monitoring* requirements refer to establishment of quality system performance (i.e. customer satisfaction, internal audits), measurement of processes, product and/or services.
- *Control of nonconformity* relates to identification, recording and reviewing of products and/or services, which do not conform to requirements.
- *Analysis of data for improvement* sets demands on data analysis for determination of effectiveness of the quality management system, and for identification of possible improvements.
- *Improvement* relates to establishment of a system procedure that describes use of quality policy, objectives, internal audit results, analysis of data, corrective and preventive action and management review to facilitate continual improvement.

The ISO 9001:2000 can to a certain extent be tailored to a company by omitting requirements that do not apply to that organisation (ISO, 1999).

Revised ISO-9004:2000

ISO 9004:2000 is the other part of the consistent pair of quality management systems, the other being ISO 9001:2000. The two standards were designed as a pair, but can be used as stand-alone documents as well.

ISO 9004 provides comprehensive guidelines for achieving an excellent organisation through continuous improvement of business performance. The standard is focused on improving internal processes of the organisation. Furthermore, the eight management principles have been integrated in this standard. These quality principles, which have been established by ISO for achieving Total Quality Management, include:
- *customer focused organisation*, i.e. focus the organisation on sustained customer satisfaction while giving benefits to other interested parties as well
- *leadership*, i.e. provide a visible personal leadership
- *involvement of people*, i.e. encourage employee efforts
- *process approach*, i.e. consider activities as processes with an input which is transferred to an output
- *system approach to management*, refers to consideration of the *total* system and not sub-systems
- *factual approach to decision making*, i.e. make decisions based on facts and not on feelings
- *continuous improvement*, relates to development of a culture for continuous improvement

- *mutually beneficial supplier relationships,* i.e. create a valued relationships with suppliers (NNI, 1999)

The structure of topics in the new ISO 9004 is similar to ISO 9001, which facilitates extension of the quality system to Total Quality Management. However, the scope of ISO 9004 is different. In ISO 9004 those elements that are necessary for improvement of performance of the organisation are extensively considered, while ISO 9001 provides quality management system requirements to demonstrate its capability to fulfil customer requirements. The revised ISO 9004 version is thus *not* a guideline to implement the revised ISO 9001:2000. For details of ISO 9004:2000 guidelines is referred to the document of ISO/TC 176 (1998).
A comparison of both standards is shown in Table 7.3 (ISO, 1998; NNI, 1999).

Table 7.3. Comparison of ISO 9001:2000 and ISO 9004:2000 (NNI, 1999).

ISO 9001:2000	ISO 9004:2000
Requirements	Guidelines
Focused on quality assurance (QA)	Focused on quality management (QM)
Minimum level	Directed towards excellent organisation
Focus on customers	Focus on all interested parties (customers, shareholders, employees, etc.)
Keywords: • provide confidence • comply with requirements • specify requirements • prove capability of organisation	Keywords • strive for advantages • continuous improvement of processes • permanent customer satisfaction • leads to Total Quality Management

Finally, organisations are not obliged to align the structure of their quality management system and/or documentation to the ISO structure. In fact, their quality management system should be configured in a manner that is appropriate to its unique activities (ISO, 1998).
In addition, the structure of the new ISO 9000 family aligns with the structure of ISO 14000 to facilitate compatibility with the series of Environmental Management System Standards (ISO, 1999).

7.4 (Inter) national quality systems specific for the agri-food sector

Considering the food production chain, a distinction can be made between manufactured products and unprocessed (bulk) products. Typical examples of the latter type of products are fresh fruits and vegetables, fresh milk, meat, fish and poultry, but also flowers and indoor plants. These products often haven no brand name, therefore, responsibility for product quality is not as clear as for brand products.
For manufactured (brand) products, producing companies take care of the quality system. Often they also consider previous (e.g. raw material production) and next steps (e.g. distribution, consumption) in the production chain. In fact, HACCP is obliged for each company that is

involved in preparation, processing, handling, packaging, and transport and/or trade of food. Although HACCP principles are also suitable for primary production, it is not yet widely applied.

In the Netherlands, Commodity Boards take care of development of specific quality systems for their sectors in primary production. Development of these specific quality systems has been stimulated by

- the need to improve image of specific sectors
- increasing critical attitude of consumers towards production methods and food safety
- increasing concern on environmental impact of food, flower and indoor plant production
- saturation of markets and raised competitive markets
- a number of food crises which had their origin in primary production

Although each sector has its own characteristics, primary production sectors have also some characteristics in common, i.e.

- often bulk production
- high number of small suppliers in different steps of the production chain
- each step in the production chain has its own methods of working
- frequent supply, e.g. for milk each day
- restricted view on needs and requirements of consumers because they are located at initial stages of production chains

Some major quality systems, which have been developed by Product and Commodity Boards in the Netherlands, are shortly described below.
Moreover, also in the retail sector initiatives were taken to ensure food safety and product quality. In England the British Retail Consortium (BRC) has been established, which combined elements of HACCP and ISO in BRC guidelines for retailers. Also this quality system is described in more detail in this section.

7.4.1 Quality systems in the animal production sector

Besides common characteristics of primary production, the animal production faces some typical quality affecting factors such as use of veterinary medicines for growth and animal health. Also, living conditions for cattle and poultry such as light, ventilation, accommodation and hygienic conditions are important quality factors, which should be controlled. Another typical aspect of animal production involves animal feeding production. Animal feeding is mainly composed of (undefined) waste streams of food industry, which complicates quality control of animal feeding production.

Good Manufacturing Practices for animal production
Also in animal production sector GMP codes have been developed. The FEFAC in Brussels published three GMP codes regarding, i.e. *Salmonella* control, compound and medicated feed, and undesirable substances and products and the negative list of raw materials. The GMP codes covered following topics:

- General production conditions and related requirements
- Purchase and receipt of raw materials, premixes, additives and veterinary medicines and their storage

- The manufacture with comprehensive indications for:
  - mixture of additives(?)
  - Manufacturing process with a view to avoiding contamination
  - Requirements for scales and measuring instruments
  - Requirements for mixers (maintenance, control, inspection)
  - Organisation of transport
  - Stability of additives and medicines
  - Handling of returns
- Storage and supply (Namur and Vanbelle, 1994).

Also in The Netherlands, GMP codes for primary production have developed such as GMP for animal feed production. The aim of this GMP is manufacturing of products (services) in such a way that suppliers of these products comply with legal requirements with respect to safety for humans, animals and environment (i.e. basic quality). Moreover, products must comply with requirements set by integrated quality assurance programmes for cattle, meat, egg and diary sector (Commodity Board for animal feeding, 1998). The GMP code is structured based on the ISO-9002. It provides requirements for development of a quality system for animal feed production, referring to:

- Management responsibility
- Quality system procedures and quality plan
- Norms and requirements
- Document and record-keeping
- Purchase
- Receiving control
- Identification and traceability
- Process control
- Inspection and control
- Control of measuring and inspection equipment
- Control of non-conformance products
- Evaluation and handling of non-conformance products
- Corrective and preventive measures
- Storage and packaging
- Control of quality data registration
- Quality audits (intern)
- Education and training
- Service
- Statistical techniques

Typical aspects for animal feed production that are covered in these quality system requirements are: addition of animal medicines, undesired compounds in raw materials, hygienic requirements for production, microbial criteria for raw material quality (e.g. Salmonella), requirements for mixing equipment and transport systems of animal feeding.

Integrated chain control

In the animal production sector several integrated chain control systems (In Dutch: "Integrale Keten Beheersing" IKB) have been developed, like IKB for slaughter poultry, IKB for eggs and

IKB for cattle (Van den Berg, 1993; Daniels, 1989; Product boards for livestock, meat and eggs, 1996). Integrated chain control has been defined as the entity of activities, which cover the whole production chain and tuning of units in the chain, in order to control quality-determining factors including those factors that affect labour conditions and environment (Boesten, 1995).

IKB systems aim at covering all steps in the production chain. Typical topics that are established in IKB systems are:
Communication: exchange of information between chains must be facilitated by appropriate registration of relevant data, such as results of chemical and bacterial analyses, age of animals, and vaccination schedules. These data from previous steps can be used to anticipate further in the chain.
Traceability: tracing and tracking of products must be organised, by adequate administration, to simplify removal of non-conformance products.
Veterinary medicines must be controlled and veterinary service must comply with Good veterinary practice (GVP)
Animal feeding may only be purchased from GMP-acknowledged companies
Specific hygienic measures with respect to plant design, personal hygiene, and transport must be taken to prevent microbial contamination and animal illness
Independent audits/inspections will be carried out to inspect performance of IKB
Sanctions will be executed if companies do not comply with IKB requirements

Application of an IKB system is not (yet) compulsory. However, the number of companies with IKB systems is raising because quality demands within production chains and by consumers are still increasing.

Chain quality of milk (KKM)
Chain quality of milk (In Dutch "keten kwaliteit melk", KKM) is another Dutch quality system in the animal production sector, specifically for milk production. The system is clear and simple, and farmers are supported and advised by the KKM-foundation. Chain quality of milk consists of six modules as shown in Figure 7.11. The modules cover all relevant topics for milk production and indicate control points and norms (Sol, 1998). There are several similarities between KKM and IKB-cattle especially for use of veterinary medicines, animal health and animal feeding.

7.4.2 Quality systems for vegetable and fruit production

Typical quality aspects of vegetable and fruit production relate to seed quality and illness resistance, besides product properties like colour, flavour, nutritional value and shelf life. Also in the sector of vegetable and fruit production the need of quality systems has been recognised.

Integrated quality assurance system (IKZ)
In this sector integrated quality assurance systems have been developed (In Dutch: "Integrale KwaliteitsZorg", IKZ). These systems are tailored for products e.g. IKZ-sprout, IKZ apple. The systems aim at covering all steps in the whole production chain, with consumers as point of departure (Boesten, 1995). These consumer needs must be translated to requirements in the different units of the production chain such as, sowing of varieties, cultivation, harvesting,

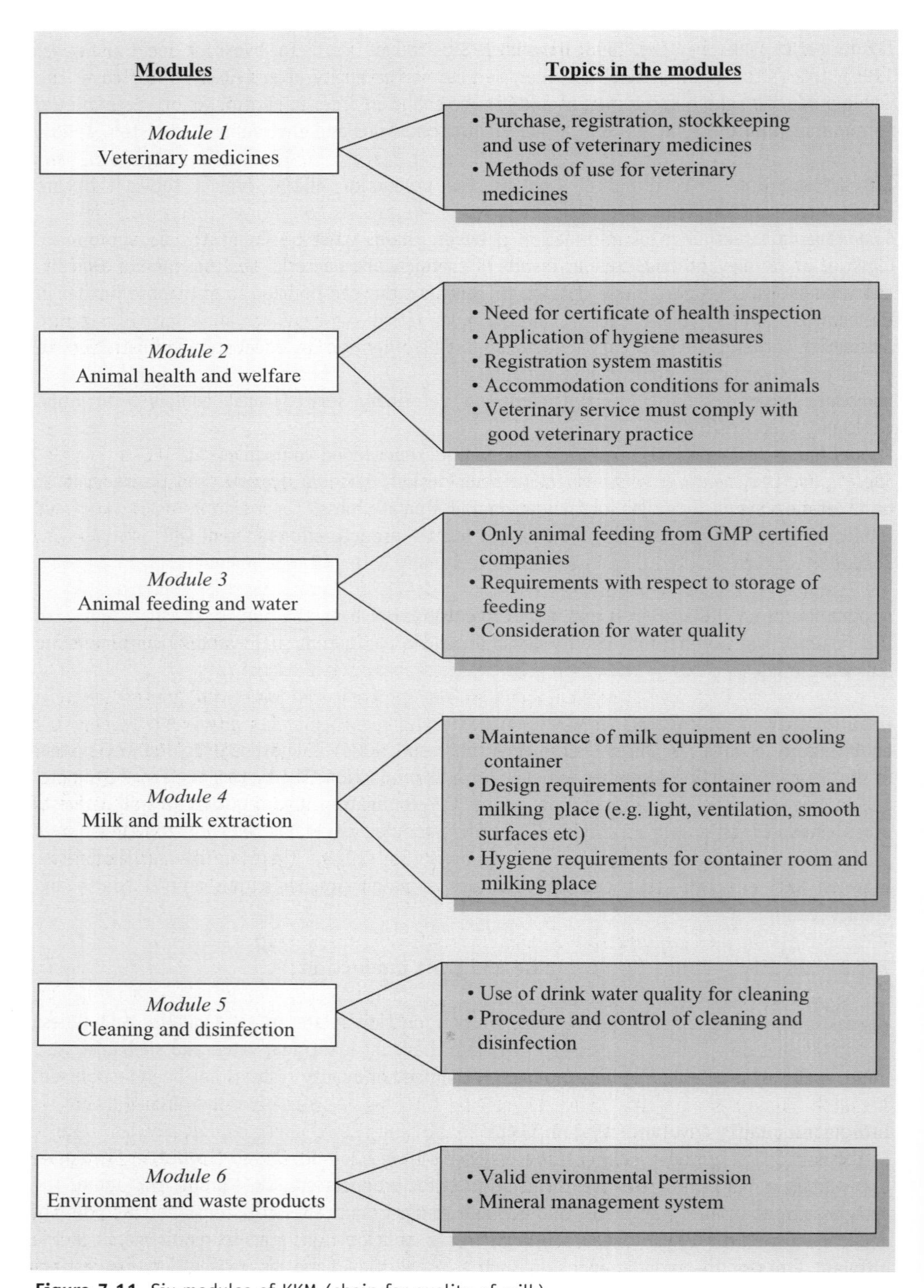

Figure 7.11. Six modules of KKM (chain for quality of milk).

handling and packaging. Communication between units in the chain is an elementary aspect of IKZ. Furthermore, existing methods to assure quality are being integrated in IKZ systems, like
Seed quality control, i.e. prior to marketing seeds varieties must be tested on purity and germinative capacity
Practical value studies, i.e. varieties are extensively tested on their illness resistance and product properties
Quality attributes inspection by auctioneers and, randomly, by other inspection authorities.
Residue analyses are carried out in greenhouses to check their compliance with legal norms.
Established storage and distribution *temperatures* for fruits and vegetables to improve shelf life
Labelling of packed fruits and vegetables to provide information about shelf life
Also environmentally sound production is an important aspect of quality in vegetable and fruit production chains. Several projects have been initiated to stimulate sound production (Boesten, 1995).

Eurep-GAP
Due to several problems and crisis in the agri food production chain (like BSE, dioxin but also use of genetic crops) new initiatives have been taken to assure quality of primary products. In 1998 retailers started with the initiative Eurep-GAP, i.e. European Retailer Working Group - Good Agricultural Practice. Eurep-GAP puts demands on the safe production of vegetable produce. The following aspects are described in this system:

- a minimum demand is the assurance of safe production
- growers are stimulated to reduce use of pesticides
- growers must use an Integrated Crop Management (ICM), whereby land use and production must be balanced while considering environmental impact
- growers must take care of natural resources, environment, employees and human health
- a tracing system must be implemented which enables distribution channels to trace back origin of vegetable products and fruit (Fresh Produce Traceability Project, FPTP).

To obtain an Eurep-GAP certificate a grower must comply with 150 control points and 100 other action points on the list. The certificate must be continued yearly. The ultimate objective of Eurep-GAP is to grow to a world-wide acknowledged standard, therefore it will comply as much as possible with other initiatives like the Global Food Safety Initiative (GFSI) (Grootenhuis, 2001). In addition, the system will be extended for animal produce as well.

7.4.3 Quality systems for flower and indoor plant production

The emphasis in this sector is mainly focused on environmental impact of production. One specific system developed in the Netherlands is shortly described.

Environmental project ornamental plant cultivation MPS
A Dutch quality system has been developed for this sector, i.e. environmental project ornamental plant cultivation (MPS). MPS is basically a registration system but aims at reducing use of pesticides, fertiliser and energy in ornamental plant production. Companies are classified depending on extent of use of pesticides, fertilisers and energy.
In conclusion, whereas for in the food manufacturing industry a few internationally acknowledged quality systems are applied, in the primary sector a wide variety of specific systems have been developed, which anticipate to specific quality determining factors for each

sector. However, the HACCP approach is receiving increasing attention in the primary sector (Noordhuizen, 1996, Noordhuizen and Welpelo, 1998).

7.4.4 Quality system for the retail

British Retail Consortium (BRC)
The British Retail Consortium (BRC) or the International Technical Standard for Food Suppliers (ITS Food) is in principle a technical standard for companies supplying private label food products. In the nineties, the BRC was founded as a reaction to the increased number of discount retailers (using private labels) in combination with poor economic perspectives in the UK. Since profit margins were reduced there was less money for quality management activities. Therefore, the BRC

- developed clear criteria for the assessment of private label suppliers
- contracted out auditing of these suppliers to a third party inspection
- so aimed at cost reduction for retailers as well as suppliers

A major advantage of the BRC guidelines is that there is more clarity for suppliers of private labels, there is only one inspection standard and a supplier can use that standard for all retailers and so reduce auditing costs (Smit, 2000, Kranghand, 2001).

BRC is a checklist that combines the HACCP principles with specific parts of GMP (e.g. pest control and facility lay-out) and parts of ISO (system control). The BRC inspection, to obtain a certificate, is very comprehensive and as a consequence less attention is paid to system aspects as for a HACCP or ISO certificate. In fact, the BRC-certificate shows an awareness of responsibility of manufacturers for food safety and evaluates the reliability of potential suppliers (Damman, 1999, 2001; Loode, 2000, Smit, 1999, 2000, Grootenhuis, 2001). Table 7.4 gives an overview of differences between HACCP and the BRC standard (Smit, 1999).

The BRC standard must be evaluated and certified by the European standard EN 45004. The BRC inspection is a relatively quick scan (1/2 a day), whereas for HACCP certification a profound evaluation of the HACCP system is executed (3 days). BRC has a broader scope whereas HACCP puts much attention on a profound analysis of safety hazards. Although, HACCP and BRC complement each other well, it is expected that in the future within the BRC system more attention will be focused on performance of the HACCP system (Damman, 2001).
Also BRC has the ultimate objective to become an international acknowledged system. Representatives of retailers from different countries in America and Europe are working on the International Technical Standard for Food suppliers (ITS Food), which is based on the BRC standard .

Global Food Safety Initiative (GFSI)
Last but not least we would like to mention the Global Food Safety Initiative (GFSI). The task force has been initiated as a reaction to the loss of consumers trust in food production. The task force consists of 38 quality managers from corresponding retailers, whereby all continents are represented. They have defined four priorities including

1. Criteria to evaluate available and new standards, i.e. the key elements
2. Development of an international Early Warning System
3. Communication to all involved parties including consumers
4. Development of partnerships between government and food industry.

Table 7.4. Differences between HACCP and BRC.

HACCP	BRC
HACCP system requirements	1. HACCP system
1. Responsibility of management with respect to food safety	2. Quality management system 2.1 Quality policy statement 2.2 Quality manual [...] 2.13 Supplier performance monitoring
2. Product information 2.1 Characteristics of product 2.2 Characteristics of user	3. Factory environment 3.1 Location 3.2 Perimeter and grounds 3.3 Lay-out and product flow [...] 3.12 Transport
3. Process information 3.1 Flow charts 3.2 Lay-out 3.3 Control and verification process information	
4. Hazards, risks and control measures 4.1 Hazard identification 4.2 Risk analysis 4.3 Control measures	4. Product control 4.1 Product design and development 4.2 Product packaging 4.3 Product analysis [...] 4.8 Control of non-conforming product
5. Critical control points 5.1 Evaluation of CCP	5. Process control 5.1 Temperature-time control [...] 5.5 Specific handling requirements

With respect to the evaluation of current standards the following key elements have been identified, quality management systems, good practices (both agricultural, technological and distribution practices) and HACCP. The latter is in fact part of a quality management system but it is considered as so important for food safety that is it is mentioned separately. Further exploration of the key elements is under progress and is expected to be finished by the end of 2002.

The Task Force is developing two additional documents one including a protocol for audits and another including guidelines with respect to certification institutes based upon ISO/IEC 65. The certification institutes will be under supervision of accredited authorities (Byrnes, 2001)

7.5 Managing quality assurance

Unlike computers the human-being is neither well suited to exact repetition of methods nor to application of consistent standards when dealing with measurement or other criteria (Smith and

Edge, 1990). It is therefore necessary to provide written standards and procedures in order to work a quality system effectively. Since quality standards and procedures are for day-to-day use they must be concise and readable. A typical and effective approach is to develop a hierarchy of procedures so that general principles and rules are described in a top level document (quality manual), which refers to subsequent levels of procedures and work instructions that provide the detail (Figure 7.12).

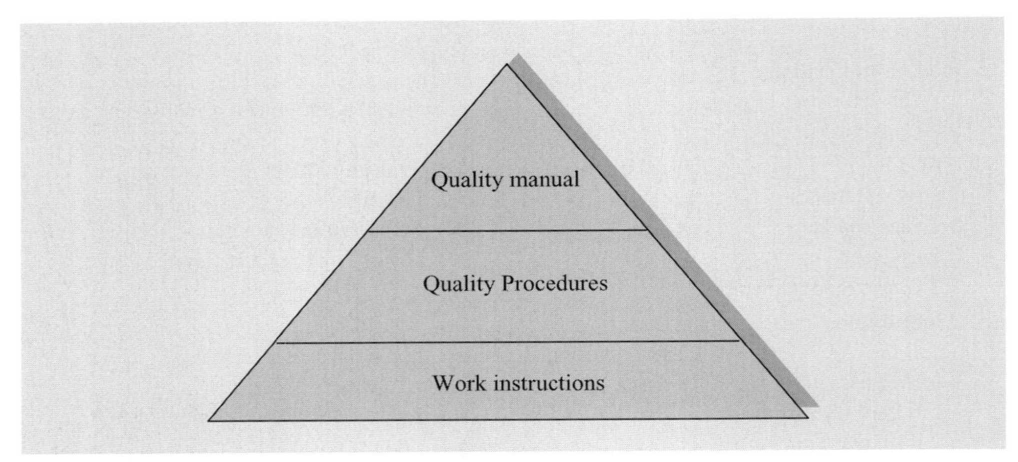

Figure 7.12. Quality system documentation pyramid.

A quality manual describes general quality policy and organisation of the company, together with responsibilities for quality. It contains outlines for specific areas for control and lower level document needed to carry out quality control are defined. Listed in the quality manual, quality procedures provide detailed instructions. The type of information given includes actions to be taken, routing of forms (such as defect sheets, purchase orders, engineering change notes and so on), information to be recorded, responsibilities for each action, etc. Work instructions deal with a lower level of activities than quality procedures. Whereas procedures describe who does what, work instructions provide specific information about standards to be achieved in work execution (such as work methods, hygiene instructions, use of equipment).

7.5.1 Quality assurance process

Quality assurance can be described as a control process (Figure 7.13). By measurement of the physical process and the quality system, information is collected on how quality is realised and to what extent quality expectations of the customer are met. As for each control process, also the quality assurance process consists of three basic steps: measurement, evaluation and corrective action.

Measurement involves evaluation of the system, which can be carried by auditing. An audit is a means by which management can determine whether people in the organisation are carrying out their duties and whether the organisation is effective in meeting goals. Audits are described in more detail in the next section.

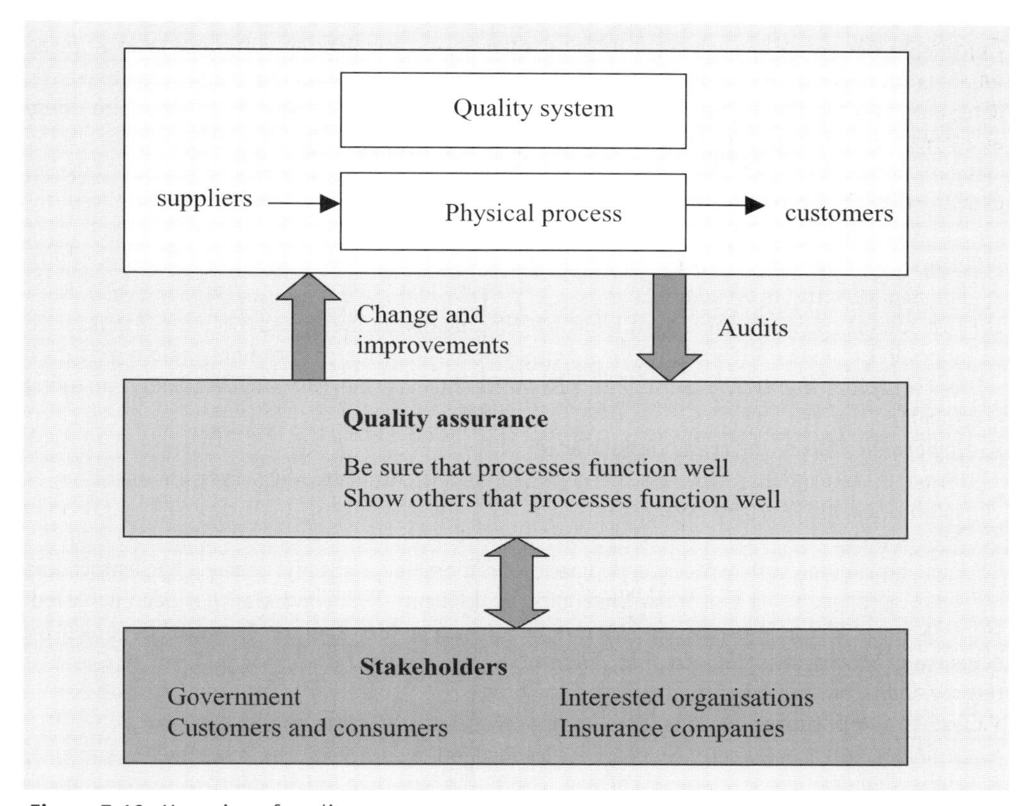

Figure 7.13. Managing of quality assurance.

Evaluation includes comparing to standards. These standards can either be derived from the company's quality policy and/or can be provided by external standards such as HACCP, ISO and BRC. Quality assurance is primarily aimed at giving both management and stakeholders guarantees about quality and quality systems. So quality assurance shows two faces: towards the own company it initiates corrective actions and towards external stakeholders it provides confidence and ensures credibility (Van den Berg and Delsing 1999).
Corrective action includes improving and changing the quality system in order to meet future quality requirements. In Chapter 6, improvement and organisational change processes are discussed. Note that external pressure can force change-processes when companies feel obliged to meet external standards, such as ISO, in a relatively short time period. Choosing an adequate change strategy is very important in order to get control on these processes.

In most companies the quality control department is responsible for quality assurance. It keeps knowledge about standards actual, it organises audits, it reports about evaluation results, it organises corrective actions and it communicates with external stakeholders. Normally quality assurance will be only a part of quality department's activities. However, when quality control already has been delegated to production departments, quality assurance becomes the main task of the quality department. It is then often is called Quality Assurance (QA) department, and is hierarchically positioned just below top management level.

The Quality Assurance department maintains intensive communication with departments in the organisation having assurance responsibilities for other aspects, for instance for environmental care, work conditions and personal welfare. Many companies are trying to integrate these assurance tasks in one integrated care/assurance system.

7.5.2 Quality auditing and certification

Auditing is a measurement process where the auditor is evaluating three aspects (Staples 1990):

1. *Compliance with the standard.*
   The auditor considers the extent to which the apparently prescribed system conforms with appropriate standard (ISO 9001, for example) by examination of documents before execution of the actual audit. The prescribed system defined by the company becomes the ruling document for the audit.
2. *Implementation of the prescribed system.*
   Examination of what people are doing will reveal the extent to which they are operating as established in their own documents.
3. *Effectiveness.*
   Each procedure, both written and practice, is there for a specific reason. Each procedure must produce some result and there must be a means of showing what has been achieved.

Information will come to the auditors in many ways. The information must be objective evidence otherwise it is not useful for auditing.
Quality audits provide several advantages, such as:

- audits give feedback on completeness and effectiveness of the quality system,
- periodic audits reveal seriousness of continuous improvement for the company, to everyone,
- audits indicate those elements of the quality system that are inadequate or need to be improved,
- audits provide a permanent record of progress in achieving goals of the quality system (Evans and Lindsay, 1996).

For auditing it is essential to establish a procedure and audit criteria. A general *procedure* for auditing involves collection of information, verification of information, establishing objective evidence, summarising audit findings and preparation of report. *Audit criteria* are procedures, norms and/or requirements against which the auditor compares objective evidence about the specified process, management system, product and/or service and/or information about these matters.

Three types of audits can be distinguished, one of them being an internal and two of them being an external audit (Staples, 1990):

- The first party audit is an internal audit and is carried out by the company for management purposes such as improvement of the own quality system.
- The second party audit is an external audit and is carried out to audit a potential supplier against proposed requirements when a contract must be made. These requirements can be prescribed in suppliers' internal documents and they may include requirements established in a quality system standard.
- The third party audit is an external audit and implies an independent audit organisation (the third party) assessing organisations against a quality system standard. It provides a certificate, which records the results of the audit that the organisation at time of

assessment had management systems, which complied with the standards. This third-party method is considered to provide users of organisations' products with confidence and therefore reduce the need for second-party audits.

Internal audit or quality system audit

Audit criteria for internal audits relate to typical issues of the quality system, such as:

- Management involvement and leadership, which refers to what extent all levels of management, are involved.
- Product and process design, relates to compliance of products with customer needs.
- Product and process control, concerns the extent of defect prevention.
- Customer and supplier communication includes to what extent customers and suppliers communicate with each other.
- Quality improvement programmes refer to presence of quality programmes and the results of these programmes.
- Employee participation relates to active involvement of employees in quality improvement.
- Education and training concerns activities that are carried out to ensure appropriate skills and experience and training in quality improvement techniques.
- Quality information relates to the collection of and feedback on quality results (Evans and Lindsay, 1996).

External audit and certification

In external audits different approaches are used to establish supplier quality, which ranges from comprehensive audits to short quality evaluations and awareness programmes. Short quality evaluations intend to see if the supplier is following appropriate quality control procedures. Awareness programmes only try to create awareness about customers' requirements, among employees of the supplier, by providing appropriate information.

In anticipation to these efforts of purchasers, suppliers try to prevent that each purchaser individually judges their quality system by obtaining a certificate. Certification is defined as a procedure by which a third party gives a written assurance that a product and/or service, process or quality management system conforms to specified requirements (ISO, 1998). Certification can refer to a person, a product/service, a process or a system and must be carried out by authorised agencies (Van den Berg, 1993).
For *certification of person* requirements and criteria are set for the specific institute or course. For example, CERKOOP is a foundation raised by the Dutch Association for Quality Experts, which certifies persons educated in quality and quality courses.
Products can be certified for specific quality attributes, for example, the Dutch EKO certification for environmental sound production or certification for free-range meat.
Process certification can be applied when appropriate qualification of the final product is not possible. For example, if the number of pathogens must be extremely low, then it is not feasible to perform a proper judgement based on control of the final product. In the dairy sector of the Netherlands, the Dutch Controlling Authority has introduced a process certification for milk and milk products (COKZ). Companies that are eligible for COKZ-certification must prove by their quality assurance system that from a hygienic point their process is under control and their products and environment are systematically inspected (Mors, 1996).

System certification is focused on certification of the entire quality system. The most common approval is certification of the ISO-based quality systems by a third independent institute. In the Netherlands, also HACCP-based quality systems can be certified. (Mors, 1996)

Auditing the auditor

Before certification institutes are allowed to assess quality systems they are audited as well, which is called accreditation. Accreditation has been defined as a procedure by which an authoritative body gives formal recognition that a body or person is competent to carry out specific tasks (ISO, 1998).

In the Netherlands, the Council for Accreditation supervises institutions in the private sector that assess products, processes and measuring instruments on behalf of government, companies and consumer associations. The Council for Accreditation can screen e.g. certification institutes based on established criteria and if the institute proved competence accreditation is granted (Figure 7.14). Figure 7.14 shows that the ministry has two lines of influence, i.e. the first by supervising the council for accreditation and the other by inspection and investigation in the company and on marketplace.

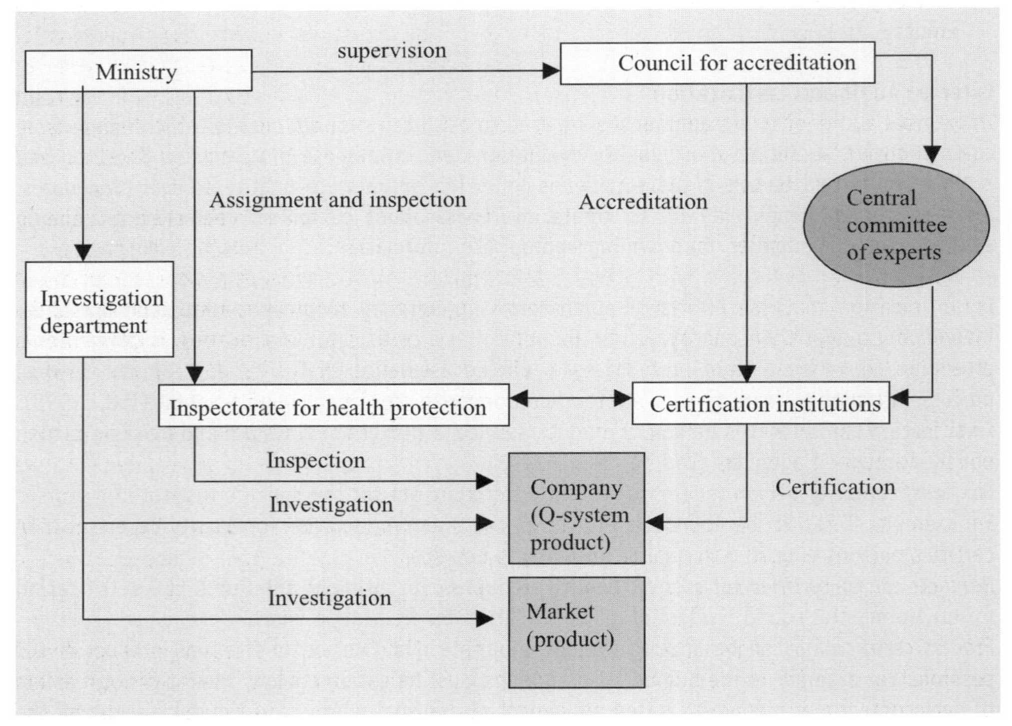

Figure 7.14. Relation between auditing and certification institutions.

7.6 Quality assurance in the food industry

Quality assurance systems are commonly applied in the agribusiness and the food industry. It started in the eighties with Good Manufacturing Practices Codes. The codes provided guidelines that were aimed at assuring minimum acceptable standards and conditions for processing and storage of agri and food products to assure basic quality and safety. In the nineties the popularity of the ISO 9000 family increased. The 9000-series provided a framework for quality management and quality assurance. The specified quality system requirements could be used when a contract between two parties required the demonstration that the supplier was capable to prevent non-conformance at all relevant stages of that supplier (e.g. ranging from design to service). Since 1996 the application of HACCP increased quickly also because it was a legal requirement. More recently, retailers have developed a new system: BRC. In this QA-system, the principles of GMP, HACCP and ISO 9000 series are combined. It is expected that this system will be extended to an international acknowledged QA-system.

Although the objective of the QA-systems is to guarantee quality by standardised guidelines or principles, in practice the interpretation and implementation of the QA-systems appeared to be distinctively different. For example, the guidelines of the ISO-9000 series have been elaborated differently. It ranges from focus on establishment of procedures to focus on quality improvement. In the first situation, comprehensive procedures are made for all types of handling to control and assure quality. In the second situation, the attention is focused on improvement, whereas procedures are only prepared where control is essential for realising quality. In the latter case a more lean system is achieved which aims at controlling and improving critical points with respect to quality.
Also in the application of HACCP principles different interpretations have been noticed in practice. Some companies have allocated many CCP's to be sure that safety of their products is assured. Others have the opinion that a high number of CCP reflects a poor condition of equipment and machines. Firstly, according to them, technological innovations should be introduced to improve the safety level, so reducing the number of CCP's. Moreover, international differences exits between acknowledgement of HACCP certificates, which hinders international trade. Therefore, ISO will develop a norm with requirements for food safety.
Another bottleneck that is experienced by the food industry is the ease by which a high number of requirements are set by government, retailers and customers. The sets of requirements are not only different but they also change regularly.
In addition, it should be noticed that a certificate is often considered as an ultimate goal. A certificate is then used as tool to protect against external interference. It is used for as a defence instrument instead of a consumer-oriented quality assurance tool.

From the techno-managerial perspective, QA-systems should be developed by a profound technical and technological analysis of product, process and process conditions to assess critical control points. Consensus must be reached within the company about the selection of critical control points. For this purpose, predictive modelling can be used to identify potential hazards and crucial quality parameters, whereas the Taguchi principles can be applied to determine social well-considered tolerances. In addition, application of the HACCP principles and risk assessment support in selecting technological critical control points, which are scientifically underpinned. At the same time they provide a scientific basis for the non critical

points. It is important that identification of these critical points is executed in collaboration with the people involved in the relevant processes.
The restriction to critical points enables the development of a lean QA-system, which serves as a tool for the people involved instead of being an enforced system. Such a lean QA-system consists of a hard core based on crucial control points and modules, which can be easily modified or extended depending on customers' demands.
From the techno-managerial approach a single focus on procedures is not advisable because it has its limitations in directing human behaviour. In the next chapter it is explained how the concept of Total Quality Management (TQM) can overcome this shortcoming. Likewise, for quality assurance in the whole agri-food chain, the lean QA-systems may be more suitable and easier to agree upon by partners involved.

8. Quality policy and business strategy

One of the most important tasks of top management is, with a vision of the company's future as a basis, to formulate long-term goals and state strategies on how to achieve these goals. In modern management, quality issues are of great significance in this process. To make this clear many companies have formulated a particular quality policy. In other companies the quality values are completely integrated into other strategic documents. Anyhow, it is important that quality values are firmly established in the company's strategy process (Bergman and Klefsjö, 1994).

Strategy refers to the determination of mission and long-term objectives of a company, and adoption of courses of action and allocation of resources necessary to achieve these aims. Therefore, objectives are a part of strategic formulation.

Policies are general statements or understandings that guide managers' thinking in decision-making. They ensure that decisions fall within certain boundaries. They usually do not require action but are intended to guide managers in their commitment to the decision they ultimately make.

The essence of policy is vision, whereas strategy concerns the direction in which human and material resources will be applied in order to achieve selected objectives (Weihrich and Koontz, 1993).

In practice, the strategic management process involves taking information from the environment and deciding on business strategy, which includes anticipating or reacting to that environment.

According to Noori and Radford (1995) the marketplace after World War II resulted in a "bigger is better" philosophy and mass production as a consequence. Mass production makes sense in an economy characterised by seemingly unlimited supply of resources, ever expanding markets and consumer acceptance of standardised products. Today, however, customers are more discriminating and competition is much more sophisticated. The following trends result in changing market strategies and policy (Figure 8.1.):

- Rapid advances in technology, biological and physical sciences, are providing industries with new materials, new products and new market needs. Over 90 percent of all recorded scientific advances have taken place in the last 30 years.
- Escalation of new product introduction is shortening the life span of most products. Electronic and other high technology products now have life spans as short as several months.

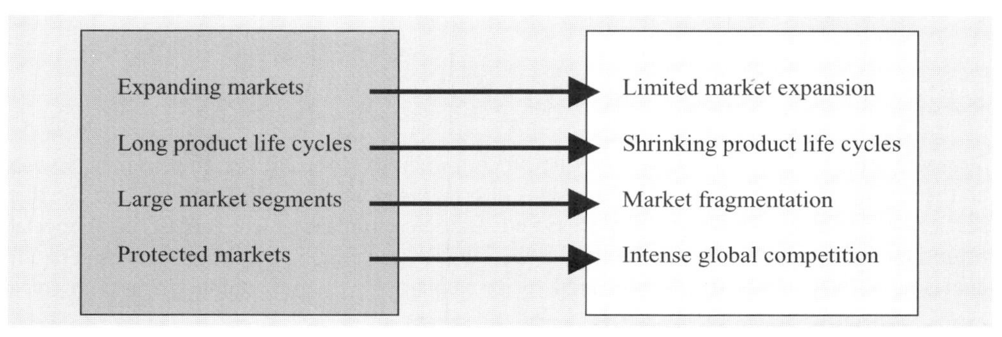

Figure 8.1. New realities of the market place.

- Continuous reductions in transportation and communication costs are increasing global competition.
- Global competition is intensifying as countries are moving to eliminate protectionism and create large free-trade zones.
- Markets in general are fragmenting, not expanding. Competition is intense. Customers are demanding high-quality products, at a reasonable price, that meet their individual needs. Unsatisfied customers are switching from one product to another.
- Growing international concern about the environment is being translated into stricter legislation. Conventional energy supplies are not limitless, nor is the world's ability to absorb disposals of products and process by-products.

As a consequence of these trends, the market will not necessarily absorb large volumes of standardised products. New and revised products must be developed quickly and brought to the market in increasing numbers. Both the increased risk of quality decrease during storage and expansion of product lines make holding inventories much too expensive. Smaller facilities will arise, which can produce a variety of products in various volumes and from random orders.
This flexibility allows a firm to be more responsive across a number of market segments and to reach different customers within each segment. The product demand should increase, while the firm maintains low inventory levels. Operating costs are being reduced through process simplification and through design of products that are easier to manufacture. Greater flexibility and lower costs are forcing firms to co-operate with suppliers and customers in a broader chain perspective.
These developments can be summarised in a strategy of mass customisation, which combines the advantages of mass production with the advantages of offering a broad product assortment corresponding to individual consumer wishes. Mass customisation and other developments in the market have far going consequences for quality and quality policy. Anticipating these developments includes quality decisions in strategic management processes and a quality policy that fits with the company's strategic objectives.

The strategic management process will be described in the next section. The concept of quality policy will be elaborated in section 8.2. Then the following two central themes in food quality management will be described: Total Quality management and strategic alliances in the food chain.
In order to verify the performance of quality policy, sets of criteria have been developed. These sets are enclosed in different awards and prices as explained in section 8.5. Finally quality policy with respect to typical aspects of the food industry will be discussed.

8.1 Strategic Management

Strategic management is a decision making process wherein analysing the environment as well as the competences of the company provides alternative strategic choice possibilities. In this section, the strategic management process and strategic alternatives will be described. Special attention will be paid to benchmarking, a well known technique in quality management for analysing a company's position compared to its competitors or firms doing the best in some area.

8.1.1 Strategic management process

Strategic management is the process of formulating and implementing strategies to accomplish an organisation's mission and objectives and ensure competitive advantage. The essence of strategic management is to look ahead, understand the environment, and effectively position an organisation for competitive success in changing times (Schermerhorn, 1999).
The strategic management process entails both formulation activities, which are the major focus of planning and implementation activities. These activities will actually realise the devised strategy (Gatewood *et al.*, 1995; Schermerhorn 1999). Both activities are shown in Figure 8.2 and will be discussed in more detail.

The **first step** in strategic management is the careful assessment and classification of organisational mission and goals.
The *mission* or purpose of any organisation may be described as its reason for existence. In today's quality-conscious and highly competitive environment, the sense of purpose must be directly centred on serving needs of customers or clients. After all, their satisfaction and continued support are the ultimate keys to organisational survival. In general the firm's mission reflects such information as what types of products or services it produces, who its customers tend to be and what important values it holds. Values are broad beliefs about what is or is not appropriate. The corporate culture is the predominant value system for the organisation as a whole. In strategic management the presence of strong core values helps build corporate identity and supports the mission statement.
Goals direct activities toward key and specific results. Typical goals of a firm are profitability, market share, product quality level and social responsibility.

The **second step** is the evaluation of the firm's internal strengths and weaknesses, and opportunities and threats (SWOT) associated with the firm's environment. SWOT analysis emphasises that the fit between firm and its environment is of paramount importance. A firm's strategy should be built around this match. *Strength* is an ability or attribute which gives the firm a distinctive competence, which may be reflected in, for example, the firm's product quality offered, or cost advantages due to the firm's overall quality performance. Conversely, a *weakness* is an attribute or skill that a firm lacks or has not developed and at which it performs poorly. An *opportunity* is an environmental circumstance that is potentially beneficial for a firm, such as convenience food due to the fact that households want to spend less time on cooking. A

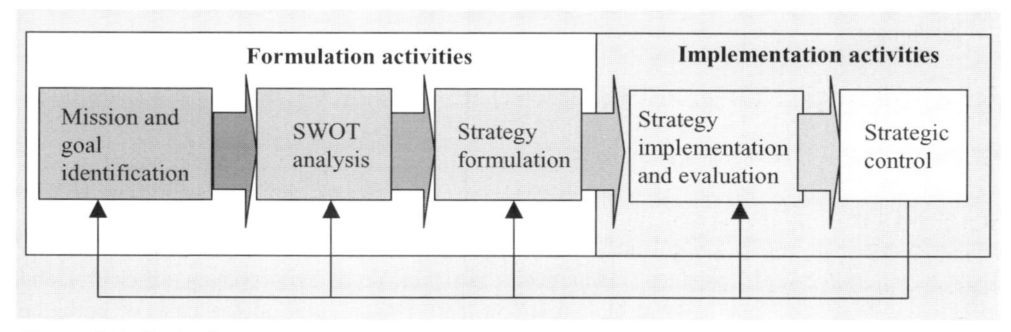

Figure 8.2. Strategic management process.

threat on the other hand, is an environmental factor that could be potentially harmful to the firm, such as a law act on abandoning battery housing systems for chickens.

The **third step** is to formulate the strategy. As a rule, a firm will formulate its strategy on its greatest strengths and keep its weaknesses from being exploited by competitors and other factors in the environment. Opportunities and threats are often the same for all firms in an industry. However, a successful strategy will better exploit the opportunities and limit the business environmental threats. One can use two criteria to select potential strategies, *i.e.* its content and the followed procedure. The *content criteria* consists of elements such as:

- Suitability, to what extent does the strategy fit in meeting the strategic problem and objectives,
- Feasibility, to what extent is the strategy economically and technically feasible,
- Acceptability, to what extent are relevant partners, departments, management and workers willing to commit to the strategy.

Procedural criteria relate to the way the selection of strategies is carried out. In other words the conditions under which the decisions are made, such as knowledge of involved persons, availability of information, and authority of decision-makers (Johnson and Scholes, 1997).

Then, the strategy has to be implemented. This requires involvement of the entire organisation. It means initiating and directing activities to ensure that all activities are focused on implementing the firm's strategy. Finally, strategic control is the feedback mechanism in the strategic management process. It compares the firm's actual performance to its intended performance. Strategic control enables evaluation of the way strategy was planned, formulated and implemented. As a result, the firm may find that a potentially successful strategy was formulated well but implemented poorly, or the other way around.

The importance of step 2, the SWOT analysis, is becoming clear with literature analysis. It offers a range of strategic analysis models and techniques that can be applied to analyse the firm's position in its environment. Besides general techniques for analysing business environment (like PEST analysis: political, economical, social and technical trends) and general techniques for market and consumer research, several specific techniques are frequently used in strategic analysis:

1. the portfolio analysis for determining the firm's product position in the market;
2. the five forces analysis for determining the firm's competitive position;
3. the value chain analysis and
4. the core competence analysis, both for determining the relative strength of internal resources.

The *portfolio analysis* (Boddy and Paton, 1998) developed by the Boston Consulting Group (BCG), is a useful model for examining market share of individual products or services in relation to growth rate in their particular markets (Figure 8.3.)

The model's assumptions are underpinned by the concept of the *experience curve*, which sustains that companies already in a particular market for a significant period of time have learned from their experience how to produce and deliver their product or service more efficiently and effectively. They are likely to have built a strong market share, and on the basis of both the associated economies of scale and their 'experience', to have achieved relatively low unit costs.

Competitive position (market share)

Market growth	High	Low
High	**Stars** (develop)	**Question marks** (investigate)
Low	**Cash cows** (milk)	**Dogs** (divest)

Business type	Profitability	Growth opportunities	Typical net cash flow
Star	High	Yes	Balanced to negative
Cash cow	High	Limited	Positive
Dog	Low	Limited	Balanced to negative
Question mark	Low	Yes	Very negative

Figure 8.3. The Boston portfolio matrix.

Thus the matrix tends to associate a high market share with an ability to generate large, positive cash flows. The matrix descriptors are explained as follows:

- *Cash cows* have high market share in a mature market. Probably 'stars' of the past now in maturity, they occupy a leading position with little need for investment in production capacity or marketing. They are likely to be cash generators, which can be 'milked', contributing to development of future stars.
- *Stars* have high market share within a growing market but expenditure may be high so that the product or service both generate and use cash. They are likely to be the cash cows of the future.
- *Question marks* are also found in growing markets, but have a low market share. They are typically products or services that are at an early stage in life requiring substantial amounts of cash to be spent in an effort to increase market share. However, the payback from high spending is uncertain.
- *Dogs* experience low market-share in low-growth markets and are likely to drain the organisation of cash by using a disproportionate amount of management time.

A portfolio of products and/or services, which is balanced in terms of this matrix is crucial to the survival of the business because investment for growth must be internally generated, which is a role of the cash cow. The star and perhaps the question mark are essential to future success. Although the dog is often regarded as a candidate for speedy elimination, Johnson and Scholes (1997) defended against immediate action without further consideration. A product should be considered, such as whether it is a necessary element in a complete product range, or whether dropping a service will be politically acceptable. In other words, there might be a good reason to retain a loss-making product or service.

The *five forces model* of Porter (Porter, 1980; Dessler, 2001; Certo, 2000) is a well known tool for analysing the firm's competitive position. Essentially, Porter's model outlines the primary forces that determine competitiveness within an industry and illustrate how those forces are related. The model suggests that in order to develop effective organisational strategies, managers must understand and react to these forces within that industry. Porter's model is presented in Figure 8.4. According to the model, the following factors determine competitiveness within an industry:

- products that might act as a substitute for goods or services that companies within the industry produce;
- new companies entering the market threatening market share of the company;
- ability of suppliers to control issues like costs of materials that industry companies use to manufacture their products;
- bargaining power that buyers possess within the industry;
- general level of rivalry or competition among firms within the industry.

According to the model, buyers, product substitutes, suppliers and potential new companies within an industry, all contribute to the level of rivalry among industry firms.

The *concept of the value chain*, introduced by Porter (1985), is derived from an established accounting practice of calculating the value added to a product by individual stages in a manufacturing processes (Figure 8.5.). Porter applied this idea to the activities of an organisation as a whole, arguing that it is necessary to examine activities separately in order to identify sources of competitive advantage. (Boddy and Paton, 1998). The usefulness of value chain analysis is that it recognises that individual activities in the overall production process play a part in determining *e.g.* cost, quality and image of the end-product or service. That means that each individual can contribute to a firm's relative position and create a basis for

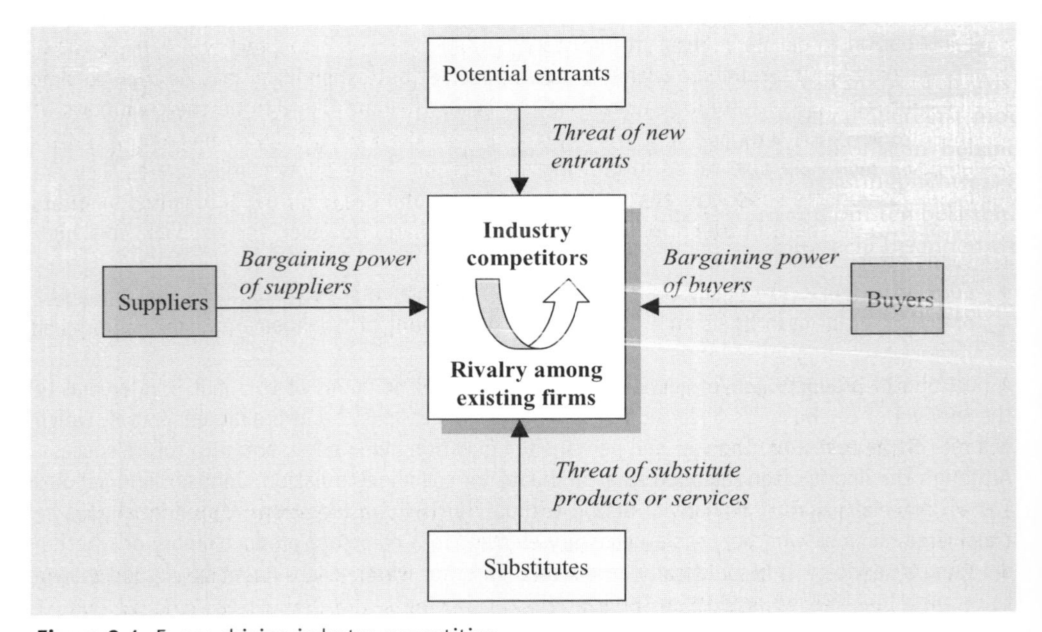

Figure 8.4. Forces driving industry competition.

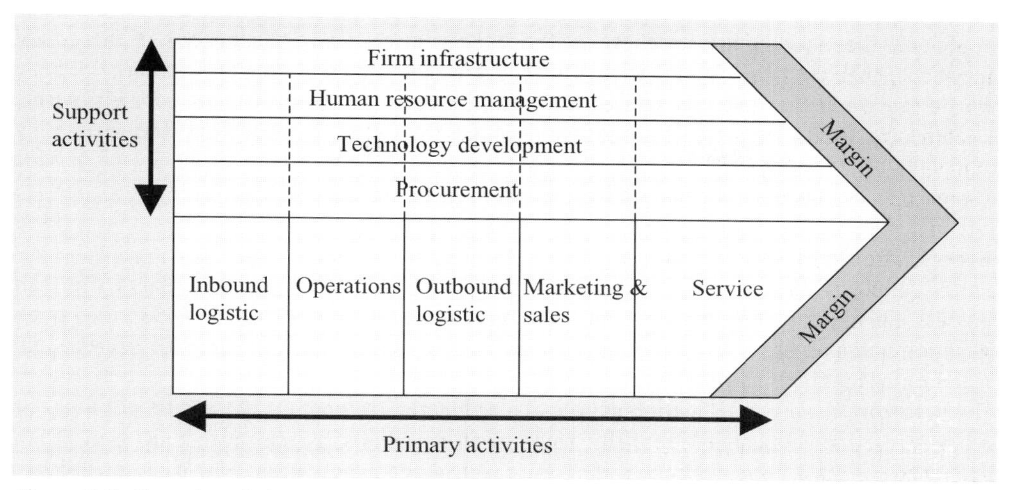

Figure 8.5. The value chain.

differentiation' (Porter, 1985). Porter argues to identify key factors, which compel cost efficiency and value, the so-called cost drivers and value drivers.

Hamel and Prahalad (1994) have developed the core competence philosophy. They argued that the difference in performance of organisations in the same 'industry' is rarely fully explainable by differences in their resource base *per se*. Superior performance is also determined by the way in which resources are *deployed* to create competences in the organisation's separate activities. Moreover, it is determined by the processes of linking these activities together to sustain excellent performance. Although the organisation needs to achieve a threshold level of competence in all of its activities, only some will be of core competence. These are the competences that underpin the organisation's ability to surpass competition -or demonstrably to provide better value for money. Core competences must be difficult to imitate, otherwise it will not provide long-term advantage. They may also be the basis on which new opportunities are created.
Johnson and Scholes (1995) derived from both models the bases on which an organisation's core competence can be built:

- cost efficiency;
- added value, *i.e.* effectiveness in matching customer requirements;
- managing linkages within the organisation's value chain, and linkages into the supply and distribution chains;
- robustness, *i.e.* competences must be difficult to imitate.

8.1.2 Strategic alternatives

In considering possible strategies to achieve its objectives, management faces three choices (Johnson and Scholes, 1997; Boddy and Paton, 1998):

- On what basis do they wish to compete with, or differentiate themselves, from their competitors? This is sometimes referred to as choosing a *positioning* or *generic* strategy.

- Should they develop new products or services, new markets or a combination of those two, should they drop certain activities? This concerns choosing a strategic direction.
- Having made the first two sets of decisions, how should they develop the new services, products and/or markets? This refers to choosing the means of developing strategy.

The choices available can be categorised as shown in Figure 8.6

Generic strategies

Porter (1980, 1985) identified two basic types of competitive advantages that a firm can possess, *i.e.* low costs or differentiation. From this concept, Porter (1980, 1985) classified three generic strategies or means by which companies can identify their position to develop and maintain competitive advantage.

The first, *cost leadership*, is a strategy whereby a firm aims to produce its product or service at a lower cost than its competitors. This could mean developing a large market share to attain economies of scale in production costs. Cost leadership does not necessarily mean low price.

The second generic strategy is *differentiation*, whereby a company aims to offer a product or service that is distinctive - and valued as such by customers - from those of its competitors. Porter argued that differentiation is 'something unique beyond simply offering a low price'. It allows firms to demand a premium price or to retain buyer loyalty during cyclical or seasonal downturns in the market. For example, firms that develop a reputation for reliability may be

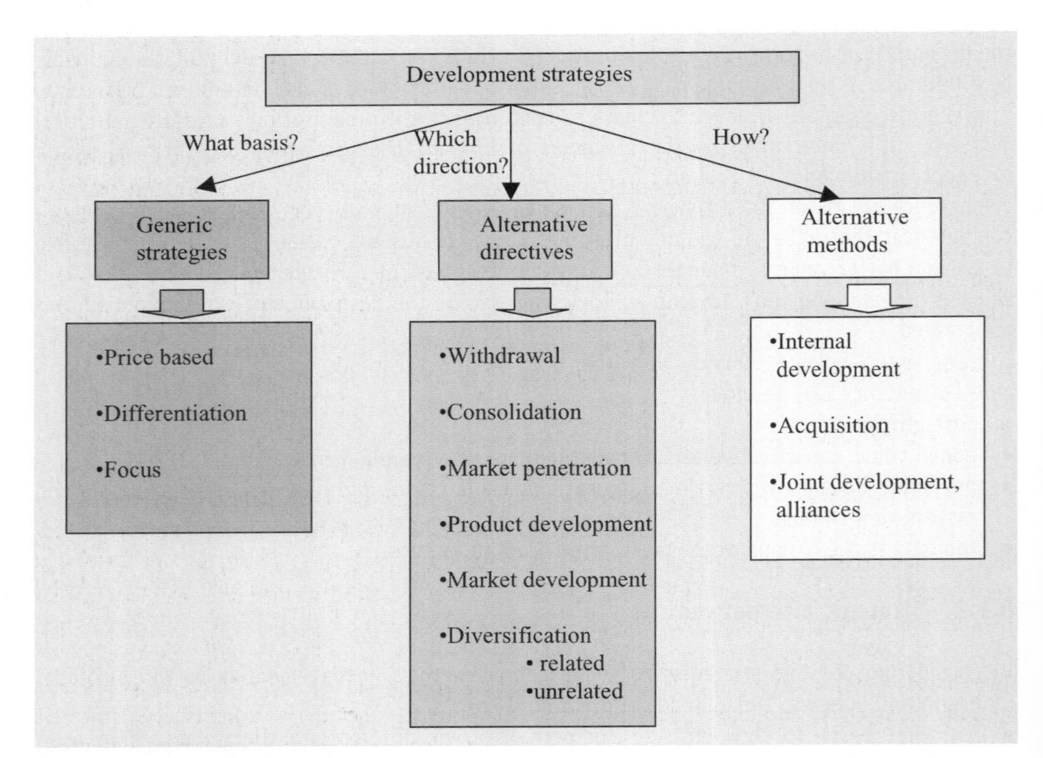

Figure 8.6. Strategic alternatives.

able to enhance their market share or to achieve higher margins by charg
prices, which customers are willing to pay.
The third, *focus strategy*, is a strategy whereby management applies eithe
differentiation to a narrow marker segment, rather than to a whole market
variants, cost focus and differentiation focus.

Strategic directions

The second set of choices that are important for organisations concern the direction of strategy development. Options can be classified in four categories:

1. *Existing markets and existing products.*
   Alternative directions are consolidation in a growing market, market penetration in a mature market in order to increase market share and withdrawal in a declining market.
2. *Existing markets and new products*
   A strategy of product (or service) development allows a company to retain the relative security of its present markets while altering products or developing new ones.
3. *New markets and existing products*
   Market development in the strategy of seeking new markets for existing products or services. This could mean new demographic markets or new groups of customers, by age, socio-economic grouping, education, lifestyle, interests, etc.
4. *New markets and new products*
   The strategy is diversification, which means bringing new products in new markets. Product/market combinations can be related. Then there is some link between existing and new activities. Completely new activities are called unrelated diversification.

Alternative Methods

A final set of decisions concerns the means by which an organisation pursues the chosen strategy. Internal development means that all aspects of the development of new products are totally under the organisation's control and under the influence of its own management and expertise. Acquisitions allow rapid entry into new product/market areas. They may also be appropriate when the acquiring company lacks necessary skills and resources.
Joint development and alliances means turning to partners to co-operate in developing products or services. Arrangements can vary from highly formal contractual relationships to looser forms. Examples are franchising, licensing and subcontracting.

8.1.3 Benchmarking

Benchmarking is the process through which a company learns how to become the best in some area by carefully analysing practices of other companies that already do extremely well in that area (best-practices companies). Benchmarking helps a company to learn its strengths and weaknesses, and those of other industrial leaders, and incorporate the best practices into its own operations (Dean and Evans, 1994).

Two major types of benchmarking are *competitive* and *generic*. Competitive benchmarking usually focuses on products and manufacturing of a company's competitors. Generic benchmarking evaluates processes or business functions against the best companies, regardless of their industry.

e benchmarking process can be described as follows:

1. Determine which functions to benchmark. These should have a significant impact on business performance and should be key dimensions of competitiveness. If fast response is an important dimension of competitive advantage, then processes that might be benchmarked would include order processing, purchasing, production planning and product distribution. There should also be an indication that the potential for improvement exists.
2. Identify key performance indicators to measure. These should have a direct link to customer needs and expectations. Typical performance indicators are quality, performance and delivery.
3. Identify the best-in-class companies. For specific business functions, benchmarking might be limited to the same industry. For generic business functions, it is best to look outside one's own industry. Selecting companies requires knowledge of which firms are superior performers in the key areas. Such information can be obtained from published reports and articles, industry experts, trade magazines, professional associations, former employees, or customers and suppliers.
4. Measure the performance of the best-in-class companies and compare the results to your own performance. Such information might be found in published sources or might require site visits and in-depth interviews.
5. Define and take actions to meet or exceed the best performance. This usually requires changing organisational systems. Simply to imitate the best is like shooting at moving target - their processes will continually improve. Therefore, attempts should be made to exceed the performance of the best (Dean and Evans, 1994).

8.2. Quality Policy

In this section quality policy will be analysed. Firstly a summary of the quality guru's philosophies will be given. Then, considering the costs of quality, their focus on prevention will be analysed. This section will end with Crosby's maturity-grid, which relates quality policy to the stage of development the company is in.

8.2.1. Quality Philosophies

As stated earlier, policies or general statements are the understandings of the guide manager's thinking in decision-making. A quality policy must be firmly established in the decision-making processes and be the guiding principle for all activities whenever quality is concerned, with respect to external as well as internal customers and also towards suppliers to the company (Bergman and Klefsjö, 1994).

Quality policy in most companies is more or less influenced by the ideas of the four major quality gurus Deming, Juran, Feigenbaum and Crosby (see Chapter 3 for a description).

Their ideas about Total Quality Management can be illustrated by the more or less philosophical approach of Deming. With this approach he believed that a commitment to quality requires transformation of the entire organisation. The Deming philosophy is based on a system known as the "Fourteen Points":

1. Create and publish for all employees a statement of the aims and purposes of the company or other organisation. The management must demonstrate constantly their commitment to this statement.

2. Learn the new philosophy, top management and everybody.
3. Understand the purpose of inspection, for improvement of processes and reduction of cost.
4. End the practice of awarding business on the basis of price tag alone.
5. Improve constantly and forever the system of production and service.
6. Institute training.
7. Teach and institute leadership.
8. Drive out fear. Create trust. Create a climate for innovation.
9. Optimise toward the aims and purposes of the company the efforts of teams, groups, and staff areas.
10. Eliminate exhortations for the workforce.
11. (a) Eliminate numerical quotas for production. Instead, learn and institute methods for improvement.
    (b) Eliminate MBO (Management by Objective). Instead, learn the capabilities of processes and how to improve them.
12. Remove barriers that rob people of pride and workmanship.
13. Encourage education and self-improvement for everyone.
14. Take action to accomplish the transformation.

All gurus agree that the ultimate focus should be on quality, customer orientation and on a company-wide approach in prevention of quality problems. However, they differ in the way to reach this. These differences can be illustrated with the aid of the administrative concepts (see also Chapter 3).

Management approaches are described in a matrix with two axes: centralisation versus decentralisation and high versus low degree of formalisation. In Figure 8.7, the gurus are positioned in the same matrix, indicating that Feigenbaum and Crosby, more than Deming and Juran, start their programs at the top of the company. Feigenbaum and Deming believe, more than Crosby and Juran, in a formalised way of doing things, whereas the latter strongly focus on everyone's attitude towards quality, an attitude that refuses to accept today's quality levels as good enough. Note that these differences between the quality philosophies are not as explicit in practice. Deming, for instance, has also stressed the importance of attitude. They all agree upon the fact that quality policy implies an obligation for management (Bergman and Klefsjö, 1994). Unless the management acts according to the quality policy, no other employee in the company will take it seriously.

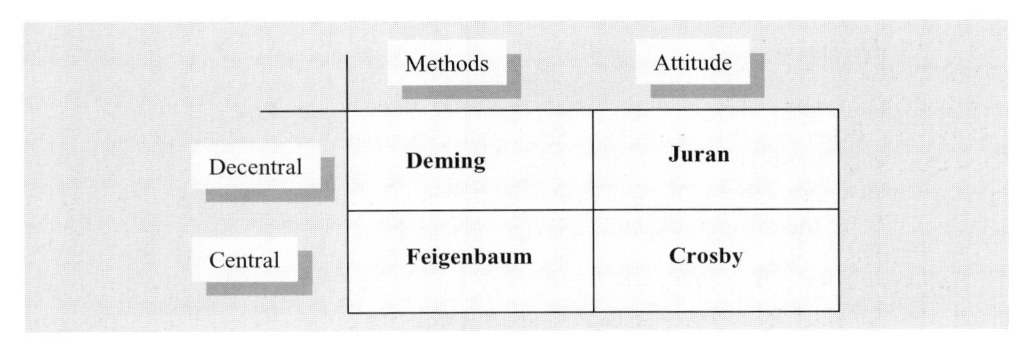

Figure 8.7. Approaches of different quality philosophies.

8.2.2 Quality costs

In most firms cost accounting is an important function. All organisations measure and report costs as a basis for control and improvement. Quality cost information serves a variety of purposes. To provide insight in costs of poor quality, those costs are associated with avoiding poor quality or incurred as a result of poor quality. To help management must be evaluated the relative importance of quality problems, thus identifying opportunities for cost reduction. Budgeting and cost control activities can aid. Finally, it can serve as a scoreboard to evaluate the organisation's success in achieving quality objectives with regard to the firm's strategy. To establish the cost of quality program, one must identify the activities that generate cost, measure them, report them in a way that is meaningful to management and analyse them to identify areas for improvement.
Quality costs (Figure 8.8) can be organised into four major categories: prevention costs, inspection costs, internal failure costs and external failure costs (Evans and Lindsay, 1996).

Prevention costs are expended in an effort to keep non-conforming products from occurring and reaching the customer, including the following specific costs:

- Quality planning costs, such as salaries of individuals associated with quality planning and problem-solving teams, the development of new procedures, new equipment design and reliability studies in order to improve quality or to prevent poor quality.
- Process control costs, which include costs spent on analysing production processes and implementing process control plans.

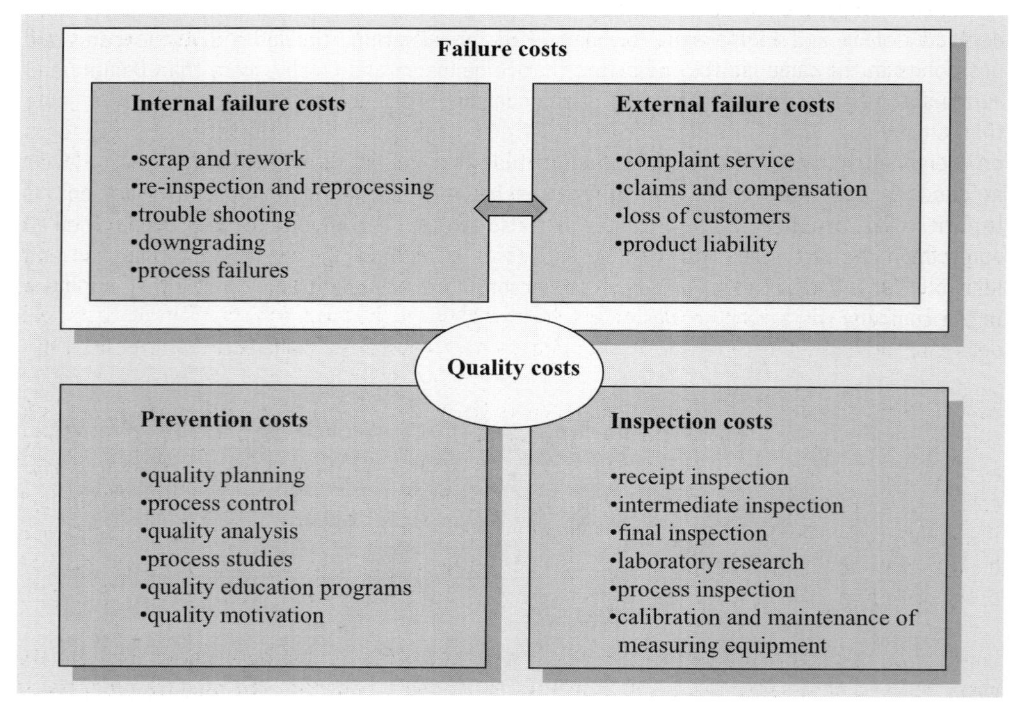

Figure 8.8. Quality costs.

- Information systems costs expended to develop data requirements and measurements.
- Training and general management costs, including internal and external training programs, clerical staff expenses and miscellaneous supplies.

Inspection costs are expended on ascertaining quality levels through measurement and analysis of data in order to detect and correct problems. Categories of inspection costs include:
- Test and inspection costs associated with incoming materials, work in process and finished goods, including equipment costs and salaries.
- Instrumental maintenance costs, due to calibration and repair or measuring instruments.
- Process measurement and control costs, which involve the time spent by workers to gather and analyse quality measurements

Internal failure costs are a result of unsatisfactory quality detected before delivery of a product to the customer; some examples include:
- Scrap and rework costs, including material, labour and overhead.
- Costs of corrective action, arising from time spent determining the causes of failure and correcting production problems.
- Downgrading costs, such as revenue lost when selling a product at a lower price because it does not meet specifications.
- Process failures, such as unplanned machine downtime or unplanned equipment repair.

External failure costs occur after poor quality products reach the customer, specifically:
- Costs due to customer complaints and returns, including rework on returned items, cancelled orders, freight premiums, and loss of customers.
- Product recalls costs and warranty claims, including the cost or repair or replacement as well as associated administrative costs.
- Product liability costs, resulting from legal actions and settlements.

Experts estimate that 60 to 90% of total quality costs are the result of internal and external failure. These costs are not easily controllable by management. In the past, management reacted to high failure costs by increasing inspection. Consequently inspection costs increased, while very often the overall result little improved. In practice, an increase in prevention usually generates larger savings in all other categories. Better prevention of poor quality clearly reduces internal failure costs, since fewer defective items are made. Also external failure costs will decrease. In addition less inspection is required, since products are made correctly the first time.

Juran (1951) was the first one to give managers the means of answering the question on how much quality is enough (Figure 8.9). He divided costs of achieving a given level of quality into avoidable and unavoidable costs. Failure costs are avoidable, whereas prevention and inspection costs are unavoidable. Juran stated that there is an optimum where a rise in prevention and inspection costs is not anymore compensated by the parallel decline in failure costs. There has been criticism on Juran's approach with respect to the optimum. Crosby (1979) argued that the optimum is at a 100% quality level (= zero defects), as is illustrated in Figure 8.9 at the right.

Considering costs from the perspective of agri and food products, the optimum of total quality costs will not necessarily be at 100% quality. It is almost impossible to produce products of constant and perfect quality because of the nature of agri and food products, such as biological

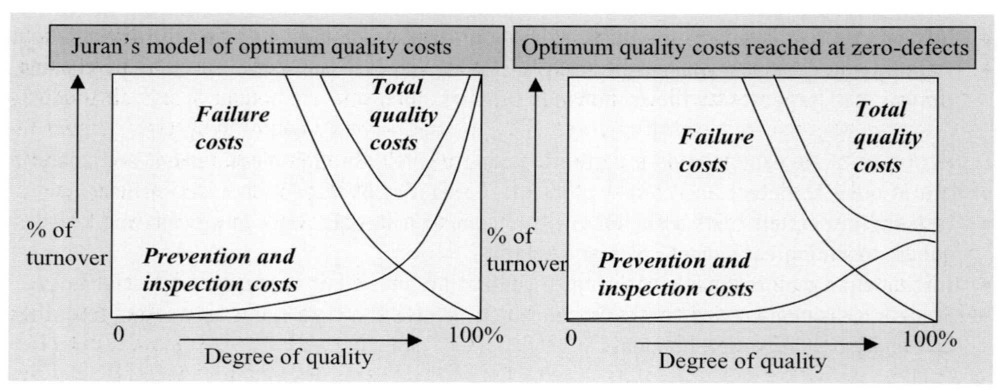

Figure 8.9. Two quality costs models.

variation, climate fluctuations and limited shelf life qualities. This is in accordance with Juran's approach of the proven reality that poor quality is uneconomical and the unproved theory that perfect quality is uneconomical. Although, supporters of the "Zero-Defects" program of Crosby are arguing that with a careful stepwise approach the level of zero-defects can be accomplished. However, nowadays higher and higher demands are put on agri and food production in terms of safety, freshness, health and sustainability. Furthermore, firms are confronted with an increasing competition and quickly changing consumer markets. Therefore, firms have to improve their products and processes continuously. As a consequence the optimum of total quality costs will decrease, while the general quality level will increase (Zuurbier *et al.*, 1996).

Quality costs themselves provide little information, since they may vary due to factors such as production volume and season effects. Thus, index numbers can more effectively analyse quality cost data. Some common measurement indicators are labour, sales, manufacturing cost and unit of products. By using these indices one should be aware of the time horizon included, the possible need to weigh to a standard, price fluctuations and interaction with other factors. This illustrates that all of these indices, although used extensively in practice, have a fundamental problem. A change in the indicator can appear to be a change in the level of quality or productivity alone. For instance, if selling price is decreased through increased competition, the direct sales-based index will increase even if quality does not change. Nevertheless, use of such indexes is widespread and useful for comparing quality costs over time. Generally, sales bases are the most popular, followed by cost, labour and unit bases. It has to be mentioned that this approach to quality costs is limited to efficiency and effectiveness and does not fit in a TQM-philosophy. For instance, if direct labour is decreased through managerial improvements, the direct labour-based index will increase even if quality does not change. However, the overall firm's competitiveness (total quality) has increased because it succeeded through improvements to reduce direct labour costs. Thus it succeeded to drop its total costs. Thus a different approach is required (Evans and Lindsay, 1996).

Quality and activity based costing

Traditional accounting systems focus on promoting efficiency of mass production, particularly production with few standard products and high direct labour. Today's products are characterised by much lower direct labour and many other activities that consume resources and are unrelated to volume of units produced. In some cases, due to automation, direct labour typically is only 15% of manufacturing cost. Meanwhile, overhead costs such, as purchasing, marketing and quality costs, have grown to 55% or more, and are spread across all products using the same formula. Because of these changes, traditional accounting systems present an inadequate picture of manufacturing efficiency and effectiveness. The weakness of these systems is that they are not capable of allocating the expenses of these support resources to individual products sufficiently; however, Activity Based Costing (ABC) does (Figure 8.10).

ABC organises information about the work (or activity) that consumes resources and delivers value in a business. People consuming resources in work ultimately achieve the value that customers pay for. Examples of activities might be inspecting, receiving, shipping and order processing. Preparation of flowcharts of each business process might help to map all activities involved. ABC allocates direct and overhead costs to those products and services that use them. Knowing the costs of activities supports efforts to improve processes. Once activities can be traced to individual products or services, then additional strategic information is made available. The effects of delays and inefficiencies become readily apparent. The firm can then focus on reducing these hidden costs.

Activity based costing (ABC)		
Insight in behaviour of overhead costs •What are the causes ? •How are they built up ?	**Starting-point** •Activities cause costs •Products and customers cause activities	**ABC system focuses on** •Which factors are responsible for activities •What are the costs of these activities •How do activities proportion to products?

Figure 8.10. Activity based costing.

8.2.3. Stages of development

Quality policy evolves during the course of time. A systematic approach is policy deployment, which is a top-down and bottom-up process in order to align objectives through all levels (Bergman and Klefsjö, 1994). Objectives are identified at each level by top-down and bottom-up consultation and communication, thus translating overall goals of the company into specific targets for departments and groups. The outcome of policy deployment processes strongly depends on the stage of development of quality understanding and management.
Like Deming and Juran, Crosby viewed the way to quality as a never-ending journey (Crosby, 1979; Melnyk and Denzler, 1996). Crosby typified stages of development in quality management in his maturity grid (Table 8.1). In this grid, corporate quality awareness and quality maturation from a level of uncertainty to one of certainty is traced. The stages of development are:

- *Uncertainty*
  The firm remains unaware of the importance of quality as a strategic necessity.
- *Awakening*
  Managers understand the importance of quality but have not yet acted to improve it.
- *Enlightment*
  Managers openly acknowledge the importance of quality and the corporate need for improvement; they started to take concrete action to improve overall quality by establishing a formal program.
- *Wisdom*
  Managers have established a quality management program that is working well, identifying problems early and routinely pursuing corrective actions. The program emphasises prevention rather than inspection.
- *Certainty*
  Quality has become an unavoidable component of the current management structure. The entire system is designed to ensure that the firm can attain its goal of zero defects. Any problems are both infrequent and truly random events.

Referring to quality costs, Crosby states that quality costs can decline form 20% of sales to 2,5% of sales.

8.3 Total quality management

Total Quality Management is a company policy that fits with the later stages of Crosby's maturity grid. In this section Total Quality Management will be explained by using the concept of different policy orientations. Implementation of this policy is shortly discussed.

8.3.1. Policy orientations

For quality policy various orientations are used, depending on the organisation's culture and individual beliefs of top management. In companies where technology plays an important role, managers tend to first introduce technological solutions for arising problems, such as automation of processes and redesign of products and processes. In other companies managers prefer changing culture and attitude of employees, because they strongly believe that this is the only way to accomplish improvements in future. Generally speaking three policy orientations can be distinguished (Figure 8.11):

1. *Product/process orientation*.
   Technological solutions are preferred in order to make product quality, processes, and materials predictable and controllable.
2. *Procedure orientation*
   Solutions are looked for within the quality system. The underlying belief is that a clear description of responsibilities, accompanied with a broad description of procedures and work instructions, form the basis for controlling quality performance.
3. *People orientation*
   Problems are analysed in the field of human behaviour. Solutions are searched in changing people by means of organisational development, training and empowerment.

Table 8.1. Quality management maturity grid of Crosby.

Quality management maturity grid					
Measurement Categories	Stage I Uncertainty	Stage II Awakening	Stage III Enlightment	Stage IV Wisdom	Stage V Certainty
Management understanding and attitude	No comprehension of quality as a management tool. Tend to blame quality department for 'quality problems'	Recognised that quality management may be of value, but not willing to provide money or time to make it all happen	While going through quality improvement program, learn more about quality management, becoming supportive and helpful	Participating. Understand absolutes of quality management. Recognise their role in continuing emphasis	Consider quality management an essential part of company system
Quality organisation status	Quality is hidden in manufacturing or engineering departments. Inspection is probably not part of organisation. Emphasis on appraisal and sorting	A stronger quality leader is appointed but main emphasis is still on appraisal and moving the product. Still part of manufacturing or other department	Quality department reports to top management, all appraisal is incorporated and manager has role in management of company	Quality manager is an officer of company; effective status reporting and preventive action. Involved with customer affairs and special assignments	Quality manager on board of directors. Prevention is main concern. Quality is a true leader
Problem handling	Problems are fought as they occur; no resolution; inadequate definition; lots of yelling and accusations	Teams are set up to attack major problems. Long-range solutions are not solicited	Corrective action communications established. Problems are faced openly an resolved in an orderly way	Problems are identified early in development. All functions are open to suggestion and improvement	Except in the most unusual cases, problems are prevented
Cost of quality as % of sales	Reported: unknown, actual 20%	Reported: 3%, actual 18%	Reported: 8%, actual 12%	Reported: 6.5%, actual 8%	Reported: 2.5%, actual 2.5%
Quality improvement actions	No organised activities. No understanding of such activities	Trying obvious 'motivational' short-range efforts	Implementation of the 14-step program with thorough understanding and establishment of each step	Continuing the 14-step program and starting. Make certain	Quality improvement is a normal and continued activity
Summation of company quality posture	We don't know why we have problems with quality'	'Is it absolutely necessary to always have problems with quality?'	Through management commitment and Q-improvement problems are identified and solved	Defect prevention is a routine part of our operation	We know why we do not have problems with quality

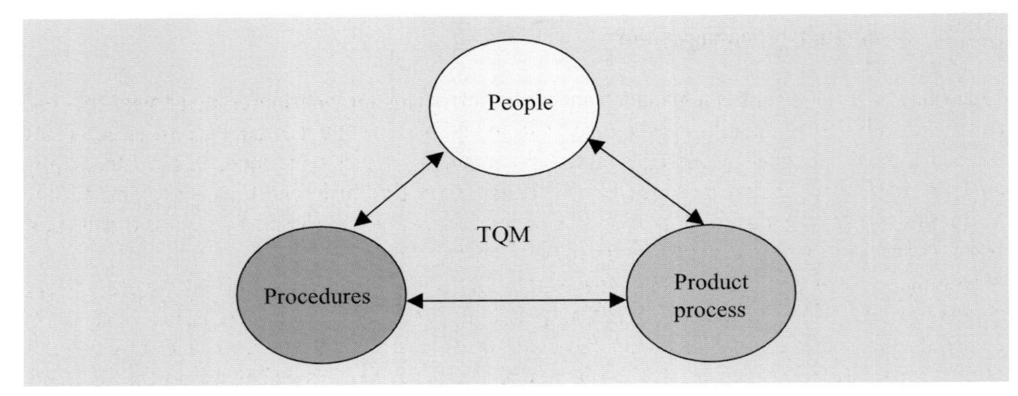

Figure 8.11. Different quality policy orientations.

Total quality management (TQM) in this book is considered as a policy, and is further elaborated in the next section. TQM embraces the three policy orientations, product/process and procedure and people at the same time, thus striving for the broadest range of analysis methods and possibilities for solutions.
Experience from food companies in the Netherlands revealed that integrating the three policy orientations is very difficult. Only a few companies succeed in doing so. Some companies combine people and procedure orientation, but more companies integrate product/process and procedure orientation. Most of the companies only use a single orientation in resolving quality problems and controlling the quality system.
Frequently, using one policy orientation won't be sufficient for solving problems (Edge, 1990). For instance, introduction of new technology, such as computer aided design and manufacturing, will give problems in traditionally organised companies, unless they question long-held belief and practices; as is the case when you replace an old car by a helicopter. New manufacturing can also shock commercial departments because automated machine tools can produce parts with more narrow specifications, which in turn requires another market approach. Another example is the tightness of the procedures that are required for automated machine operation. These automated operations might magnify harmful effects that defect upstream processes might have on downstream processes. Without machine operators, who are physically handling parts, there is no one to realign them or compensate for small machine or operation error. Process engineering might try to provide expert systems to replicate an operator's talent for recognising errors or to build in sensors and programmed control. However, there always remains the input of pure craftsmanship.

New manufacturing systems will best work in an organisation where technical and managerial aspects are integrated through contributions of various functional staffs. Such organisations have flat structures (Edge, 1990). They are often self-managing within a framework of general corporate objectives. The organisation comprises many units, which are charged with specific research and development, engineering and manufacturing tasks. The units are highly motivated and their personnel is educated and often specialist. Managers must think like cross-disciplinary generalists, encourage corporate learning, harmonise efforts of specialists and be able to respond efficiently and quickly to changing market demands.

8.3.2 Total Quality Management

Total Quality Management is a management view that strives for creation of a customer-focused culture, which defines quality for the organisation and establishes the foundation for activities aimed at attaining quality-related goals. TQM is not merely a technique, but a philosophy anchored in the belief that long-run success depends on a uniform and firm-wide commitment to quality, which includes all activities that a firm carries out. TQM can become the core of a firm's strategy to remain competitive (Gatewood *et al.*, 1995).

According to Dean and Evans (1994) the term *total quality management* (TQM) covers a total, company-wide effort including all employees, suppliers and customers, and that seeks continuously to improve quality of products and processes to meet the need and expectations of customers. TQM has become the basic *business strategy* for firms that aspire to meet and exceed needs of their customers.

There probably are as many different approaches to TQM as there are businesses. Although no TQM program is ideal, successful programs share many characteristics. According to Dean and Evans (1994) the basic attributes of TQM are (1) customer focus, (2) strategic planning and leadership, (3) continuous improvement, and (4) empowerment and teamwork (Figure 8.12).

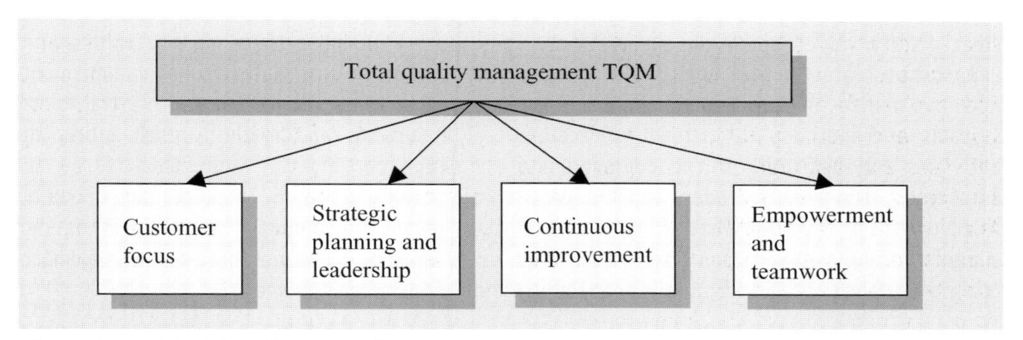

Figure 8.12. Principles of Total Quality Management (TQM).

Customer focus (1)

The customer is finally judging the quality. Quality systems must address all product and service attributes that provide value to the customer, which lead to customer satisfaction and loyalty. Many factors influence value and satisfaction throughout the customer's overall purchase, ownership, and service experiences. This includes the relationship between the company and customers, *i.e.* the trust and confidence in products and services, which leads to loyalty. This concept of quality includes not only product and service attributes that meet basic requirements, but also those features that enhance the product and differentiate it from competing offerings. This is often referred to as the "service bundle" or "service package". A business can achieve success only by understanding and fulfilling the needs of customers.

TQM is an important element of business strategy that is directed toward market-share gain and customer retention. It demands constant sensitivity to emerging customer and market requirements, and measurement of the factors that drive customer satisfaction. TQM also requires an awareness of developments in technology and rapid and flexible response to customer and market needs. Such requirements extend beyond merely reducing defects and complaints or meeting specifications. Nevertheless, defect and error reduction and elimination of causes of dissatisfaction contribute significantly to the customers' views of quality and are thus important parts of TQM. In addition, the company's approach to recovering from defects and errors is crucial to its improving both quality and relationships with customers.

Crucial to assuring quality for external customers are also the internal customers, such as the next department in a manufacturing process. Failures in meeting needs of internal customers will likely affect external customers as well. Employees must view themselves as customers of other employees and suppliers to others.

Strategic planning and leadership (2)
Achieving quality and market leadership requires a long-term strategy. Improvements do not occur overnight. Planning and organising of improvement activities require time and major commitments of all members of the organisation. Strategies, plans and budget allocations need to reflect long-term commitments to customers, employees, stockholders and suppliers. They also must address training, employee development, supplier development, technology evolution and other factors that support quality. A key part of the long-term commitment is regular review and assessment of progress in long-term goals.
Leadership for quality is the responsibility of top management. Senior leadership must create clear quality values and high expectations. They must implement these quality values and expectations into the way companies operate. This requires substantial personal commitment and involvement of senior management. The leaders must take part in creation of strategies, systems and methods for achieving excellence. The systems and methods must guide all activities and decisions of the company and encourage participation and creativity by all employees. The senior leaders must serve as role models through their regular personal involvement in visible activities, such as planning, reviewing company quality performance, supporting improvement teams and recognising employees for quality achievement. They must reinforce the values and encourage leadership in all levels of management.

Continuous improvement (3)
Continuous improvement is part of the management of all systems and processes. Achieving the highest levels of quality and competitiveness requires a well-defined and well-executed approach to continuous improvement. Such improvement needs to be part of all operations and all work unit activities in a company. Improvements may be of several types:

- enhancing value to the customer through new and improved products and services;
- reducing errors, defects, and waste;
- improving responsiveness and cycle time performance; and
- improving productivity and effectiveness in the use of all resources.

Thus, improvement is driven not only by the objective to provide better quality, but also by the need to be responsive and efficient. Both activities result in additional marketplace advantages. To meet all these objectives, the process of continuous improvement must contain regular cycles of planning, execution and evaluation. This requires a, preferably quantitative, basis for assessing progress and for deriving information for future cycles of improvement. Quality must be *measured*.
A company should select data and indicators that best represent the factors that determine customer satisfaction and operational performance. A system of indicators tied to customer and

company performance requirements will provide a clear and objective basis for directing all activities of the company toward common goals.

Empowerment and teamwork (4)

All functions at all levels of an organisation must focus on quality to achieve corporate goals. Teamwork can be viewed in three ways:

1. *Vertical view, i.e.* teamwork between top management and lower-level employees. Employees are empowered to make decisions that satisfy customers without a lot of bureaucratic hassles; barriers between levels are removed.
2. *Horizontal view, i.e.* teamwork within work groups and across functional lines. An example is a cross functional product development team that might consist of designers, manufacturing personnel, suppliers, salespeople, and customers.
3. *Interorganisational view, i.e.* organisational partnerships with suppliers and customers. Rather than dictating specifications for purchased parts, a company might develop specifications jointly with suppliers to take advantage of the suppliers' manufacturing capabilities.

The person who best understands his or her job is the one performing it. Also everyone must participate in quality improvement efforts. Thus employees must be empowered to make decisions that affect quality and to develop and implement new and better systems. This often requires a radical shift in the philosophy of senior management. In the traditional philosophy the workforce should be "managed" to comply with existing business-systems, which is contrary to what is required in TQM.

Companies can encourage participation by recognising team and individual accomplishments. So they share success stories throughout the organisation and encourage risk taking by removing the fear of failure. Moreover, they encourage the formation of employee involvement teams, implementing suggested systems that act rapidly, provide feedback and reward implemented suggestions. Moreover, they provide financial and technical support to employees to develop their ideas.

Employees need training in quality skills related to performing their work, and understanding and solving quality-related problems. Training brings all employees to a common understanding of goals and objectives and the means to attain them. Training usually begins with awareness of quality management principles and is followed by specific skills in quality improvement. Training should be reinforced through on-the-job applications of learning, involvement and empowerment.

Firms applying TQM successfully are realising a decrease in long run costs by building a more satisfied customer base (Gatewood , 1995). However, because of short run costs for new equipment, materials and additional labour are sometimes required to implement TQM processes and many firms are hesitating to adopt TQM. Time is another bottleneck to adoption of TQM. Many firms do not want to wait the two to ten years typically required to obtain the initial benefits from TQM.

8.3.3 Implementing TQM

In most companies it will be evident that implementing a TQM program implies a long-term fundamental change process. These organisational change processes and strategies have been

previously discussed in Chapter 6. Shared power strategies and, in particular, Organisation Development (OD) were proposed to be adequate for complex, long-range change efforts, involving improvement of internal problem-solving capacities.
At implementation of these change efforts, senior management, middle management and the workforce each have a critical role to play. Senior managers must ensure that their plans and strategies are successfully executed within the organisation. Middle managers provide leadership by which the vision of senior management is translated into the operation of the organisation. In the end, it is the workforce that delivers quality and must have not only empowerment, but also a true commitment to quality for TQM to succeed (Dean and Evans, 1994).
Senior managers must ensure that the organisation is focusing on needs of the customer. They must promote the mission, vision and values of the company throughout the organisation. Senior managers must identify critical processes that need attention and improvement. They also have to recognise the resources and trade-offs that must be made to fund the TQM activity. They must improve the progress and remove barriers to implementation. Finally, they must improve the processes in which they are involved (strategic planning, for example), both to improve performance of the process and to demonstrate their ability to use quality tools for problem solving (Tenner and De Toro, 1992).
Middle management has been viewed by many as a direct obstacle to creating a supportive environment for TQM. Middle managers are often seen as feeding territorial competition, stifling information flow, not developing and/or preparing employees for change and feeling threatened by continuous improvement efforts. However, middle management's role in creating and sustaining a TQM culture is critical. Middle managers improve the operational processes that are the foundation of customer satisfaction; they can make or break co-operation and teamwork. They are the principal means by which the workforce prepares for change.
Transforming middle managers into change agents requires a systematic process that dissolves traditional management boundaries and replaces them with an empowered and team-oriented state of accountability for organisational performance (Samuel, 1992).
The workforce must develop ownership of the quality process. Ownership and empowerment gives employees the right to have a voice in deciding what needs to be done and how. It is based on a belief that what is good for the organisation is also good for the individual and vice versa. Training, recognition and better communication are the key success factors for transferring ownership to the workforce.

8.4 Strategic alliances

Quality management in the agri- and food industry is strongly depending on co-operation, due to the mutual dependencies between elements of the food chain. In this section the concept of the value chain is followed by an analysis of alliances in the food chain.

8.4.1 The value system

In section 8.1.1 the concept of the value chain is mentioned. One of the key features of most industries is that a single organisation almost never undertakes all of the value activities from product design to delivery of the final product or service to the final consumer. There is usually specialisation and any organisation is part of the wider *value system,* which creates a product or service (see Figure 8.13)

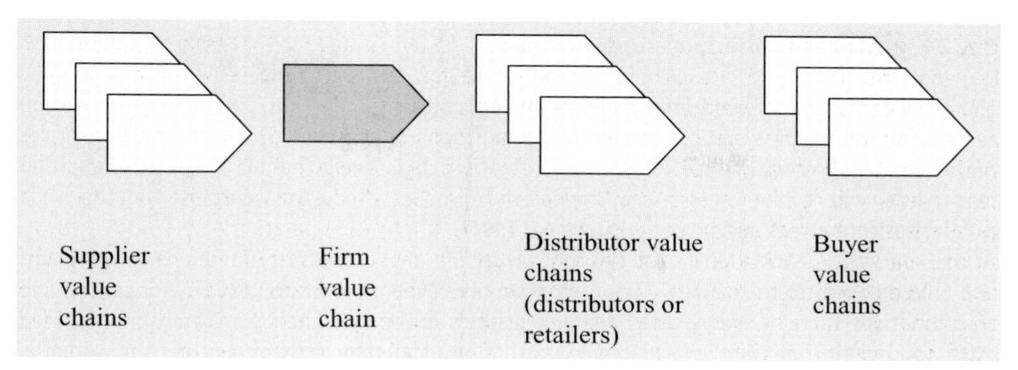

Figure 8.13. The value system.

In understanding the basis of an organisation's strategic capability, it is not sufficient to consider the organisation's internal position alone. Much of the value creation will occur in the supply and distribution chains, and this *whole process* needs to be analysed and understood. The quality of a food product when it reaches the final purchaser is not only influenced by the activities within one manufacturing company itself. It is also determined by *e.g.* quality of raw materials, half products, ingredients and performance of the distributors and retailers. The ability of an organisation to influence performance of other organisations in the value chain is crucially important for competence and a source of competitive advantage (Porter, 1985, Johnson and Scholes, 1997).

In order to be able to assure the highest quality levels on product quality and food safety, food companies and other food chain members feel a strong incentive for co-ordination. Highly co-ordinated food and agri product systems have special challenges (Downey, 1996). Co-ordination, whether it is voluntary or vertically integrated, means that the success of each party or business unit is dependent on the ability of other partners in the system to perform as expected. This is a risk that many entrepreneurs are not willing to take.

In some cases individual business entities value independence and autonomy so highly that they are unwilling to allow the needs of the system to dictate their decisions. Some agribusiness firms feel they are better off by retaining the ability to change sources at their discretion or find new outlets for their products without worrying about implications for a channel relationship. Even in a single vertically integrated firm there are often significant challenges of getting each level of the chain to consistently perform at required levels. Clearly, the longer the controlled chain, the greater the management challenges to make it work effectively. There are also external factors that affect the degree of co-ordination that may exist in any industry. Public desire to protect small independent farms or agri businesses sometimes results in the development of legal barriers to contractual or vertically integrated co-ordination.
In some cases, the savings or market advantages of co-ordination may not offset the costs associated with co-ordinating the activities. Here, a more traditional system of independent suppliers bringing commodities to an open market, sorting product into various categories, and allowing prices to be negotiated based on available supply and demand serves the needs of the market quite adequately (Downey, 1996).

8.4.2 Alliances in the food supply chain

There is a continuum of ways for organising the agricultural production/processing/distribution system. All alternatives along this continuum are forms of alliances where different processes are bound together with some form of co-ordination. It is likely that no single co-ordination mechanism will evolve across the agricultural products industries because of the wide differences in markets and products (Downey, 1996).

At one end of the continuum is the outright *ownership* of the necessary functions to bring agri and food products to the market. This alternative offers the greatest opportunity for control and co-ordination. There are many cases where companies in the food chain have built or purchased other food chain functions and placed them totally under their control. But such a system is capital intensive and requires high levels of management capability. As many large vertically integrated organisations can confirm, ownership does not automatically ensure trust, easy flow of information, and co-operative activity within the organisation.

In *joint ventures*, two or more companies form a separate business entity to accomplish mutually beneficial functions. These joint ventures have become relatively common, but are not always successful. The long-term commitment of joint venture partners often faces significant differences in philosophies and strategies that might result in a less successful relationship.

Contractual relationships between agribusiness entities can also accomplish the objective of co-ordination if they are carefully negotiated with proper motivation. An infinite set of contractual arrangements is possible. Contractual arrangements are usually short-term with less commitment than joint ventures and thus more attractive to a larger number of agri and food businesses.

Informal agreements or simply developing a "preferred supplier" relationship with a buyer are the least formal type of buyer/seller alliances and possibly the most common. "Preferred supplier" arrangements are totally based on informal relationships and proven dependability. With no legally binding commitment, this alliance is the least restrictive, but many will argue that it is still a very effective business arrangement.

8.4.3 Strategic alliances

As the complexity of markets has evolved, a new form of alliance is emerging, one that is thought to generate mutual benefit for both buyer and seller. It is based on trust and disclosure. This strategic alliance goes beyond contractual arrangement and generates close working relationships that can enhance the operating efficiency and marketing effectiveness of the combined business entities (Downey, 1996).

This business-to-business relationship more closely resembles a "partnership". It is a long-term relationship with common goals, mutual understanding and a depth of commitment that goes beyond what can easily be negotiated into a contract. To realise this strategic alliance, there must be a non-adversarial attitude in all parties involved in the alliance. Moreover, all parties must perceive the relationship as beneficial. True strategic alliances generally exceed simple single person relationships, so these alliances will survive personal changes. The details of the alliance have to be negotiated with several levels of management involved in order to seek areas of mutual benefit. A major advantage is that joint solutions to system problems can be found more easily and the benefits can be shared among all parties. Often these benefits are improvements in quality and result in competitive advantage in the marketplace that is reflected in customer satisfaction.

Paperwork is often reduced and transactions between partners are streamlined because there is mutual understanding and agreement. Usually companies can be managed more efficiently. Generally the level of customer service is enhanced because of a system that operates both more efficiently and effectively. The number of strategic alliances is usually limited because of the intensity of the relationship. There must be considerable compatibility of goals, philosophy and operations, for true strategic alliances to work effectively. Open and trusting communications are the hallmark of successful strategic alliances and partnerships. While a formal contract is typical in a strategic alliance, much of what makes it really work is intangible and difficult to commit to paper (Downey, 1996).

8.5 Quality policy evaluation

A company's quality policy should be evaluated from a strategic point of view. Mostly this is executed within the normal strategy evaluation processes, which are related to overall business performance. It should be noticed, however, that frequently business performance evaluation is restricted to financial criteria. A limited number of frameworks exist that are meant for evaluating (total) quality management. The most prominent frameworks are management awards like the United States' Malcolm Baldrige Award, the Japanese Deming's price and the European Foundation for Quality Management Award (Evans and Lindsay, 1996).

The Malcolm Baldrige Award

The criteria of the Malcolm Baldrige Award are built upon a set of core values and concepts that include:

- customer driven quality
- leadership
- continuous improvement
- employee participation and development
- quick response
- design quality and prevention
- long-range view of the future
- management by fact
- partnership development
- corporate responsibility and citizenship
- results orientation

These core values are embodied in seven categories, which form the basis for the assessment (Figure 8.14.):

The seven categories include leadership, information and analysis, strategic planning, human resource development and management, management of process quality, business results and customer focus and satisfaction.

Leadership (1): This category examines senior executives' leadership and personal involvement in setting strategic directions, building and maintaining a leadership system conductive to high performance, individual development and organisational learning.

Information and analysis (2): This category addresses the information and analysis requirements for performance improvements based upon the improvement of key processes.

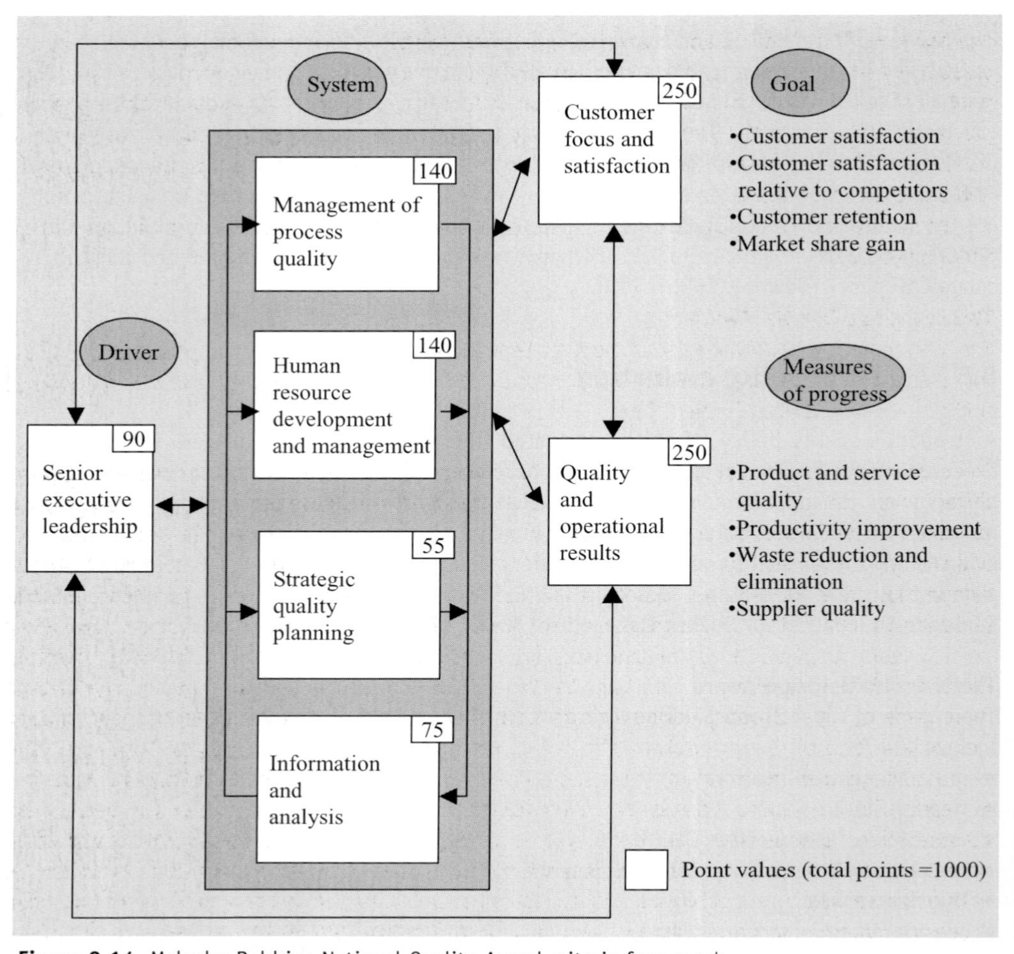

Figure 8.14. Malcolm Baldrige National Quality Award criteria framework.

Strategic quality planning (3): This category focuses on the firm's strategic and business planning and development of plans, along with the firm's attention to the customer and operational performance requirements.

Human Resource Development and management (4): The areas addressed in this category are concerned with how well human resource practices tie into and are aligned with the firm's strategic directions.

Management of process quality (5): In this category the key issues of process management - effective design, a prevention orientation, evaluation and continuous improvement, linkage to suppliers and overall high performance are assessed.

Business results (6): The company's performance in key business areas. It provides an appropriate orientation for all processes and process improvement activities.

Customer focus and satisfaction (7): The final category addresses how the company determines current and emerging customer requirements and expectations, provide effective customer relationship management and determines customer satisfaction.

The framework has four basic elements. Senior executive leadership sets (1) directions, (2) creates values, (3) goals and (4) systems. It guides the pursuit of customer value and firm performance improvement. The system comprises a set of well-defined and well-designed processes for meeting the firm's customer and performance requirements. Measures of progress establish a results-oriented basis for channelling actions to delivering ever improving customer value and company performance.
The basic aims of the system are delivery or ever-improving value to customers and success in the marketplace.

The European Quality Award
The European Quality Award assessment is based on customer satisfaction, business results, processes, leadership, people satisfaction, resources, people management, policy and strategy and impact on society (Figure 8.15).

In comparing these elements with the Baldrige Award framework, many similarities and several differences are apparent. Results, including customer satisfaction, people (employee) satisfaction and impact on society, constitute a higher percentage of the total score. Leadership and the quality system drive these results like in the Baldrige Award. The three criteria of people satisfaction, the impact on society and business results are somewhat different. People satisfaction refers to how employees feel about the organisation, including the working environment, perception of management style, career planning and development and job security. Unlike category 4 of the Baldrige Award (human resource development and management), the category people satisfaction is an independent result. The category impact on society focuses on perception about the company by the community at large and the company's approach to quality of life, environment and the preservation of global resources. The European Quality Award criteria put more emphasis on this category than on the public responsibility item in the Baldrige Award criteria. The business results criterion explicitly addresses the financial performance of the firm, its market competitiveness and its ability to satisfy shareholders' expectations. Other non-financial areas of performance, such as order processing time, new product design lead-time, and time to break-even are also considered.

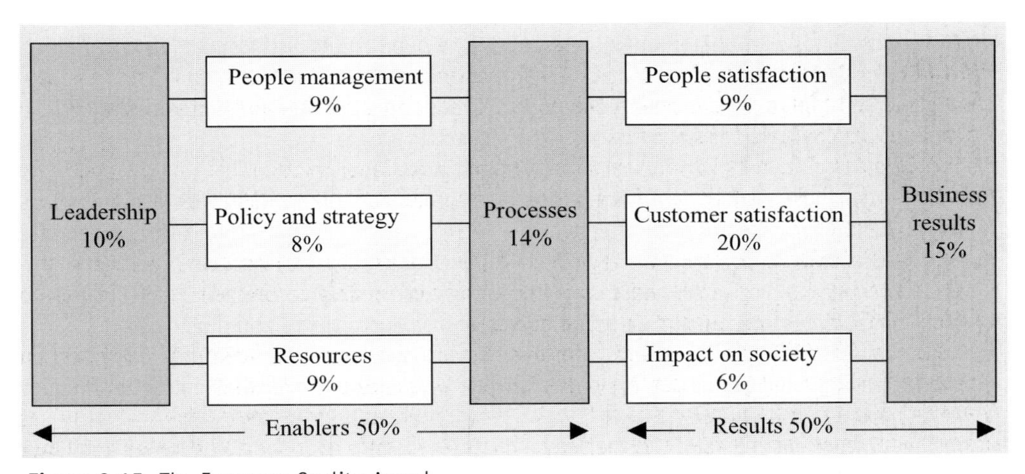

Figure 8.15. The European Quality Award.

8.6 Quality policy in the food industry

Nowadays the food industry faces the challenge to meet the broad range of demands from the market and environment (like other companies, legislation etc). Requirements are not only put on product quality, cost and availability, but also on flexibility, service and dependability of the business as reflected in the extended quality triangle. In practice, it appears to be difficult for agri-business and food companies to comply with all these different requirements.

Considering the different policy orientations reveal that in food production the product-process orientation is predominant. As a matter of fact, the food industry is characterised by a technological focus and most employees have a technical and/or technological education. Next to the product-process orientation is the procedure-orientation. The latter orientation is also widely accepted in the food industry because procedures are considered as a powerful tool to provide assurance in technological processes. However, in practice the procedure-orientation often gives complications because the effect of human behaviour and decision making processes on the accurate execution of procedures is underestimated. The people-orientation is not widely applied in food production because it is often considered as a soft and intangible approach.

Considering the policy orientations in agri business and food companies, it is not surprising that QA-systems are commonly perceived as technological supporting tools with a predictable outcome. This might explain why these systems (like HACCP, ISO) are rather appreciated by the food industry. Nevertheless, it is clear that these QA-systems have their restrictions. In practice, employees and operators experience that the QA-systems are pushed and that benefits are not clear. Systems that are developed by operators and employees themselves and which are considered as useful, are most successful and well accepted. This is one of the reasons why systems should be as simple and practical as possible. Too complex QA-systems will not be effective in practice. Pressing operators to comply with the enforced guidelines will be experienced as unpleasant and can result in a discrepancy between operators and quality department or QA managers. Moreover, external certification will increase the enforcement idea.

Another typical problem for the food industry originates from the demands from the market situation. Considering the administrative concepts shows that the market situation put conflicting demands as reflected in Figure 8.16. On the one hand quality demands are set which make a strong appeal to the bureaucratic administrative concept like:

- demands on high and advanced technological product quality, like for functional foods
- legal obligations and market demands with respect to safety
- increased requirements with respect to costs and price level

This set of demands requires a consistent quality approach at a high level with a high degree of assurance. In practice, this is reflected in a high level of centralisation and bureaucratic systems with many rules based on legislative obligations. However, on the other hand the continuous changing market and environment require a more flexible concept:

- high demands are put on service accompanied with a customer-oriented attitude
- high demands on flexibility due to *e.g.* unexpected happenings, seasonal influences, etc.
- the increasing reduction of life cycle of products, which leads to much attention on product development
- increased mass customisation in the food industry leads to a enormous broadening and large changes in product assortments

Al these aspects require a responsive quality approach that puts responsibilities low in the organisation and that is not hindered by detailed rules and guidelines.

From the techno-managerial view point, the above mentioned challenge should be approached using the following starting points:

- The technological discussion must be focused on the technological core competence of the business. If a company has a competitive position in the market for specific products (and the corresponding technology) then it can not allow any failures.
- Technological core competence should be supported by suitable QA-systems, especially GMP and HACCP and to a lesser extent ISO.
- Integration of the three policy orientations in a concept of Total Quality Management (TQM), which is build upon strategic consumer values and core competence of the company.
- The QA-systems must be a tool for all employees and is subservient to the TQM principles.
- The core competence of each unit in the production chain should take advantage of intensive collaboration and chain development. This process can be started by extensive exchange of information between units of the production chain.

In conclusion, the total food production chain needs to focus on the technological core competence coupled to the principles of Total Quality Management in order to comply with customer value and consumer satisfaction.

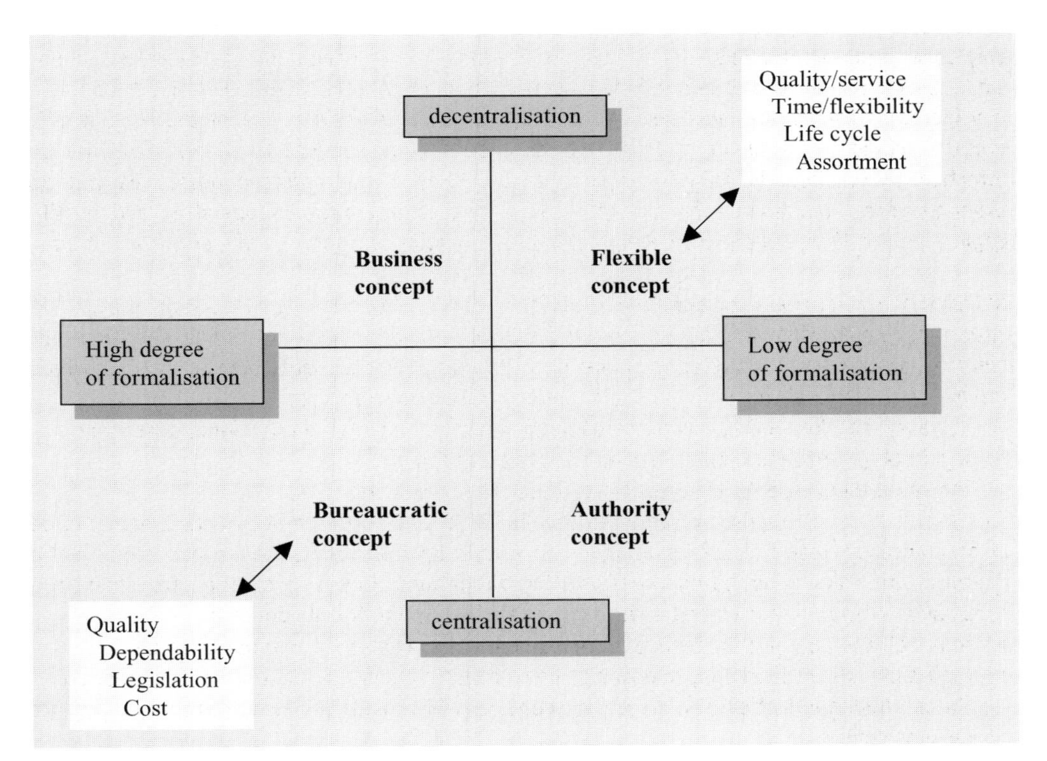

Figure 8.16. Conflicting requirements.

9. Food quality management in perspective

The current situation in the food markets and the agri-food production chain is very turbulent due to changing consumer requirements, increased competition, environmental issues and governmental interests. In this chapter will be described the current dynamics of food quality. The techno-managerial approach gives finally, considering the dynamics, a concise picture of the core developments for the future of food quality management.

9.1 Food quality dynamics

Food quality is not a static concept. As a matter of fact, the agri- business and food industry are in a transition phase from a productivity/costs focus, towards a customer focus, resulting in a dynamic situation in the agri-food chain.

From a chain oriented perspective, three major cycles of change can be distinguished (Figure 9.1). The first cycle encompasses developments in the market. A decreasing life cycle of products and rapidly changing preferences of consumers lead to more impulsive behaviour by consumers who have become a moving target for product developers. The second cycle refers to the technologies associated with processing and production systems. In general, innovation in technologies is slower than changes in the market situation. The third cycle deals with breeding and primary production and is actually the slowest cycle even with the use of modern biotechnology. Short-term changes in the market are impossible to follow. From the chain perspective it is of utmost importance for those actors that are active in the primary production cycle or the processing cycle to have a strategic view on market developments and to identify market niches where they can be strong and ahead of their competitors. On the other side, it is of vital importance to have a clear view on technological possibilities and limitations from the processing and the primary production perspective to be able to follow the most effective quality design strategy.

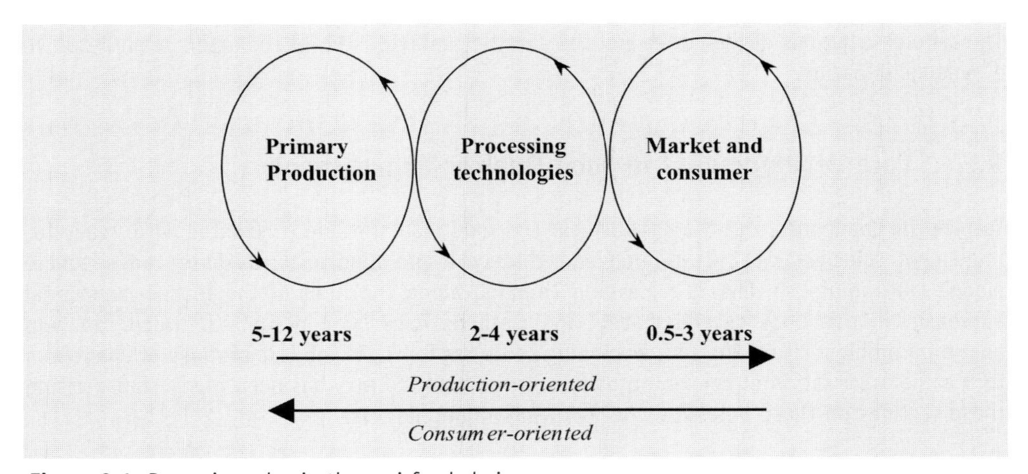

Figure 9.1. Dynamic cycles in the agri-food chain.

One other development relevant in this context is the globalisation of food production systems. Raw materials are taken from all over the world, which is similar for finished food products. In Western countries consumers are used to the fact that all kind of food products are available all over the year. Trade barriers are disappearing, reinforcing this type of development. From this perspective it is obvious that strategic collaboration within the supply chain is of utmost importance

Considering the cycles in more detail reveals that each actor in the chain faces different requirements and environmental influences. On the consumer side, the many changes in the market for new food products call for a repositioning of existing food production systems and raise the question whether the concepts currently used can survive the challenges of the future. Apart from market saturation, a number of other developments have a large influence on market conditions. Generally consumers are becoming better educated and more demanding. They have become less predictable in their purchase behaviour; they eat more outside and are more conscious about health-related aspects. As a result there is a continuous need for new products and a more differentiated food product assortment. One typical example is the current developments in the functional food area. Related to these developments product life cycle becomes shorter, and efficiency and flexibility of food production systems become even more important.
Besides the continuous changing consumer demands, retailers have become more powerful and put high demands on food companies (with respect to safety, quality, profits, and circulation rate). Retailers are in fact the major link between consumers and food companies. Their knowledge about consumer purchases is even increased due to the use of information technology, which gives them a very powerful position.
Food companies attempt to anticipate these dynamics by concentrating on product development, intensifying their knowledge potential, geographic expansion and differentiation by brands.
The primary sector also faces the dynamic demands. Due to several severe food safety problems, which had their origin in the primary sector, much attention has to be paid to improvement of quality systems. Besides, they have increased costs due to environmental regulations, labour and soil prices and there is pressure on prices due to liberalisation of the market.
The different rhythms of the cycles and the specific characteristics of the actors really result in a dynamic situation.

9.2 Core developments in Food Quality Management

To face the challenges that the agribusiness and food industries are confronted with, ready-to-implement solutions can't be offered immediately. A profound analysis, based on adequate models and points of view, is necessary. Thinking along the lines of the techno-managerial approach, in this book several models and methods have been presented, which can help analysing problem situations and developing solutions for it. In the last sections of chapters 4 to 8 a characterisation of common problems in the food industry has been given. Summarising these sections on those problems, the following highlights arise:

a. The *functional organisation,* which is common in the food industry, provides too much of a basis for departments to meet their own interests and responsibilities, for centralisation of knowledge and for a top-down culture.
b. Although progress is clear, food companies and agribusiness lack *responsiveness and market orientation*. The conflicting requirements, flexibility at one hand and assurance at the other hand, tend to be solved in the direction of assurance. This seems to favour the formal quality assurance systems.
c. The food industry does not handle *knowledge* with care. Much of the knowledge remains at the suppliers of basic materials and equipment and is not available for the users. Internal knowledge is also badly stored and feedback to production and other places in the organisation, where quality can be improved, is as well not organised. Moreover, this is a problem when realising that knowledge plays an important role in developing core competence of the organisation.
d. *Chain collaboration* is still difficult due to lack of mutual trust and believes in the advantages of co-operations. In addition, the fact that actors consider quality from different viewpoints results in a poor understanding of the concept of product quality.

In the previous chapters it has been made clear that each of these problems is a driver for poor fulfilling of the different quality management functions. Moreover, since these problems occur simultaneously, negative effects are enhanced. Considering the food dynamics as mentioned before, it can be concluded that the situation is becoming worse. Modern developments like customer-orientation, market competition, chain collaboration and using advanced technology, typically require flexible organisations and extensive use of knowledge.

In this book possible solutions have been proposed from the techno-managerial viewpoint. These different directions of solutions have been summarised in Figure 9.2, which is a simplified representation of the core model of this book (Figure 1.4).
In Figure 9.2, the core developments in food quality management for the near future have been depicted based on considerations described in the book and experiences of the authors.

The three core developments as depicted in Figure 9.2 are:

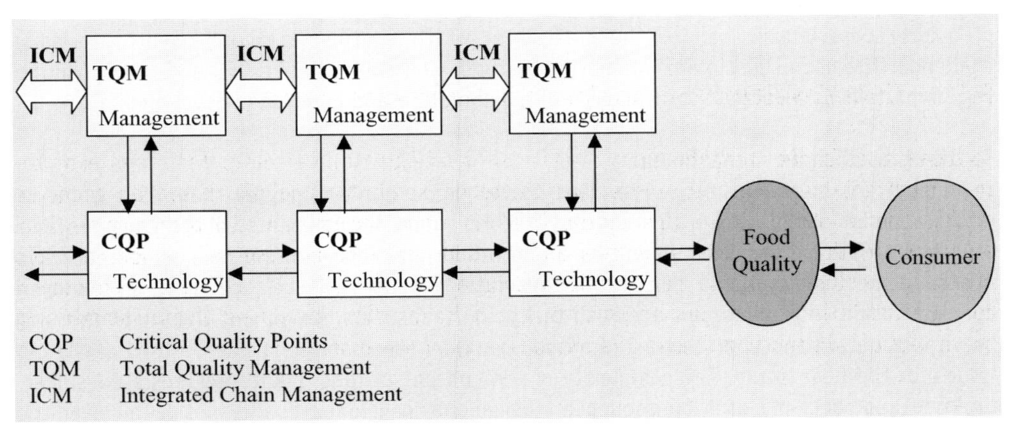

Figure 9.2. Core developments in food quality management.

1. Focus on critical quality points

In order to comply with the conflicting requirements: costs and dependability on the one hand and responsiveness the other hand, those points in the process should be assessed that have decisive influence on final product quality. These points are often related to the technological core competence of the company.

These critical quality points require both an extensive analysis with respect to necessary investment, and research and control activities based on detailed consideration of costs, environmental impact and meeting consumer demands. The methods and principles of risk analysis are expected to play a major role. Moreover, the use of predictive modelling can accelerate and optimise design processes, which will support the technological core competence of the company. Development in information and communication technology will support the above developments.

2. Total quality management

The principles of Total Quality Management should be embraced in the agribusiness and food industry.

To avoid the disadvantages of the functional organisation, the principles of empowerment in combination with customer focus will result in a transformation from a vertical into a horizontal organisation. A potential risk of this transformation is that knowledge development and management do not receive sufficient attention and will be dispersed throughout the organisation. Consequently the senior management has to create an organisational structure, in which knowledge-intensive departments, like for instance the quality department, the engineering department and the maintenance department, will have a strong position. Of course, good co-operation between these departments and the production departments is necessary. Advanced knowledge management systems are useful tools in achieving this.

Teambuilding will be more important for responsiveness of the company as well as knowledge integration in product design and business development. Therefore, it will become an important tool in management decision making. In this context the use of current quality systems such as ISO is no longer a top-down assignment of procedures but a tool that assists and guides everyone involved in quality management. As a result systems will become lean and bureaucratic overkill will be avoided.

Commitment of the senior management will be crucial and they have to take the initiative to integrate the technological and managerial aspects (Kramer and Van den Briel, 2002). A climate of quality consciousness should be created in which continuous improvement is a daily challenge. The use of discipline-integrating methods and concepts like QFD and HACCP will be very important as a catalyst for this type of collaboration and knowledge development.

3. Integrated chain management

Food quality is more and more dependent on the production and supply chain as a whole to create superior quality. Using this concept, technological developments will be triggered in a way that each actor in the chain will be supported in improving areas of core competence and at the same time superior quality of the final product will be achieved. In addition communication with consumers about all technological aspects throughout the food chain will be vital to obtain and improve trust. Transparency from the chain perspective is one of the big issues for the near future. The complexity of the concept of food quality requires a consumer-oriented approach and as a consequence a commonly developed quality notion all over the chain. Collaboration is crucial and will be extended towards innovation and design processes.

Quality systems will have to be redesigned into chain quality systems from the perspective of efficiency as well as assurance as a future consumer demand.

The combined use of these core elements offers opportunities for policy making in food quality. Policy-making is very complex and it is submitted to severe judgement. Policy-making requires both fundamental and policy-relevant research. Both types of research require profound analysis and we hope that the methods, concepts and models as proposed in this book might contribute to this process.

Literature references

Abram, I.; Baumbach, F.; Curiel, G.J.; Harrison, D.C.; Pederson, T.; Peschel, P.; Quente, S.; Sondergaard, B.; Tuuslev, T. 1994. Hygienic design of valves for food processing. Trends in Food Science and Technology, 5,169-171.

Ahlmann, H., 1989. Quality strategies for survival and increased prosperity. The Quality Meeting at Linköping University (in Swedish).

Alwitt, L.; Berger, I. 1993. Understanding the link between environmental attitudes and consumer product usage; Measuring the moderate role of attitude strength. In: Advances in Consumer Research.

Banens, P.J.A.; Does, R.J.M.M.; Dongen van G.B.W.; Engel, J.; Hasselaar, M.M.A.; Lieshout, R.A.J.M.; Praagman, J.; Schriever, B.F.; Trip, A. Veen van der H. 1994. Regelkaarten en stabiliteit. In: Industriële statistiek en kwaliteit. Banens P.J.A. et al. Kluwer Bedrijfswetenschappen. BV. Deventer. The Netherlands, 432 pp. (in Dutch).

Barendsz, A.W. 1994. A logic sequence to design a HACCP quality management system. Voedingsmiddelen Technologie, 8, 13-17 (in Dutch).

Barendsz, A.W. 1997. Quantification of risks by quantitative and semi-quantitative risk assessment. Voedingsmiddelen Technologie, 13, 11-13(in Dutch).

Bech, A.C.; Hansen, M.; Wienberg, L. 1997. Application of the house of quality in translation of consumer attributes into sensory attributes measurable by descriptive sensory analysis. Food Quality and Preferences, 8, 329-348.

Bergman, B.; Klefsjö, B, 1994. Quality, from customer needs to customer satisfaction. McGraw-Hill, New York.

Bergmeyer, H.U.1983. Methods of Enzymatic analysis. Bergmeyer, H.U. Academic press, New York.

Blake, R.R. , Mouton, J.S.,Barnes, B.L. and Greiner, L.E., 1964. Breakthrough in Organization Development. Harvard Business Review, Nov.-Dec. 1964.

Blanchfield, J.R. 1988. How the new food product is designed. Food Sci. Technol. Today, 2, 54.

Boddy D.; Paton, R. 1998. Management, an introduction. Prentice Hall, London.

Boesten, M.E.W. 1995. Integrated food-chain security (IKZ) must result in surplus value for the vegetable and fruit sector. Voedingsmiddelen technologie, 8, 11-13. (in Dutch).

Boulding, K. 1956. General systems theory, the skeleton of science. Management Science 1956.

Bouman, H.E. 1996. The origin and contents of HACCP. In: HACCP in meat, poultry and fish processing. Pearson A.M.; Dutson, T.R. (eds.), Chapman & Hall, Glasgow, pp 1-7.

Bounds, G. ,Yorks,L., Adams, M. And Ranney, G., 1994. Beyond Total Quality Management. McGraw-Hill, New York.

Buchanan, R.L. 1995. The role of microbiological criteria and risk assessment in HACCP. Food Microbiology, 12, 421-424.

Buchanan, R.L.; Damert, W.G.; Whiting, R.C.; Van Schothorst, M. 1997. Use of epidemiological and microbial food survey data to develop a 'worst-case' estimate of the dose-response relationship between *Listeria monocytogenes* levels and the incidence of foodborne listeriosis. Journal of Food Protection, 60, 918-922.

Burggraaf, W.N.A. 1998. An approach to come to hygienic design. Voedingsmiddelentechnologie, 8, 44-45 (in Dutch).

Burns, T.; Stalker, G. 1961. The management of innovation. Tavistock, republished by Oxford University Press, London, 1994.

Buzzell, R.D.; Gale, B.T., 1987. The PIMS principles: Linking strategy to Performance. The Free Press. New York.

Byrnes, H. 2001. World-wide norm for food safety. Voedingsmiddelentechnologie (VMT), 9, 20-21. (in Dutch).

Callis, J.B.; Illman, D.L.; Kowalski, B.R. 1987. Process analytical chemistry. Anal. Chem., 59, 104-116.

Certo, S.C. 1997. Modern management. Prentice Hall, New Yersey.

Certo, S.C., 2000. Modern Management. Prentice Hall, New York.

Charteris, W.P. 1993. Quality function deployment: a quality engineering technology for the food industry. Journal of the Society of Dairy Technology 46,12-21.

Codex Alimentarius Commission. 1995. Report on the 9th session of the Codex Committee on residues of veterinary drugs in Foods. Washington D.C. USA, 5-8 December 1995. Alinorm 97/31.

Codex Alimentarius commission; a joint FAO/WHO food standards programme. 1997. Codex Alimentarius, general requirements (food hygiene) supplement volume 1b, 2nd ed. Rome, Italy.

Codex Alimentarius commission; a joint FAO/WHO food standards programme. 1998. Report of 31st session of the codex committee on food hygiene. Alinorm 99/13A. Rome, Italy.

Cohen, J.C. 1990. Applications of qualitative research for sensory analysis and product development. Food Technology, 41, 164.

Cohen, L. 1995. Quality function deployment. How to make QFD work for you. Engineering Process Improvement Series. Addison-Wesley Publishing Company. Massachusetts, USA, pp. 348.

Commodity Board for animal feed production of the Netherlands. 1998. GMP codes for the animal feed production, Den Haag, The Netherlands (in Dutch).

Cooper, R.G. 1993. Winning at new products: accelerating the process from idea to launch, 2nd edition Reading, Addision-Wesley.

Cooper, R.G.; Kleinschmidt, E.J. 1990. New Products: the key success factors. America Marketing Association. Chicago.

Costa, A.I.A.; Dekker, M., Jongen, W.M.F. 2001. Quality function deployment in the food industry: a review. Trends in Food Science and Technology, 11, 306-314.

Crosby, P.B., 1979. Quality is Free. McGraw Hill, New York.

Curiel, G.J.; Hauser, G.; Peschel,P.; Timperley, D.A. 1993. Hygienic equipment design criteria. Trends in Food Science and Technology, 4, 225-229.

Cyert, R.M.; March, J.G. 1963. A behavioral theory of the firm. Prentice Hall. Englewood Cliffs.

Dale, B.; Boaden, R.; Wilcox, M.; McQuater, R. 1998. The use of quality management techniques and tools: an examination of some key issues. Int. J. Technology Management, 16, 305-325.

Dalen, G.A. 1996. Assuring eating quality of meat. Meat Science, 43, 21-33.

Damman, H. 2001. 'BRC-standard zet industrie op het verkeerde been'. Voedings-middelentechnologie, 33 (9): 15 (in Dutch).

Damman, J. 1998. Codex Alimentarius will become increasingly important. Voedingsmiddelen Technologie , 14, 16-18 (in Dutch).

Damman, J. 1999. CBL-akkoord over BRC-standaard moet aantal audits reduceren. *Voedings-middelentechnologie*, 31 (22): 15-17 (in Dutch).

Damman, H. 2000. EFSN co-ordinates and make uniform; SAFE will inform EVA. Voedingsmiddelentechnologie, 6, 17 and 19.

Daniëls, H.P. 1989. Integral quality assurance system for slaughter poultry IKB. Commodity board for Poultry and Eggs, 186 pp.

Dean, J.W.; Evans, J.R. 1994. Total Quality. West Publishing Company, New York.

Dekker, M. Linnemann, A.R. 1998. Product development in the food industry. In: Innovation of food production systems. Jongen W.M.F.; Meulenberg, M.T.G. (eds.). Wageningen Pers, Wageningen, The Netherlands. 67-86.

Deming, W.E. 1993. The new economics for industry, government, education. MIT Centre for Advanced Engineering study. Cambridge MA.

Deming, W.E. 1986. Out of the Crisis. MIT Centre for advanced Engineering Studies, Massachusetts.

Dessler, G., 2001, Management. Prentice Hall, New York.

Downey. W.D., 1996. The challenge of food and agri product supply chains. In: Proceedings of the 2nd International Conference chain management in agri-and food business. J.H. Trienekes; P.J. Zuurbier (eds.) Wageningen Agricultural University, Wageningen.

Dufey, P.A. 1995. Fleisch- und Fettqualitat bei Schweinemast mit Weidegang. Agrarforschung, 2, 453-456.

Early, R. 1997. Putting HACCP into practice. International Journal of Dairy Technology, 1, 7-13.

Edge J. 1990. Quality Improvement. In: Handbook of Quality management. Lock D (ed). Gower Publishing Company, Aldershot, Chapters 18,19 and 20..

Edwards, S.A.; Casabianca, F. 1997. Perception and reality of product quality from outdoor pig systems in Northern and Southern Europe. In: Livestock farming systems. More than food production. Proceedings of the 4th International Symposium on Livestock farming systems. EAAP Publication, 145-156.

EHEDG, 1995. European Hygienic Equipment Design Group. In: Trends in Food Science and Technology. Elsevier Science Publishers Ltd., Cambridge,UK, 134 p.

Ellis, M.J. 1994. The methodology of shelf life determination. In: Shelf life evaluation of foods. Man, C.M.D.; Jones, A.A. (eds.) Blackie Academic & Professional, Chapman & Hall, Glasgow, 27-39.

European Commission Directorate-General III Industry. 1996. Guide for the introduction of an HACCP system in pursuance of Article 3 of Directive 94/43/EEC on the hygiene of foodstuffs in small and medium-sized business in the food industry. III/5087/96-5087EN1.

European Union. 1990. Council Regulation (EEC). No. 2377/90 of 26 June 1990 laying down a Community procedure for the establishment of maximum residue limits of veterinary medicinal products in foodstuffs of animal origin.

Evans, J.R., Lindsay, W.M. 1996. The management and control of quality. Evans, J.R., Lindsay, W.M. (eds.) West publishing Company. St Paul.

Feigenbaum, A. 1961. Total Quality Control. McGraw-Hill, New York.

Feil, B.; Stamp, P. 1993. Sustainable Agriculture and product quality: a case study for selected crops. Food Reviews International, 9, 361-388.

Fennema, O.R.; Tannenbaum, S.R. 1985. Introduction to Food Chemistry. In: Food Chemistry. Fennema, O.R. (ed.), Marcel Dekker, Inc., New York., 1-23.

Food and Drug Administration FDA. 1994. Title 21, Code of Federal Regulations, Part 110 Current Good Manufacturing Practice in manufacturing, packing or holding human food.

Forsythe, S.J.; Hayes, P.R. 1998. Chapter 3: Food Spoilage. In: Food Hygiene, Microbiology and HACCP. Forsythe, S.J.; Hayes, P.R. (eds). Aspen Publishers, Inc. Gaithersburg, Maryland.86-149.

Fortin, A. 1989. Pre-slaughter management of pigs and its influence on quality (PSE/DFD) on pork. Proc. 35. ICoMST, Vol. III, 981-986.

Fortuna, R.M. 1988. Beyond quality: taking SPC upstream. Quality Progress, 21, 23-28.

Fox, P.F. 1991. Food Enzymology, Vols 1 and 2. Fox, P.F. (ed.) Elsevier, New York.

French, J.; Raven, B. 1960. The bases of social power. In Group Dynamics: Research and Theory, D. Cartwright and A. Zanders (eds). Harper and Row, New York.

Frewer, L.J.; Howard, C.; Hedderley, D.; Shepherd, R. 1997. Consumer attitude towards different food-processing technologies used in cheese production. The influence of consumer benefit. Food Quality and Preferences, 8, 271-280.

Fuller, G.W. 1994. New food product development. From concept to marketplace. CRC Press, Boca Rotan, Florida, USA, pp. 275.

Galbraith, J.R. 1973. Designing complex organizations. Addison Wesley, Reading.

Gardner, S. 1995. Food safety: An overview of international regulatory programs. European Food Law Review 2, 123-149.

Garret, E.S.; Jahncke, M.L.; Cole, E.A. 1998. Effects of Codex and GATT. Food Control, 9, 177-182.

Garrett, E.S.; Hudak-Roos, M. and Ward, D.R. 1995. Implementation of the HACCP program by the fresh and processes seafood industry. In: HACCP in meat, poultry and fish processing. Advances in meat research series 10. Pearson, A.M.; Dutson, T.R. (eds.). Blackie Academic & Professional. London, 109-133.

Garvin, D.A. 1984. What does product quality really mean? Sloan Management Review, 26, 25-43.

Gatewood, R.D.; Taylor, R.K., Ferrell, O.C., 1995. Management: Comprehension, Analysis and Application, Austen Press, London.

Gerats, G.E.C. 1990. Working towards quality (in Dutch with a summary in English). Dissertation, University of Utrecht.

Gerwen van, S.J.C.; Wit de, J.C.; Notermans, S.; Zwietering, M.H. 1997. An identification procedure for foodborne microbial hazards. International Journal of Food Microbiology, 38, 1-15.

Gobbi, C. and M. Mufatto. 1999. A reference model for extend enterprise. In: Productivity and Quality Management. W.Werther, J. Takala, D.J.Sumanth. (eds.) MCB University press, Bradford.

Goodfellow, S.J. 1995. Implementation of the HACCP program by meat and poultry slaughterers. In: HACCP in meat, poultry and fish processing. Advances in meat research series vol. 10. Pearson, A.M.; Dutson, T.R. (eds.). Blackie Academic & Professional. London, 58-71.

Gould, G.H. 1996. Methods for preservation and extension of shelf life. International Journal of Food Microbiology, 33, 51-64.

Gould, W.A.; Gould, R.W. 1993. Total quality assurance for the food industries 2nd edition. CTI Publications, Inc. Maryland USA, 464 pp.

Govers, C.P.M. 1996. What and how about quality function deployment (QFD). Int. J. Production Economics, 46-47, 575-585.

Graf, E.; Saguy, I.S. 1991. Food product development: from concept to marketplace. Graf, E.; Saguy, I.S. (eds.) Van Nostrand reinhold, New York, Chapter 3.

Grootenhuis, A. 2001. BRC-inspectie is momentopname: uniformiteit rond BRC-standaard nog ver weg. Voedingsmiddelentechnologie, 33 (6): 44-45. (in Dutch).

Grunert, K.G.; Baadsgaard, A.; Larsen, H.H.; Madsen, T.K. 1996. Market orientation in food and Agriculture. Kluwer Academics Publishers, Boston.

Haard, N.F. 1985. Characteristics of edible plants tissues. In Food Chemistry. Fennema, O. (ed). Marcel Dekker, Inc., New York, 857-911.

Hamel, G. and Prahalad C.K., 1994. Competing for the future. Harvard Business Schol Press, Boston.

Hammer, M. and J. Champy, 1993. Re-engineering the corporation: a manifesto for business revolution. Harper Collins, New York.

Harding, F. 1995. Hygienic quality. In: Milk quality. Harding, F. (ed.) Blackie Academic & Professional. London, 40-59.

Heeschen, W.H. 1998. Milk hygiene and milk safety in the European and International markets. Kieler Milchwirtschaftliche Forschungsberichte, 50, 53-77.

Heeschen, W.H. Harding, F. 1995. Contaminants. In: Milk quality. Harding, F. (ed.) Blackie Academic & Professional. London, 133-150.

Heizer, J.; B. Render, 1993. Production and Operations Management, strategies and tacties. Allyn and Bacon, Boston.

Hellriegel, D. and Slocum, J.W. 1992. Management. Addison-Wesley, New York.

Henning, G.J. 1988. Research guidance as aid for product development. Voedings-middelentechnologie, 6 and 8, 13-16 and 13-17(in Dutch).

Hersey, P. and Blanchard, K. 1988. Management and organizational behavior.Prentice Hall, Englewoods Cliffs.

Hird, D.W. 1987. Review of evidence for zoonotic listeriosis. J. Food Protection, 50, 429-433.

Hofmeister, K.R. 1991. Quality function deployment: market success through customer-driven products. In: Food Product Development. Graf, E.; Saguy, I.S. (eds.) Van Nostrand Reinhold, New York, 189-210.

Hogarth, R.H., Kahneman, D. And Tversky, A., 1994. In: Judgement in managerial decision making, M. Bazerman. Wiley, New York.

Holah, J.T.; Venema-Keur, B.M.; Trägårdh,C.; Illi, H.; Lalande, M.; Cerf, O. 1992. A method for assessing the in-place cleanability of food-processing equipment. Trends in Food Science and technology, 3, 325-328.

Hoogland, J.P.; Jellema, A.; Jongen, W.F.M., 1998. Quality assurance. In: Innovation of food production systems, eds. W.M.F. Jongen and M.T.G. Meulenberg. Wageningen Pers, Wageningen.pp 139-158.

Horwitz, W. 1988. Sampling and preparation of sample for chemical examination. J. Assoc. Off.Anal. Chem., 71, 241-245.

Hughes, D. 1994. Breaking with tradition: building partnerships and alliances in the European Food Industry. Wye College Press. Wye.

Hultin, H.O. 1985. Characteristics of muscle tissue. In: Food Chemistry. Fennema, O. (ed.). Marcel Dekker, Inc., New York., 726-790.

Hurst, W.C.; Schuler, G.A. 1992. Fresh produce processing- An Industry perspective. Journal of Food Protection, 55, 824-827.

IFST: Institute of Food Science and Technology of the U.K., 1991. Food and Drink Good Manufacturing Practice: A guide to its responsible management, London.

Imai, M. 1986. Kaizen, The Key to Japan's Competitive Success. Random House, New York.

Inspectorate for Health protection, Food Commodities and Veterinary Affairs. 1995. First safety of food . Information on the Commodity Regulation on Food Hygiene article 20 and 31. Rijswijk, The Netherlands (in Dutch).

Institute of Food Science and Technology (IFST). 1998. IFST publication 'Food and Drink - Good Manufacturing Practice: A Guide to its Responsible Management' 4th Edition.

International Commission on Microbiological Specifications for Food (ICMSF) 1989. Micro-organisms in food (vol. 4). Application of the hazard analysis critical control point (HACCP) system to ensure microbiological safety and quality. Blackwell Scientific Publications, London, UK. 357 pp.

International Commission on Microbiological Specifications for Foods (ICMSF) working group on microbial risk assessment. 1998. Potential application of risk assessment techniques to microbiological issues related to international trade in food and food products. Journal of food protection, 61, 1075-1086.

International standardisation Organisation (ISO) 1987. Quality systems - model for quality assurance in design/development, production, installation and servicing (ISO:9001); Quality management and quality systems elements (ISO:9004).

International standardisation Organisation (ISO) 1994. Quality systems - model for quality assurance in design/development, production, installation and servicing (ISO:9001); Quality management and quality systems elements (ISO:9004).

International standardisation Organisation (ISO) 1998a. Quality management systems-concepts and vocabulary ISO 9000:2000. ISO/TC 176/SC1/N 185.

International standardisation Organisation (ISO). 1998. Quality management systems- Requirements (9001:2000); Quality management systems- Guidelines (9004:2000). Document ISO/TC 176/SC 2/N 415.

International standardisation Organisation ISO/CD2 9001:2000. 1999. Quality management systems-Requirements. Document ISO/TC/SC 2/N 434.

Ishikawa, K., 1985. What is Total Quality Control? The Japanese Way. Prentice Hall, Englewood Cliffs, New Yersey.

Ivancevich, J.M. ,Lorenzi, P. and Skinner, S., 1994. Management, quality and competitiveness. Irwin, Boston.

Janssen, F.J.A. 1991. Process description (in Dutch). In: Handbook Integrated Quality Assurance. Schuurman, F.J.H. (ed.). ISBN 90 B 987 2004.

Jay, J.M. 1996. Modern Food Microbiology, 5th edition. Heldman, D.R. (ed.). Chapman and Hall, New york. 661 pp.

Johnson G. and K. Scholes, 1997. Exploring corporate strategy Prentice Hall, London.

Jongen W.M.F.; Meulenberg, M.T.G. 1998. Introduction. In: Innovation of food production systems. Jongen W.M.F.; Meulenberg, M.T.G. (eds.). Wageningen Pers, Wageningen, The Netherlands. 1-3.

Jongen, W.M.F. 1998. Technological innovations in a changing environment. In Proceedings.

Proceedings Agri Food Quality II, Turku, Finland (1998).

Juran, J.M. 1951. Quality Control Handbook. McGraw Hill, New York.

Juran, J.M. 1964. Managerial Breakthrough. McGraw Hill, New York.

Juran, J.M. 1988. Juran on planning for quality. The Free Press, New York.

Juran, J.M. 1990. Juran on leadership for Quality. The Free press, New York.

Juran, J.M. 1992. Juran on quality by design. The new steps for planning quality into goods and services. New York: The Free Press Juran, J.M. (ed.), pp. 538.

Micossi, S. 1998. Perspectives on European Food Legislation. Paper prepared for the Mentor Group 1. The forum for US-EU legal- Economic Affairs. Helsinki, September, 1998, 1-9.

Kamann, D. 1989. Actoren in netwerken, in: F. Boekema en D. Kamann. Sociaal economische netwerken. Wolters-Noordhoff (in Dutch).

Kampfraath, A.A. ; Marcelis, W.J. 1981. Management and Organization. Kluwer, Deventer (in Dutch).

Kaulio, M.A. 1998. Customer, consumer and user involvement in product development: a framework and a review of selected methods. Total quality management, 9, 141-149.

Kehoe, D.F. 1995. The fundamentals of quality management. Kehoe, D.F. (ed.). Chapman and Hall. .

Keuning, R. 1994. Product development in the food industry. Lecture notes. Department of Food Technology and Nutrition Sciences, Wageningen University, The Netherlands.

Koeferli, C.S.; Schwegler, P.P.; Hong-Chen, D. 1998. Application of classical and novel sensory techniques in product optimisation. Lebensm.-Wiss. U.- Technol., 31, 407-417.

Kogure, M.; Akao, Y.; 1983. Quality function deployment and CWQC in Japan. Quality Progress, 16, 25-29.

Kolarik, W.J. 1995. Creating Quality: concepts, systems, strategies and tools. London, McGraw-Hill.

Korteweg, L.I., 1991. Complaints: compliment and warning. Voedingsmiddelentechnologie VMT, 5. (in Dutch).

Kotler, P. 1994. Marketing management: analysis, planning, implementation and control, 8th edition. Englewood Cliffs, Prentice Hall.

Kotter, J.P. 1992. "What leaders really do", in J.J. Gabarro (ed), Managing People and Organizations. Harvard Business School Press, Boston.

Kramer, M. and Van den Briel, S. 2002. Total Quality Management in the food industry. MSc-thesis Wageningen University, The Netherlands, pp. 90.

Kranghand, J. 2001. CBL-BRC code inspection standard for retailers of food. Voedingsmiddelentechnologie VMT, 9, 16-18. (in Dutch).

Krajewski, L. ; Ritzman, L. 1999. Operations Management, strategy and analysis. Addison, Wesley, Reading, Massachusetts.

Krajewski, L.J.; Ritzman, L.P. 1999. Operations management: strategy and analysis. Addison Wesley, New York.

Lancaster, G. And L. Massingham, 1999. Essentials of marketing. Mc-Graw-Hill, London.

Lawrence, P.R. and Lorsch, J.W. 1967. Organisations and environment. Harvard University, Boston.

Leaper, S. 1997. HACCP: A Practical Guide. Technical Manual 38. HACCP Working Group. Campden food & drink research association. Gloucestershire, UK, 59 pp.

Lelieveld, H.L.M. 1993. The EC Machinery Directive and food processing equipment. Trends in Food Science and Technology, 4, 153-154.

Lelieveld, H.L.M.; Hugelshofer, W.; Jepson, P.C.; Lalande, M. Mostert, M.A.; Nassaur, J.; Ringström, R. 1992. Microbiologically safe continuous pasteurisation of liquid foods. Trends in Food Science and Technology, 3, 303-307.

Leniger; Beverloo 1975. Food Processing Engineering. D. Reidel Publishing Company, Dordrecht, The Netherlands, p.546.

Lewin, K., 1952. Group decision and social change. In: Readings in Social Psychology, Swanson G.E. (ed.) Holt Rinehart, New York.

Loode, M. 2000. Van HACCP naar BRC: veiliger voedsel door 'ketenbeïnvloeding'. *Specifiek, (19): 4-5*. (in Dutch).

Lyon, D.; Francombe, M.A.; Hasdell, T.A.; Lawon, K. 1992. Guidelines for sensory analysis in food product development and quality control.

Maas, H.F.A.; Barendsz, A.W. 1997. Nomogram as simple and successful to for classification of risks. Voedingsmiddelen Technologie, 18/19, 37-39 (in Dutch).

Marcelis, W.J., 1984. Maintenance management development (in Dutch). Kluwer, Deventer.

Mayes, T., 1998. Risk analysis in HACCP: burden or benefit? Food control, vol. 9, 2-3, 171-176.

Mc. Alister, L.; Rothschild, M.L.(eds.). UT. Association for Consumer Research.

Meilgaard, M; Civille, G.V.; Carr, B.T. 1991. Sensory Evaluation Techniques, CRC Press, Inc., Boca Raton, Florida, USA.

Melnyk, S.A. and Denzler, D.R., 1996. Operations management, a value-driven approach. Irwin, Chicago.

Microbiology and Food Safety Committee of the National Food Processor Association (MFSCNFPA). 1993a. Implementation of HACCP in a food processing plant. Journal of Food Protection, 56, 548-554.

Microbiology and Food Safety Committee of the National Food Processor Association (MFSCNFPA). 1993b. HACCP implementation: A generic model for chilled foods. Journal of Food Protection, 56, 1077-1084.

Mintzberg, H. et al. 1976. The structure of "unstructured" decision processes. Administration Science Quaterly, 21, June 1976.

Mintzberg, H. 1993. The structuring of Organisations. Prentice-Hall, Englewood Cliffs.

Montgomery, D.C. 1985. Introduction to statistical quality control. New York. John Willey.

Mors, Y. 1996. Uniekaas: 'Good can always become better'. Quality in the dairy chain. Zuivelzicht 88.35 (in Dutch).

Moskowitz, H.R. 1994. Food concepts and products. Just-in-time development. Food and Nutrition Press, Inc. Connecticut, USA, pp. 502.

Mostert, M.A.; Shiessl, S.; Rysstad, G.; Weber, W.; Wilke, B.; Larsen, L.S.; Harvey, P.C.; Reinecke, G.; Delaunay. 1993. Hygienic packaging of food products. Trends in Food Science and Technology, 4, 406-407.

Namur, A.P.; Vanbelle, M. 1994. Animal feed industry, process control. In: Quality control and requirements of food of animal origin. REU Technical series 40, 36-41.

National Advisory Committee on Microbiological Criteria for Foods (NACMCF) of the USA. 1998. Hazard analysis and critical control point principles and application guidelines. Journal of Food Protection, 61, 762-775.

National Advisory Committee on Microbiological Criteria of Foods (NACMCF) of the USA. 1992. Hazard Analysis and Critical Control Point System, International Journal of Food Microbiology no.16, 1-23.

Nickerson, S.C. 1995. Milk production: Factors affecting milk composition. In: Milk quality. Harding, F. (ed.) Blackie Academic & Professional. London, 3-24.

NNI, Dutch Normalisation Institute. 1999. Information workshop on changes in the ISO 9000 norms. How to assure quality after 2000? NNI-Education.

Noordhuizen, J.P.T.M. 1998. Risk analysis and preventive health care for pigs. In: Preventive health care in pig breeding. Hartog, den, L.A. Visser-Reyneveld, M.N. (eds). 1998. Foundation Post graduate education Wageningen University, The Netherlands, 101-113 (in Dutch).

Noordhuizen, J.P.T.M.; Welpelo, H.J. 1996. Sustainable improvement of animal health care by systematic quality risk management according to the HACCP concept. The Veterinary quarterly, 18, 121-126.

Noori, H.; Radford, R. 1995. Production and operations management: Total Quality and Responsiveness. McGraw Hill, New York.

Northolt, M.D.; Hoornstra, E. 1998. Quantitative analysis of companies' risk exceeds HACCP. Voedingsmiddelen Technologie, 10, 11-16 (in Dutch).

Northolt, M.D.; Sterrenburg, P.; Hoornstra, E.; Folkerts, H.; Schwarz-Bovee, E.H.G., 1998. HACCP in the food production chain; Bottlenecks and risks of raw materials. Voedingsmiddelen Technologie (in Dutch), 21, 12-16.

Notermans, S., Louve, J.L., 1995. Quantitative risk analysis and HACCP: some remarks. Food Microbiology, 12, 425-429.

Notermans, S., Mead, G.C., 1996. Incorporation of elements of quantitative risk analysis in the HACCP system. International Journal of Food Microbiology, 30, 157-173.

Notermans, S.; Nauta, M.J.; Jansen, J.; Louve, J.L.; Mead, G.C. 1998. A Risk assessment approach to evaluating food safety based on product surveillance. Food Control, 9, 217-223.

Notermans, S.; Van der Giessen, A.1993. Foodborne diseases in the 1980's and 1990's - The Dutch experience. Food Content 4.

Ohlsson, T. 1994. Minimal processing- preservation methods of the future: an overview. Trends in Food Science and Technology, 5, 341-344.

Oickle, J.G. 1990. New product development and value added. Food Development Division, Agriculture, Canada.

Otten, A.; Verdooren, L.R. 1996. Statistical process and quality control. Lecture book Agricultural University Wageningen. Wageningen, The Netherlands. Pomeranz, Y.; Meloan, C.E. (eds.) Chapman & Hall, New York, 778 pp.

Oude Ophuis, P.A.M. 1994. Sensory Evaluation of 'free range' and regular pork meat under different conditions of experience and awareness. Food Quality and Preferences, 5, 173-178.

Paine, F.A.; Paine, H.Y. 1983. Developing packs for food (chapter 14); Using barrier materials efficiently (chapter 16). In: Handbook of food packaging. Blackie Academic & Professional, London, Paine, F.A.; Paine, H.Y.(eds.), 347-356, 390-425.

Phillips, C.A. 1996. Review: Modified atmosphere packaging and its effects in the microbiological quality and safety of produce. International Journal of Food Science and Technology, 31, 463-479.

Piggott, J.R. 1988. Sensory analysis of foods. Elsevier, New York.

Pomeranz, Y.; Meloan, C.E. 1994. Food analysis. Theory and practice, 3rd ed. Chapman and Hall, New York, pp.778.

Porter, M. 1985. Competitive advantage. The Free Press, New York.

Porter, M.E. 1980. Competitive Strategy. The Free Press, New York.

Poulsen, C.S., Juhl, H.J., Kristensen, K., Bech, A.C, Engelund, E. 1996. Quality guidance and quality formation. Food Quality and Preferences, 7, 127-135.

Prins, A.; Vliet, van, T.; Walstra, P. 1994. Introduction to food physics. Lecture book. Dept. Food Technology: food physics. Wageningen University, The Netherlands.

Product Boards for Livestock, Meat and Eggs, 1996. General conditions for integrated chain control, IKB-chicken.

Radford, K.J. 1975. Managerial Decision Making. Reston Publishing Company, Inc., Reston.
Regulation Milk Quality Payment 1996. Printed by the Commodity board for Dairy Produce. The Netherlands ISBN 90-9007677-5 (in Dutch).
Richardson, T.; Hyslop, D.B. 1985. Enzymes. In: Food Chemistry. Fennema, O.R. (ed.), Marcel Dekker, Inc., New York., 371-476.
Robbins, S.P., 2000. Essentials of organizational behavior. Prentice Hall, Upper Saddle River, New Yersey.
Ross, J.E. 1999. Total quality management. St. Lucie Press. Boca Raton, USA.
Roussel, P.; Saad, K. Erickson, T. 1991. Third generation R&D, managing the link to corporate strategy. Harvard Business Presss, Boston, Massachusetts.
Salunkhe, D.K.; Bolin, H.R.; Reddy, N.R. 1991. Storage, processing, and nutritional quality of fruits and vegetables. Volume I: Fresh Fruits and vegetables. Salunkhe, D.K.; Bolin, H.R.; Reddy, N.R. (eds.). CRC Press, Inc. Florida, 195 pp.
Samuel, M., 1992. Catalysts for change. The TQM magazine.
Sanders, D.H.; Allaway, W.H.; Ohlson, R.A. 1987. Nutritional quality of cereal grains. Ohlson, R.A.; Frey, K.J. (eds.). ASA-CSAA-SSA. Madison, WI. 1-45.
Schein, E. H. 1990. Organizational Culture. Ammerican Psychologist, vol. 45.
Schermerhorn, J.R., 1999. Management. John Wiley, New York.
Schmidt, R. 1997. The implementation of simultaneous engineering in the stage of product development: A process oriented improvement of quality function deployment. European Journal of Operational Research, 100, 293-314.
Scientific Committee for Food (SCF). 1996. Principles for the development of risk assessment of microbiological hazards for the development of hygiene rules for foodstuffs as covered by the hygiene of foodstuffs directive 93/43/EEG. CS/FMH/CRIT/2 Rev 1. Submitted by the secretariat.
Senge, P. 1990. The fifth discipline. Harper, New York.
Shapiro, A.; Mercier, C. 1994. Safe food Manufacturing. The Science of Total Environment, 143, 75-92.
Shapton, D.A.; Shapton, N.F. 1991. Principles and practices for the safe processing of foods. Shapton, D.A.; Shapton, N.F. (eds.). Butterworth-Heinemann, Oxford.
Shewart, W.A. 1980. Economic Control of Quality of Manufactured Product. Milwaukee, ASQC.
Siegall, M. 1987. "The simplistic five: an integrated framework for teaching motivation". Organizational Behavior Teaching Review 12 (1987/1988).
Simon, H.A. 1960. The New Science of Management Decision. Harper & Row, New York.
Singh, R.P. 1994. Scientific principles of shelf life evaluation. In: Shelf life evaluation of foods. Man, C.M.D.; Jones, A.A. (eds.) Blackie Academic & Professional, Chapman & Hall, Glasgow, 3-26.
Sitter, H. de. 1998. Regulatory aspects of food production systems. In: Innovation of food production systems. Jongen W.M.F.; Meulenberg, M.T.G. (eds.). Wageningen Pers, Wageningen, The Netherlands, 159-175.
Slomp, J. 1999. Development of systems. Bedrijfskunde, 4, 26-33 (in Dutch).
Smit, M. 2000. Het hoe en waarom van de BRC-inspectie. *Voedingsmiddelentechnologie, 32 (11): 29-32* (in Dutch).
Smit, M.J. 1999. De retailerseisen op een rij: HACCP-certificaat en BRC Standard vullen elkaar aan. *Voedingsmiddelentechnologie, 31 (12): 11-13*. (in Dutch).
Smith, D.J.; Edge, J. 1990. Essential Quality procedures. In: Handbook of Quality Management, Dennis Lock (ed.) Gower, Aldershot.
Smulders, F.J.M. 1988. Review: lactic acid: consideration in favour of its acceptance as a meat decontaminant. Journal of Food Technology, 21, 419-436.

Snyder, O.P. 1993. The applications of HACCP for MAP and sous vide products. In: Principles of Modified-atmosphere and sous-vide product packaging. Farber, J.M.; Dodds, K.L. (eds.) Technomic publishing Co, INC, Lancaster, 325-383.

Sol, M. 1998. Chain for quality of milk (Keten Kwaliteit Melk KKM). Informative brochure of KKM foundation, Leusden, The Netherlands (in Dutch).

SPS-agreement, 1996. Understanding the world trade organisation agreement on sanitary and phytosanitary measures. WTO secretariat.

Stanley, S.E. 1998. The challenges and opportunities of ISO 90000 registration: 'Your customers are calling'. CIM Bulletin, 1018, 215-220.

Staples, G. 1990. Quality Audits and reviews. In: Handbook of Quality Management, Dennis Lock (ed.) Gower, Aldershort.

Steenkamp J.E.B.M. 1990. A conceptual model of the quality perception process. Journal of Business Research, 21, 309-333.

Steenkamp, J.E.B.M. 1989. Product quality: An investigation into the concept and how it is perceived by consumers. Van Gorcum. Assen, The Netherlands.

Stevenson, W.J., 1999. Production/operations management. McGraw-Hill, New York.

Stone, H.; Sidel, J.L. 1985. Sensory evaluation practices. Academic press, London.

Sullivan, L.P. 1986. Quality function deployment. A system to assure that customer needs drive the product design and production process. Quality progress, 39-50.

Syson, R., 1992. Improving Purchasing Performance. Pitman Publishing, London.

Tarrant, V.; Grandin, T. 1993. Cattle transport. In: Livestock handling and transport. Grandin, T. (ed.). CAB International, Oxon, UK, 109-125.

Tenner A.R.; De Toro, J.J. 1992. Total Quality Management: Three Steps to Continuous Improvement. Addison Wesley, Reading.

Thomas, K.W., 1976. Conflict and conflict management. In: handbook of industrial and organizational behavior, M.D. Dunnett, (ed.) John Wiley, New York.

Timperley, A.W.; Axis, J.; Grasshoff, A.; Hodge, C.R. Holah, J.T.; Kirby, R.; Maingonnat, J-F; Trägårdh,C.; Venema-Keur, B.M.; Cerf, O. 1993. A method for the assessment of bacteria tightness of food-processing equipment. Trends in Food Science and Technology, 4, 190-192.

Tompkin, R.B. 1995. The use of HACCP for producing and distributing processed meat and poultry products. In: HACCP in meat, poultry and fish processing. Advances in meat research series vol. 10. Pearson, A.M.; Dutson, T.R. (eds.). Blackie Academic & Professional. London, 72-108.

Traill, B. and Grunert, K.G. 1997. Product and process innovation in the food industry. Traill, B., Grunert, K.G. (eds.) Blackie Academic & Professional, London, 242 pp.

Troeger, K. 1996. Transportation of slaughter animals. Treatment during storage and its consequences for product quality. Fleischwirtschaft 76, 157-158.

Urban, G.L; Hauser, J.R. 1993. Design and marketing of new products, 2nd edition Englewood Cliffs, Prentice Hall.

Van Boekel, M.A.J.S. 1998. Developments in technologies for food production. In: Innovation of food production systems. Jongen W.M.F.; Meulenberg, M.T.G. (eds.). Wageningen Pers, Wageningen, The Netherlands. 87-116.

Van Dam, Y.K.; Hoog, de, C.; Ophen, van J.A.C. 1994. Reflecties op gemak bij voeding. In: Eten in de jaren negentig. Dam, van Y.K.; Hoog, de, C.; Ophen, van J.A.C. (eds.) Serie Economie van Landbouw en Milieu. Eburon Delft, 1-7 (in Dutch).

Van den Berg, M.G. 1993. Quality of Food products. Berg, M.G. (ed). Kluwer, Deventer, pp.360 (in Dutch).

Van den Berg, M.G.; Delsing, B.M.A. 1999. Quality of food (2nd edition). Berg, van den M.G. (ed). Kluwer, Deventer, The Netherlands, pp.360 (in Dutch).

Van den Braak, S.; Bloemen, K. 1999. HACCP and risk assessment. A conceptual HACCP plan for Dutch/French ready-to-eat meals. MSc-thesis Agricultural University Wageningen, The Netherlands, 177 pp (in Dutch).

Van der Spiegel, M.; Luning, P.A. Ziggers, G.W.; Jongen, W.M.F. *in preparation*. evaluation of performance measurement concepts on their use for measuring effectivity of food quality systems.

Van Oosterhout, , E. 2001. Application of the quality guidance model in the biological pig meat chain. MSC thesis Wageningen agricultural University, The Netherlands (in Dutch).

Van Trijp, J.C.M. 1996. Sensorisch onderzoek in relatie tot marketing onderzoek, Onderzoek, 9-10 (in Dutch).

Van Trijp, J.C.M.; Steenkamp, J.E.B.M. 1998. Consumer-oriented new product development: principles and practice. In: Innovation of food production systems. Jongen W.M.F.; Meulenberg, M.T.G. (eds.). Wageningen Pers, Wageningen, The Netherlands. 37-66.

Walker, S.J. 1994. Principles and practice of shelf life prediction for micro-organisms. In: Shelf life evaluation of foods. Man, C.M.D.; Jones, A.A. (eds.) Blackie Academic & Professional, Chapman & Hall, Glasgow, 40-51.

Wandel, M.; Bugge, A. 1997. Environmental concern in consumer evaluation of food quality. Food Quality and Preference, 8, 19-26.

Warriss, P.D.; Kestin, S.G.; Brown, S.N.; Collins, L.J. 1984. The time required for recovery from mixing stress in young bulls and prevention of dark cutting beef. Meat Science, 10, 53-68.

Weihrich H.; Koontz, H. 1993. Management, a global perspective. Mc Graw Hill, New York.

Wheelwright, S. and K. Clark, 1992. Revolutionising Product Development. The Free Press, New York.

Wijtzes, T. 1996. Modelling the microbiological quality and safety of foods. Thesis Wageningen University, The Netherlands.

Wogan, G.N.; Marletta, M.A. 1985. In: Food Chemistry. Fennema, O.R. (ed.), Marcel Dekker, Inc., New York., 689-724.

Wolf, I.D. 1992. Critical issues in food safety, 1991-2000. Food Technol. 46, 64-70.

Zeuthen, P.; Cheftel, J.C.; Eriksson, C.; Gormley, T.R.; Linko, P.; Paulus, K. 1990. Processing and Quality of Foods, Volume 1 High Temperature/short time (HTST): guarantee for high quality food with long shelf life. Elsevier Applied Science, Barking, England, 383 pp.

Zuurbier, P.J.P., Trienekens, J.H. and Ziggers, G.W. 1996. Vertical Cooperation: Concepts to Start Partnerships in Food and Agribusiness. Kluwer, Deventer (in Dutch).

Zwietering, M.H.; Wit, de C.J.; Notermans, S. 1996. The application of predictive microbiology to estimate the number of Bacillus cereus in pasteurised milk at the point of consumption. International Journal of Food Microbiology, 30, 55-70.

Index

–D–

–E–

–F–

–T–

–U–

–V–

Food q
a tech

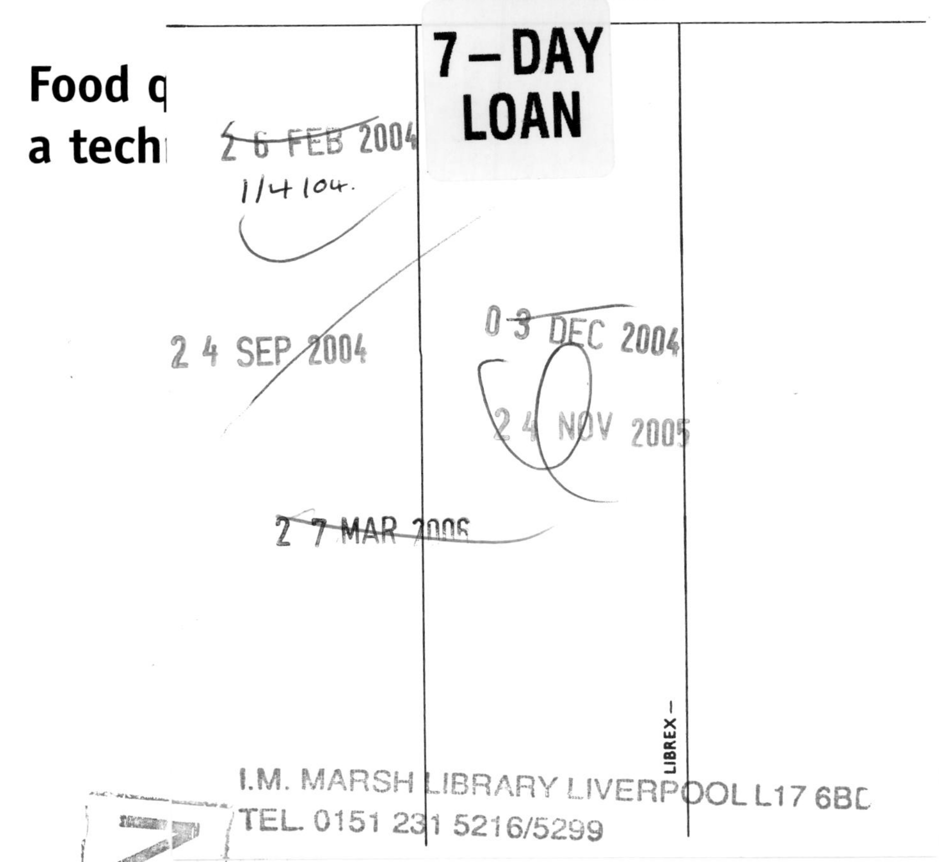
Books are to be returned on or before
the last date below.
7 – DAY
LOAN
26 FEB 2004
1/4/04.
24 SEP 2004
03 DEC 2004
24 NOV 2005
27 MAR 2006
LIBREX –
I.M. MARSH LIBRARY LIVERPOOL L17 6BD
TEL. 0151 231 5216/5299

LIVERPOOL JMU LIBRARY
3 1111 00947 8924